Springer Series in the Data Sciences

Springer Series in the Data Sciences focuses primarily on monographs and graduate level textbooks. The target audience includes students and researchers working in and across the fields of mathematics, theoretical computer science, and statistics.

Data Analysis and Interpretation is a broad field encompassing some of the fastestgrowing subjects in interdisciplinary statistics, mathematics and computer science. It encompasses a process of inspecting, cleaning, transforming, and modeling data with the goal of discovering useful information, suggesting conclusions, and supporting decision making. Data analysis has multiple facets and approaches, including diverse techniques under a variety of names, in different business, science, and social science domains. Springer Series in the Data Sciences addresses the needs of a broad spectrum of scientists and students who are utilizing quantitative methods in their daily research.

The series is broad but structured, including topics within all core areas of the data sciences. The breadth of the series reflects the variation of scholarly projects currently underway in the field of machine learning.

More information about this series at http://www.springer.com/series/13852

Mayer Alvo • Philip L. H. Yu

A Parametric Approach to Nonparametric Statistics

Mayer Alvo
Department of Mathematics and Statistics
University of Ottawa
Ottawa, ON, Canada

Philip L. H. Yu
Department of Statistics and Actuarial Science
University of Hong Kong
Hong Kong, China

ISSN 2365-5674 ISSN 2365-5682 (electronic)
Springer Series in the Data Sciences
ISBN 978-3-030-06804-2 ISBN 978-3-319-94153-0 (eBook)
https://doi.org/10.1007/978-3-319-94153-0

Softcover re-print of the Hardcover 1st edition 2018

This Springer imprint is published by the registered company Springer Nature Switzerland AG
The registered company address is: Gewerbestrasse 11, 6330 Cham, Switzerland

Preface

In randomized block designs, the Friedman statistic provides a nonparametric test of the null hypothesis of no treatment effect. This book was motivated by the observation that when the problem is embedded into a smooth alternative model to the uniform distribution over a set of rankings, this statistic emerges as a score statistic. The realization that this nonparametric problem could be viewed within the context of a parametric problem was particularly revealing and led to various consequences. Suddenly, it seemed that one could exploit the tools of parametric statistics to deal with several nonparametric problems. Penalized likelihood methods were used in this context to focus on the important parameters. Bootstrap methods were used to obtain approximations to the distributions of estimators and to construct confidence intervals. Bayesian methods were introduced to widen the scope of applicability of distance-based models. As well, the more commonly used test statistics in nonparametric statistics were reexamined. The occurrence of ties in the sign test could be dealt with in a natural formal manner as opposed to the traditional ad hoc approach. This book is a first attempt at bridging the gap between parametric and nonparametric statistics and we expect that in the future more applications of this approach will be forthcoming.

The authors are grateful to Mr. Hang Xu for contributions that were incorporated in Chapter 10. We are grateful to our families for their support throughout the writing of this book. In particular, we thank our wives Helen and Bonnie for their patience and understanding. We are also grateful for the financial support of the Natural Sciences and Engineering Research Council of Canada (NSERC) and the Research Grants Council of the Hong Kong Special Administrative Region, China (Project No. 17303515), throughout the preparation of this book.

Ottawa, ON, Canada — Mayer Alvo
Hong Kong, China — Philip L. H. Yu

Contents

Notation

$\mathbb{R}$	a set of all real numbers
$X = (X_1, \ldots, X_n)$	a random vector
i.i.d.	independent and identically distributed
θ	a parameter, possibly a vector
H_0	null hypothesis
H_1	alternative hypothesis
$f_X(x;\theta)$	the probability density function of X
$F_X(x;\theta)$	the cumulative distribution function of X
cdf	cumulative distribution function
pdf	probability density function
$\Phi(x)$	the cumulative distribution of a standard normal random variable
$\mathcal{N}_k(\mu, \Sigma)$	the k dimensional multivariate normal distribution with mean μ and variance-covariance Σ
$\text{vMF}(\boldsymbol{x}\|\boldsymbol{m}, \kappa)$	von Mises-Fisher distribution
$\pi(x;\theta)$	a smooth alternative density of X
$U(\theta; X)$	score vector
$U(X)$	score vector evaluated under the null hypothesis
$R_1, \ldots, R_n$	the ranks of $X_1, \ldots, X_n$ from the smallest to the largest
$a_n(R)$	score function evaluated at rank R
$L(\theta)$ or $L(\theta; x)$	the likelihood function of θ based on $X_1, \ldots, X_n$
$l(\theta)$ or $l(\theta; x)$	the log of the likelihood function of θ based on $X_1, \ldots, X_n$
$I_n(\theta)$	the Fisher information function of θ based on $X_1, \ldots, X_n$
$\bar{X}_n$	the sample mean of $X_1, \ldots, X_n$
S_n^2	the sample variance of $X_1, \ldots, X_n$
$X_n \xrightarrow{P} X$	the sequence of random variables X_n converges in probability to X
$X_n \xrightarrow{\mathcal{L}} X$	the sequence of random variables X_n converges in distribution to X
$X_n \xrightarrow{a.s.} X$	the sequence of random variables X_n converges almost surely to X
$E[X]$	mean of X
$\text{Var}(X)$	variance of X
CLT	central limit theorem

LAN	local asymptotic normality
MLE	maximum likelihood estimation
$T_1 \sim T_2$	the statistics T_1 and T_2 are asymptotically equivalent
$\chi_r^2(\delta)$	the chi-squared distribution with d.f. r and noncentrality parameter δ
ANOVA	the analysis of variance
VI	variational inference
MCMC	Markov Chain Monte Carlo
SIR	sampling-importance-resampling
ELBO	evidence lower bound
LTRC	left-truncated right-censored

Part I.

Introduction and Fundamentals

1. Introduction

This book grew out of a desire to bridge the gap between parametric and nonparametric statistics and to exploit the best aspects of the former while enjoying the robustness properties of the latter. Parametric statistics is a well-established field which incorporates the important notions of likelihood and sufficiency that are part of estimation and testing. Likelihood methods have been used to construct efficient estimators, confidence intervals, and tests with good power properties. They have also been used to incorporate incomplete data and to pool information from different sources collected under different sampling schemes. As well, Bayesian methods which rely on the likelihood function can be used to combine information acquired through a prior distribution. Constraints which restrict the domain of the likelihood function can also be taken into account. Likelihood functions are Bartlett correctable which helps to improve the accuracy of the inference. Additional tools such as penalized likelihood via Akaike or Bayesian information criterion can take into account constraints on the parameters. Recently, the notion of composite likelihood has been introduced to extend the range of applications. Problems of model selection are naturally dealt with through the likelihood function.

A difficulty that arises with parametric inference is that we need to know the underlying distribution of the random variable up to some parameters. If that distribution is misspecified, then inferences based on the likelihood may be inefficient and confidence intervals and tests may not lead to correct conclusions. Hence, when there is uncertainty as to the exact nature of the distribution, one may alternatively make use of traditional nonparametric methods which avoid distributional assumptions. Such methods have proven to be very efficient in several instances although their power is generally less than the analogous parametric counterparts. Moreover, nonparametric statistics rely more on intuition and the subject has developed along in a nonsystematic manner, always in an attempt to mimic parametric statistics. Bootstrap methods could also be used in most cases to provide estimates and to construct confidence intervals but their interpretation may not be easy. For example, the shape of a confidence region for a vector parameter will often appear as a cloud in space.

To act as a bridge between parametric and nonparametric statistics, Conover and Iman (1981) used rank transformations in an ad hoc manner. They suggested using parametric methods based on the ranks of the data in order to conduct nonparametric

M. Alvo, P. L. H. Yu, *A Parametric Approach to Nonparametric Statistics*,
Springer Series in the Data Sciences, https://doi.org/10.1007/978-3-319-94153-0_1

analyses. However, as mentioned by Conover and Iman (1981), such an approach has a number of limitations, for instance the severe lack of robustness for the test for equality of variances.

In a landmark paper, Neyman (1937) considered the nonparametric goodness of fit problem and introduced the notion of smooth tests of fit by proposing a parametric family of alternative densities to the null hypothesis. The type of embedding proposed by Neyman was further elaborated by Rayner et al. (2009a) in connection with goodness of fit problems. In this book, we propose an embedding which focuses on local properties more in line with the notion of exponential tilting. Hence, we obtain a new derivation of the well-known Friedman statistic as the locally most powerful test in an embedded family of distributions. In another direction, we exploit Hoeffding's change of measure formula which provides an approach to obtaining locally most powerful tests based on ranks for various multi-sample problems. This is then followed by applications of Le Cam's three lemmas in order to obtain the asymptotic distribution of various statistics under the alternative. Together, these results enable us to determine the asymptotic relative efficiency of our test statistics.

This book is divided into three parts. In Part I, we outline briefly fundamental concepts in probability and statistics. We introduce the reader to some of the important tools in nonparametric statistics such as U statistics and linear rank statistics. In Part II, we describe Neyman's smooth tests in connection with goodness of fit problems and we obtain test statistics for some common nonparametric problems. We then proceed to make use of this concept in connection with the usual one- and two-sample tests. In Chapter 6, we present a unified theory of hypothesis testing and apply it to study multi-sample problems of location. We illustrate the theory in the case of the multi-sample location problem as well as the problem involving umbrella alternatives. In Chapter 7, we obtain a new derivation of the Friedman statistic and show it is locally most powerful. We then make us of penalized likelihood to gain further insight into the rankings selected by the sample. Chapter 8 deals with locally most powerful tests, whereas Chapter 9 is devoted to the concept of efficiency. In Part III, we consider some modern applications of nonparametric statistics. Specifically, we couch the multiple change-point problem within the context of a smooth alternative. Next, we propose a new Bayesian approach to the study of ranking problems. We conclude with Chapter 12 wherein we briefly describe the application of methodology to the analysis of censored data.

2. Fundamental Concepts in Parametric Inference

In this chapter we review some terminology and basic concepts in probability and classical statistical inference which provide the notation and fundamental background to be used throughout this book. In the section on probability we describe some basic notions and list some common distributions along with their mean, variance, skewness, and kurtosis. We also describe various modes of convergence and end with central limit theorems. In the section on statistical inference, we begin with the subjects of estimation and hypothesis testing and proceed with the notions of contiguity and composite likelihood.

2.1. Concepts in Probability

2.1.1. Random Variables and Probability Functions

The study of probability theory is based on the concept of random variables. The *sample space* is the set of all possible outcomes of an experiment. A real valued random variable X is defined as a function from the sample space to the set of real numbers. As an example, suppose in an experiment of rolling two independent dice, X represents the sum of the face values of the two dice. Then the sample space consists of the ordered pairs $\{(i,j) : i,j = 1,2,\ldots,6\}$ and the range of X is $\{2,3,4,\ldots,12\}$.

The *cumulative distribution function* (cdf) of a random variable X is the probability that X takes on a value less than or equal to x, denoted by $F_X(x) = P(X \leq x)$. Suppose 30% of students have heights less than or equal to 150 cm. Then the probability that the height X of a randomly chosen student is less than or equal to 150 cm is 0.3 which can be expressed as

$$F_X(150) = P(X \leq 150) = 0.3.$$

The cdf $F_X(x)$ in general is a nondecreasing right continuous function of x, which satisfies

$$\lim_{x \longrightarrow -\infty} F_X(x) = 0, \lim_{x \longrightarrow \infty} F_X(x) = 1.$$

M. Alvo, P. L. H. Yu, *A Parametric Approach to Nonparametric Statistics*,
Springer Series in the Data Sciences, https://doi.org/10.1007/978-3-319-94153-0_2

Most random variables are either discrete or continuous. We say that a random variable is *continuous* if its cdf is a continuous function having no jumps. A continuous random variable, such as weight, length, or lifetime, takes any numerical value in an interval or on the positive real line. Typically, a continuous cdf has a derivative except at some points. This derivative denoted by

$$f_X(x) = \frac{d}{dx} F_X(x) = F'_X(x),$$

is called the *probability density function (pdf)* of X. The cdf of a continuous random variable X on the entire real line satisfies

$$f_X(x) \geq 0 \quad \int_{-\infty}^{\infty} f_X(t)dt = 1 \quad \text{and} \quad F_X(x) = \int_{-\infty}^{x} f_X(t)\, dt.$$

A random variable is called *discrete* if it takes a finite or a countably infinite number of values. Examples are the result of a soccer game (win, lose, or draw), the number of sunny days in a week, or the number of patients infected with a disease. The probability mass function (pmf) of a discrete random variable is defined as $f_X(x) = P(X = x)$. The pmf for a discrete random variable X, $f_X(x)$ should satisfy the following conditions:

$$f_X(x) \geq 0 \quad \text{and} \quad \sum_{x \in \Omega} f_X(x) = 1,$$

where Ω is the range of X.

There are occasions when one is interested in the conditional density of a random variable X given a random variable Y. This is defined as

$$f(x|y) = \frac{f(x,y)}{f_Y(y)}, f_Y(y) > 0.$$

The joint cdf for a vector of $p(\geq 1)$ random variables, $\boldsymbol{X} = (X_1, X_2, \ldots, X_p)$, is defined as

$$F_{\mathbf{X}}(x_1, x_2, \ldots, x_p) = P(X_1 \leq x_1, \ldots, X_p \leq x_p).$$

The joint density can be defined as

$$f_{\mathbf{X}}(x_1, x_2, \ldots, x_p) = \frac{\partial^p F_{\mathbf{X}}(x_1, x_2, \ldots, x_p)}{\partial x_1 \ldots \partial x_p}$$

provided the multiple derivatives exist. The p random variables $X_1, X_2, \ldots, X_p$ are said to be *independent* if their joint pdf is the product of the p individual densities, labeled marginal pdfs and denoted $\{f_{X_i}(x)\}$, i.e.,

$$f_{\mathbf{X}}(x_1, x_2, \ldots, x_p) = f_{X_1}(x_1) f_{X_2}(x_2) \cdots f_{X_p}(x_p), \text{for all } x_1, \ldots, x_p.$$

Definition 2.1. A *random sample* of size p of the random variable X is a set of independent and identically distributed (i.i.d.) random variables $X_1, X_2, \ldots, X_p$, with the same pdf as X.

Definition 2.2. The order statistics from a sample of random variables $X_1, X_2, \ldots, X_p$ are denoted $X_{(1)} \leq X_{(2)} \leq \ldots \leq X_{(p)}$ and indicate which are the smallest, second smallest, etc.

Hence, $X_{(1)} = min\{X_1, X_2, \ldots, X_p\}$ and $X_{(p)} = max\{X_1, X_2, \ldots, X_p\}$.

In probability and statistics, we are interested in properties of the distribution of a random variable. The expected value of a function $g(X)$ of a real valued random variable X, denoted by $E[g(X)]$, is defined as

$$E[g(X)] = \begin{cases} \int_{-\infty}^{\infty} g(x)f(x)\,dx & \text{if } X \text{ is continuous} \\ \sum_{x\in\Omega} g(x)f(x) & \text{if } X \text{ is discrete.} \end{cases}$$

Similarly, the conditional expectation of a function $g(X)$ given $(Y = y)$ is defined as

$$E[g(X)|Y = y] = \begin{cases} \int_{-\infty}^{\infty} g(x)f(x|y)\,dx & \text{if } X \text{ is continuous} \\ \sum_{x\in\Omega} g(x)f(x|y) & \text{if } X \text{ is discrete.} \end{cases}$$

We encounter the conditional expectation in regression problems.

The nth *moment* of a random variable X is $E[X^n]$ and the nth *central moment* about its mean is $\mu_n = E[(X - \mu)^n]$, where $\mu = E[X]$, the mean of X. Note that $\mu_1 = 0$ and $\mu_2 = Var(X)$ is the variance of X, usually denoted by σ^2.

Definition 2.3. The moment generating function of a random variable X if it exists is defined to be

$$M_X(t) = E\left[e^{tX}\right], t > 0.$$

The moment generating function as the name suggests can be used in part to obtain the moments of a distribution by differentiating it:

$$M_X^{(k)}(t)\,|_{t=0} = E\left[X^k\right].$$

As an example, the moment generating function of a normally distributed random variable with mean μ and variance σ^2 is given as

$$M_X(t) = e^{\mu t + \frac{1}{2}\sigma^2 t^2}.$$

Another important property is that the moment generating function when it exists is unique. The moment generating function of the sum of independent random variables is equal to the product of the individual moment generating functions. This fact, coupled with the uniqueness property helps to determine the distribution of the sum of the variables.

The third and fourth central moments are used to define the skewness and (excess) kurtosis as

$$\gamma = \mu_3/\sigma^3 \quad \text{and} \quad \kappa = \mu_4/\sigma^4 - 3$$

respectively. The skewness measures the "slant" of a distribution with $\gamma = 0$ for a symmetric distribution. When $\gamma < 0$, the distribution is slanted to the left (with a long tail on the left) and when $\gamma > 0$, it is slanted to the right (with a long tail on the right). The kurtosis measures the "fatness" of the tails of a distribution. A positive value indicates a heavier tail than that of a normal distribution whereas a negative value points to a lighter tail.

Knowledge of the mean, variance, skewness, and kurtosis can often be used to approximate fairly well a given distribution (see Kendall and Stuart (1979)). Table 2.1 exhibits some important pmfs/pdfs along with their mean, variance, skewness, and (excess) kurtosis. The multinomial distribution generalizes the binomial for the case of r categories.

Using the linear properties of the expectation operator, we can determine the expected value and moments of a linear combination of a set of random variables. For instance, for p random variables $X_1, \ldots, X_p$ and constants $a_i, i = 1, \ldots, p$, we have,

$$\begin{aligned} E\left[\sum_{i=1}^{p} a_i X_i\right] &= \sum_{i=1}^{p} a_i E\left[X_i\right] \\ Var(\sum_{i=1}^{p} a_i X_i) &= \sum_{i=1}^{p} a_i^2 Var(X_i) + 2\sum_{i=1}^{p}\sum_{j=i+1}^{p} a_i a_j Cov(X_i, X_j), \end{aligned}$$

where

$$Cov(X_i, X_j) = E\left[X_i X_j\right] - E\left[X_i\right] E\left[X_j\right].$$

Some useful results in connection with conditional distributions are in the following theorem.

Theorem 2.1. *Let X and Y be two random variables defined on the same probability space. Then*

(a) $E\left[Y\right] = E\left[E\left[Y|X\right]\right]$,

(b) $Var\left[Y\right] = E\left[Var(Y|X)\right] + Var(E\left[Y|X\right])$.

Example 2.1 (Symbolic Data Analysis). In symbolic data analysis, it is common to observe a random sample in the form of intervals $\{(a_i, b_i), i = 1, \ldots, n\}$. Let X be a random variable whose conditional distribution is uniform on the interval (A, B). In that case, the conditional mean and variance of X are respectively

$$\frac{(A+B)}{2}, \frac{(B-A)^2}{12}.$$

Table 2.1.: Some important discrete and continuous random variables

Name	pmf / pdf	Mean	Variance	Skewness	Kurtosis
Discrete					
Uniform on $1, 2, \ldots, m$	$\frac{1}{m}$	$\frac{m+1}{2}$	$\frac{m^2-1}{12}$	0	$-\frac{6(m^2+1)}{5(m^2-1)}$
Binomial	$\binom{n}{x} p^x (1-p)^{n-x}$ $x = 0, 1, \ldots, n,\ 0 \leq p \leq 1$	np	$np(1-p)$	$\frac{1-2p}{\sqrt{np(1-p)}}$	$\frac{1-6p(1-p)}{np(1-p)}$
Multinomial	$\frac{n!}{x_1!x_2!\ldots x_r!} p_1^{x_1} p_2^{x_2} \cdots p_r^{x_r}$, $\sum_{i=1}^r x_i = n, \sum_{i=1}^r p_i = 1$				
Continuous					
Uniform on (a, b)	$\frac{1}{b-a},\ a < x < b$	$\frac{a+b}{2}$	$\frac{(b-a)^2}{12}$	0	$-\frac{6}{5}$
Normal	$\frac{1}{\sqrt{2\pi\sigma^2}} \exp\left(-\frac{(x-\mu)^2}{2\sigma^2}\right)$, $-\infty < x, \mu < \infty, \sigma > 0$	μ	σ^2	0	0
Exponential	$\lambda e^{-\lambda x},\ x \geq 0, \lambda > 0$	$\frac{1}{\lambda}$	$\frac{1}{\lambda^2}$	2	6
Gamma	$\frac{\beta^\alpha}{\Gamma(\alpha)} x^{\alpha-1} e^{-\beta x},\ x \geq 0, \alpha, \beta > 0$	$\frac{\alpha}{\beta}$	$\frac{\alpha}{\beta^2}$	$\frac{2}{\sqrt{\alpha}}$	$\frac{6}{\alpha}$
Chi-square	$\frac{1}{2^{\frac{m}{2}} \Gamma\left(\frac{m}{2}\right)} x^{\frac{m}{2}-1} e^{-\frac{x}{2}}, x \geq 0, m > 0$	m	$2m$	$\sqrt{\frac{8}{m}}$	$\frac{12}{m}$

Table 2.1.: Continued.

Name	pmf / pdf	Mean	Variance	Skewness	Kurtosis
Discrete					
Laplace	$\frac{1}{2\sigma}\exp\left(-\frac{\lvert x-\mu\rvert}{\sigma}\right), -\infty < x, \mu < \infty, \sigma > 0$	μ	$2\sigma^2$	0	3
Logistic	$\frac{\exp\left(-\frac{x-\mu}{\sigma}\right)}{\sigma\left(1+\exp\left(-\frac{x-\mu}{\sigma}\right)\right)^2}, -\infty < x, \mu < \infty, \sigma > 0$	μ	$\frac{\pi^2\sigma^2}{3}$	0	1.2
Student's t	$\frac{\Gamma\left(\frac{\nu+1}{2}\right)}{\sqrt{\pi\nu}\Gamma\left(\frac{\nu}{2}\right)}\left(1+\frac{x^2}{\nu}\right)^{-\frac{\nu+1}{2}}$ $, -\infty < x < \infty, \nu > 0$	$0, \nu > 1$	$\frac{\nu}{\nu-2}, \nu > 2$	0	$\frac{6}{\nu-4}, \nu > 4$

From Theorem 2.1, the unconditional mean is given by

$$E\left[\frac{A+B}{2}\right]$$

which can be estimated by

$$\frac{\sum (a_i + b_i)}{2n}.$$

The unconditional variance is given by

$$E\left[Var\left[X|A,B\right]\right] + Var\left[E\left[X|A,B\right]\right] = E\left[\frac{(B-A)^2}{12}\right] + Var\left(\frac{A+B}{2}\right)$$

which can be estimated unbiasedly by

$$\frac{\sum (b_i - a_i)^2}{12n} + \frac{1}{4(n-1)}\left\{\sum (a_i + b_i)^2 - \frac{1}{n}\left[\sum (a_i + b_i)\right]^2\right\}.$$

We shall also need to make use of the distribution of the order statistics which is given in the lemma below.

Lemma 2.1. *Let $\{X_1, X_2, \ldots, X_n\}$ be a random sample of size n from a cumulative distribution $F(x)$ with density $f(x)$. Then the density of the ith order statistic is given by*

$$f_{X_{(i)}}(x) = \frac{n!}{(i-1)!\,(n-i)!}\left[F(x)\right]^{i-1} f(x)\left[1-F(x)\right]^{n-i}, i = 1, \ldots, n. \tag{2.1}$$

Proof. An intuitive proof may be given by using the multinomial distribution. The probability that $X_{(i)}$ lies in a small interval around x implies that there are $(i-1)$ observations to the left of x and $(n-i)$ to the right of x. □

2.1.2. Modes of Convergence and Central Limit Theorems

We shall be concerned with three basic modes of convergence: weak convergence (equivalently, convergence in distribution), convergence in probability, and convergence almost surely. An important tool in proving limit results is the Borel-Cantelli lemma which provides conditions under which an event will have probability zero. We shall have occasion to make use of the Borel-Cantelli lemma in Chapter 10 to prove consistency results.

Lemma 2.2 (Borel-Cantelli)**.** *Let $A_1, \ldots, A_n$ be a sequence of events in some probability space such that*

$$\sum P(A_i) < \infty.$$

Then

$$P\left(\limsup_{n\longrightarrow\infty} A_n\right) = P\left(\bigcap_{n=1}^{\infty}\bigcup_{k\geq n}^{\infty} A_k\right) = 0.$$

The notation $\limsup_{n\longrightarrow\infty} A_n$ *indicates the set of outcomes which occur infinitely often in the sequence of events.*

We shall say that a sequence of random variables X_n *converges weakly* or *converges in distribution* to X, denoted $X_n \xrightarrow{\mathcal{L}} X$, if as $n \to \infty$,

$$P(X_n \leq x) \to P(X \leq x)$$

for all points of continuity x of the cdf of X.

We shall say that a sequence of random variables X_n *converges in probability* to X, denoted $X_n \xrightarrow{P} X$, if as $n \to \infty$

$$P(|X_n - X| > \varepsilon) \to 0$$

for $\varepsilon > 0$.

We shall say that a sequence of random variables X_n *converges almost surely* to X, denoted $X_n \xrightarrow{a.s.} X$, if as $n \to \infty$

$$P\left(\lim_{n\longrightarrow\infty} |X_n - X| > \varepsilon\right) = 0$$

for $\varepsilon > 0$.

Convergence almost surely implies convergence in probability. On the other hand, if a sequence of random variables converges in probability, then there exists a subsequence which converges almost surely. As well, convergence in probability implies convergence in distribution. The following inequality plays a useful role in probability and statistics.

Lemma 2.3 (Chebyshev Inequality). *Let X be a random variable with mean μ and finite variance σ^2. Then for $\varepsilon > 0$,*

$$P(|X - \mu| \geq \varepsilon) \leq \frac{\sigma^2}{\varepsilon^2}.$$

As an application of Chebyshev's inequality, suppose that $X_1, \ldots, X_n$ is a sequence of independent identically distributed random variables having mean μ and finite variance σ^2. Then, for $\varepsilon > 0$,

$$P\left(\left|\bar{X}_n - \mu\right| > \varepsilon\right) \leq \frac{\sigma^2}{n\varepsilon^2}$$

from which we conclude that $\bar{X}_n \xrightarrow{P} \mu$ as $n \to \infty$. This is known as the *Weak Law of Large Numbers*.

There is as well the *Strong Law of Large Numbers* (Billingsley (2012), Section 22, p. 301) which states that if $X_1, \ldots, X_n$ is a sequence of independent identically distributed random variables with mean μ for which $E|X_i| < \infty$, then for $\varepsilon > 0$.

$$P\left(\lim_{n \longrightarrow \infty} |\bar{X}_n - \mu| > \varepsilon\right) = 0.$$

That is $\bar{X}_n \xrightarrow{a.s.} \mu$. The single most important result in both probability and statistics is the central limit theorem (CLT). In its simplest version, it states that if $\{X_1, \ldots, X_n\}$ is a sequence of independent identically distributed (i.i.d.) random variables with mean μ and *finite* variance σ^2, then for large enough n,

$$\frac{\sqrt{n}\left(\bar{X}_n - \mu\right)}{\sigma} \xrightarrow{\mathcal{L}} \mathcal{N}(0,1),$$

where $\mathcal{N}(0,1)$ is the standard normal distribution with mean 0 and variance 1. Since the assumptions underlying the CLT are weak, there have been countless applications as in for example approximations to the probabilities of various events involving the sample mean. As well, it has been used to approximate various discrete distributions such as the binomial, Poisson, and negative binomial. An important companion result is due to Slutsky (Casella and Berger (2002), p. 239) which can often be used together with the central limit theorem.

Theorem 2.2 (Slutsky's Theorem). *Suppose that $X_n \xrightarrow{\mathcal{L}} X$, and $Y_n \xrightarrow{P} c$ for a constant c. Then,*

$$X_n Y_n \xrightarrow{\mathcal{L}} cX$$

and

$$X_n + Y_n \xrightarrow{\mathcal{L}} X + c.$$

Moreover, if $c \neq 0$,

$$X_n / Y_n \xrightarrow{\mathcal{L}} X/c.$$

A direct application of Slutsky's theorem is as follows. Let $\{X_1, \ldots, X_n\}$ be a sequence of i.i.d. random variables having finite variance σ^2, $n > 1$ and let

$$\bar{X}_n = \frac{\sum_{i=1}^n X_i}{n}, \quad S_n^2 = \frac{\sum_{i=1}^n \left(X_i - \bar{X}_n\right)^2}{n-1}$$

be the sample mean and sample variance respectively. Then, it can be shown that

$$\begin{aligned} Var\left[S_n^2\right] &= \frac{1}{n}\left[\mu_4 - \frac{n-3}{n-1}\sigma^4\right] \\ &= \frac{\sigma^4}{n}\left[\kappa + \frac{2n}{n-1}\right] \end{aligned}$$

where $\mu_4 = E[X-\mu]^4$ and $\kappa = \frac{\mu_4}{\sigma^4} - 3$. It follows from Chebyshev's inequality that

$$S_n^2 \xrightarrow{P} \sigma^2$$

and hence, from Slutsky's theorem,

$$\frac{\sqrt{n}\left(\bar{X}_n - \mu\right)}{S_n} = \frac{\sqrt{n}\left(\bar{X}_n - \mu\right)}{\sigma} \cdot \frac{\sigma}{S_n} \xrightarrow{\mathcal{L}} \mathcal{N}(0,1).$$

An additional result known as the Delta method enables us to extend the CLT to functions (Casella and Berger (2002), p. 243).

Theorem 2.3 (Delta Method). *Let $\{Y_n\}$ be a sequence of random variables having mean θ and finite variance σ^2 and for which*

$$\sqrt{n}\left(Y_n - \theta\right) \xrightarrow{\mathcal{L}} \mathcal{N}\left(0,\sigma^2\right), as\, n \to \infty.$$

Let g be a differentiable real valued function such that $g'(\theta) \neq 0$. Then,

$$\sqrt{n}\left(g\left(Y_n\right) - g\left(\theta\right)\right) \xrightarrow{\mathcal{L}} \mathcal{N}\left(0, \sigma^2\left[g'(\theta)\right]^2\right).$$

The proof can be obtained by applying the first order Taylor expansion to $g\left(Y_n\right)$.

Example 2.2. Let $\{X_1, \ldots, X_n\}$ be a random sample from the Bernoulli distribution for which $X_n = 1$ with probability θ and $X_n = 0$ with probability $1-\theta$. Let

$$g\left(\theta\right) = \theta\left(1-\theta\right).$$

The CLT asserts that $\frac{\sqrt{n}\left(\bar{X}_n - \theta\right)}{\sqrt{\theta(1-\theta)}} \xrightarrow{\mathcal{L}} \mathcal{N}(0,1), as\, n \to \infty$. Then using the Delta method the asymptotic distribution of $g\left(\bar{X}_n\right)$ is

$$\sqrt{n}\left(g\left(\bar{X}_n\right) - g\left(\theta\right)\right) \xrightarrow{\mathcal{L}} \mathcal{N}\left(0, \theta\left(1-\theta\right)\left[g'\left(\theta\right)\right]^2\right)$$

provided $g'\left(\theta\right) \neq 0$. We note, however, that at $\theta = \frac{1}{2}$, $g'\left(\frac{1}{2}\right) = 0$. In order to determine the asymptotic distribution of $g\left(\bar{X}_n\right)$ at $\theta = \frac{1}{2}$, we may proceed by using a second-order Taylor expansion

$$\begin{aligned} g\left(\bar{X}_n\right) &= g\left(\frac{1}{2}\right) + g'\left(\frac{1}{2}\right)\left(\bar{X}_n - \frac{1}{2}\right) + \frac{1}{2}g''\left(\frac{1}{2}\right)\left(\bar{X}_n - \frac{1}{2}\right)^2 \\ &= \frac{1}{4} + 0 - \left(\bar{X}_n - \frac{1}{2}\right)^2. \end{aligned}$$

This implies that

$$4n\left(\frac{1}{4}-g\left(\bar{X}_n\right)\right)\xrightarrow{\mathcal{L}}\chi_1^2, as\ n\to\infty.$$

A weakening of the simple CLT to the case of independent but not necessarily identically distributed random variables is given by the Lindeberg-Feller theorem.

Theorem 2.4 (Lindeberg-Feller). *Suppose that $\{X_i\}$ are independent random variables with means $\{\mu_i\}$, finite variances $\{\sigma_i^2\}$, and distribution functions $\{F_i\}$. Let $S_n=\sum_{i=1}^n (X_i-\mu_i)$, $B_n^2=\sum_{i=1}^n\sigma_i^2$ and suppose that*

$$\max_k \frac{\sigma_k^2}{B_n^2}\to 0 \quad as\ n\to\infty.$$

Then for large enough n,

$$\frac{S_n}{B_n}\xrightarrow{\mathcal{L}}\mathcal{N}(0,1)$$

if and only if for every $\epsilon>0$,

$$\frac{\sum_{i=1}^n E\left[(X_i-\mu_i)^2 I\left(|X_i-\mu_i|>\epsilon B_n\right)\right]}{B_n^2}\to 0 \quad as\ n\to\infty. \tag{2.2}$$

The Lindeberg condition (2.2) is implied by the Lyapunov condition which states that there exists a $\delta>0$ such that

$$\frac{\sum_{i=1}^n E\left|X_i-\mu_i\right|^{2+\delta}}{B_n^{2+\delta}}\to 0 \quad as\ n\to\infty.$$

Remark 2.1. Note that the condition $\max_k \sigma_k^2/B_n^2\to 0$ as $n\to\infty$ is not needed for the proof of the "if" part as it can be derived from (2.2).

Example 2.3. (Lehmann (1975), p. 351) Let $\{Y_1,\ldots,Y_n\}$ be a random sample from the Bernoulli distribution for which $Y_n=1$ with probability θ and $Y_n=0$ with probability $1-\theta$. Set $X_i=iY_i$. We would like to determine the asymptotic distribution of $\sum_{i=1}^n X_i$. We note that $\mu_i=i\theta$ and $\sigma_i^2=i^2\theta(1-\theta)$. Consequently,

$$B_n^2=\sum_{i=1}^n\sigma_i^2=\frac{n(n+1)(2n+1)}{6}\theta(1-\theta).$$

On the other hand, since

$$(X_i-\mu_i)^2=\begin{cases} i^2(1-\theta)^2 & \text{with probability }\theta\\ i^2\theta^2 & \text{with probability }1-\theta\end{cases}$$

it follows that for sufficiently large n,

$$(X_i - \mu_i)^2 < n^2,$$

and consequently,

$$\lim_{n \longrightarrow \infty} \frac{\sum_{i=1}^{n} E\left[(X_i - \mu_i)^2 I\left(|X_i - \mu_i| > \epsilon B_n\right)\right]}{B_n^2} = 0.$$

Therefore, applying Theorem 2.4, $B_n^{-1}(\sum_{i=1}^{n} X_i - \theta n(n+1)/2) \xrightarrow{\mathcal{L}} \mathcal{N}(0,1)$ for large n. We note that in this example, the Lyapunov condition would not be satisfied.

2.2. Multivariate Central Limit Theorem

Definition 2.4. A random p-vector $\boldsymbol{Y}$ is said to have a multivariate normal distribution with mean vector $\boldsymbol{\mu}$ and non-singular variance-covariance matrix $\boldsymbol{\Sigma}$ if its density is given by

$$f(\boldsymbol{y}) = (2\pi)^{-p/2} |\boldsymbol{\Sigma}|^{-1/2} \exp\left[-\frac{1}{2}(\boldsymbol{y}-\boldsymbol{\mu})' \boldsymbol{\Sigma}^{-1} (\boldsymbol{y}-\boldsymbol{\mu})\right],$$

where $|\boldsymbol{A}|$ denotes the determinant of a square matrix $\boldsymbol{A}$. We write $\boldsymbol{Y} \sim \mathcal{N}_p(\boldsymbol{\mu}, \boldsymbol{\Sigma})$.

An important companion result is the Cramér-Wold device (Serfling, 2009) which states that a random vector $\boldsymbol{Y}$ has a multivariate normal distribution if and only if every linear combination $\boldsymbol{c}'\boldsymbol{Y}$ for $\boldsymbol{c} \in \mathbb{R}^p$ has a univariate normal distribution.

Proposition 2.1. *If $\boldsymbol{Y} \sim \mathcal{N}_p(\boldsymbol{\mu}, \boldsymbol{\Sigma})$ and $\boldsymbol{Z} = \boldsymbol{A}\boldsymbol{Y} + \boldsymbol{b}$ for some $q \times p$ matrix of constants $\boldsymbol{A}$ of rank $q \leq p$, and $\boldsymbol{b}$ is a constant q-vector, then*

$$\boldsymbol{Z} \sim \mathcal{N}_q(\boldsymbol{A}\boldsymbol{\mu} + \boldsymbol{b}, \boldsymbol{A}\boldsymbol{\Sigma}\boldsymbol{A}').$$

The multivariate generalization of the simple univariate central limit theorem is given in the following theorem.

Theorem 2.5 (Multivariate central limit theorem). *Let $\{\boldsymbol{Y}_1, \ldots, \boldsymbol{Y}_n\}$ be a random sample from some p-variate distribution with mean $\boldsymbol{\mu}$ and variance-covariance matrix $\boldsymbol{\Sigma}$. Let*

$$\bar{\boldsymbol{Y}}_n = \frac{1}{n}\sum_{i=1}^{n} \boldsymbol{Y}_i$$

be the sample mean. Then, as $n \to \infty$, the asymptotic distribution of $\sqrt{n}\left(\bar{\boldsymbol{Y}}_n - \boldsymbol{\mu}\right)$ is multivariate normal with mean $\boldsymbol{0}$ and variance-covariance matrix $\boldsymbol{\Sigma}$, i.e.,

$$\sqrt{n}\left(\bar{\boldsymbol{Y}}_n - \boldsymbol{\mu}\right) \xrightarrow{\mathcal{L}} \mathcal{N}_p(\boldsymbol{0}, \boldsymbol{\Sigma}).$$

Corollary 2.1. *Let $\boldsymbol{T}$ be an $r \times p$ matrix. Then*

$$\sqrt{n}\left(\boldsymbol{T}\bar{\boldsymbol{Y}}_n - \boldsymbol{T}\boldsymbol{\mu}\right) \xrightarrow{\mathcal{L}} \mathcal{N}_r\left(\mathbf{0}, \boldsymbol{T}\boldsymbol{\Sigma}\boldsymbol{T}'\right).$$

Example 2.4 (Distribution of Quadratic Forms). We cite in this example some well-known results on quadratic forms in normal variates. Let $\boldsymbol{Y} \sim \mathcal{N}_p\left(\boldsymbol{\mu}, \boldsymbol{\Sigma}\right)$ where Σ is positive definite. Then,

(a) $\boldsymbol{AY} \sim \mathcal{N}_p\left(\boldsymbol{A\mu}, \boldsymbol{A\Sigma A'}\right)$ for a constant matrix $\boldsymbol{A}$

(b) $(\boldsymbol{Y} - \boldsymbol{\mu})' \boldsymbol{\Sigma}^{-1} (\boldsymbol{Y} - \boldsymbol{\mu}) \sim \chi^2_p$

(c) $\boldsymbol{Y}'\boldsymbol{AY} \sim \chi^2_r(\delta)$ if and only if $\boldsymbol{A\Sigma A} = \boldsymbol{A}$, that is, $\boldsymbol{A\Sigma}$ is idempotent and $rank\, \boldsymbol{A} = r$. Here, A is a symmetric $p \times p$ positive definite matrix of constants,$\chi^2_r(\delta)$ is the noncentral chi-square distribution, and $\delta = \boldsymbol{\mu}'\boldsymbol{A\mu}$ is the noncentrality parameter.

2.3. Concepts in Statistical Inference

Statistics is concerned in part with the subjects of estimation and hypothesis testing. In estimation, one is interested in both the point-wise estimation and the computation of confidence intervals for the parameters of a distribution function. In hypothesis testing one aims to verify a hypothesis about a parameter. In the next section, we briefly review some important concepts related to these two topics.

2.3.1. Parametric Estimation

An important part of statistics is the subject of estimation. On the one hand, one is interested in estimating various parameters of a distribution function which characterizes a certain physical phenomenon. For example, one may be interested in estimating the average heights of individuals in a population. It is natural then to collect a random sample of such individuals and to calculate an estimate based on that sample. On the other hand, one is also interested in determining how accurate that estimate is. We begin with the definition of the likelihood function which forms the basis of parametric estimation.

Suppose that $X_1, \ldots, X_n$ are random variables having joint density $f(x_1, \ldots, x_n; \boldsymbol{\theta})$, $\boldsymbol{\theta} \in \mathbb{R}^p$. The likelihood viewed as a function of $\boldsymbol{\theta}$ is defined to be the joint density

$$L(\boldsymbol{\theta}) = f(x_1, \ldots, x_n; \boldsymbol{\theta}). \tag{2.3}$$

Most statistical inference is concerned with using the sample to obtain knowledge about the parameter $\boldsymbol{\theta}$. It sometimes happens that a function of the sample, labeled a statistic, provides a summary of the data which is most relevant for this purpose.

Definition 2.5. We say that a statistic $T(X_1, \ldots, X_n)$ is *sufficient* for $\boldsymbol{\theta}$ if the conditional density of $X_1, \ldots, X_n$ given $T(X_1, \ldots, X_n)$ is independent of $\boldsymbol{\theta}$.

The factorization theorem characterizes this concept in the sense that T is sufficient if and only if there exists a function h of $t = T(x_1, \ldots, x_n)$ and $\boldsymbol{\theta}$ only and a function g such that

$$f(x_1, \ldots, x_n; \boldsymbol{\theta}) = h(t, \boldsymbol{\theta})\, g(x_1, \ldots, x_n).$$

The concept of sufficiency allows us to focus attention on that function of the data which contains all the important information on $\boldsymbol{\theta}$.

There are some desirable properties of estimators which provide guidance on how to choose among them. An estimator $T(X_1, \ldots, X_n)$ is said to be *unbiased* for the estimation of $g(\boldsymbol{\theta})$ if for all $\boldsymbol{\theta}$

$$E_\theta[T(X_1, \ldots, X_n)] = g(\boldsymbol{\theta}).$$

Example 2.5. Let $\{X_1, \ldots, X_n\}$ be a random sample drawn from a distribution with population mean μ and variance σ^2. It is easy to see from the properties of the expectation operator that the sample mean $\bar{X}_n = \frac{1}{n}\sum_{i=1}^n X_i$ is unbiased for μ:

$$E\left[\bar{X}_n\right] = \frac{1}{n}\sum_{i=1}^n E\left[X_i\right] = \frac{1}{n}\sum_{i=1}^n \mu = \mu.$$

Also, the sample variance $S_n^2 = \frac{1}{n-1}\sum_{i=1}^n (X_i - \bar{X}_n)^2$ is unbiased for σ^2 since S_n^2 can be reexpressed as

$$S_n^2 = \frac{1}{n-1}\left\{\sum_{i=1}^n (X_i - \mu)^2 - n(\bar{X}_n - \mu)^2\right\},$$

and

$$\begin{aligned} E\left[(X_i - \mu)^2\right] &= Var(X_i) = \sigma^2 \\ E\left[(\bar{X}_n - \mu)^2\right] &= Var(\bar{X}_n) = \frac{\sigma^2}{n}. \end{aligned}$$

A desirable property of an unbiased estimator is that it should have the smallest variance

$$E_{\boldsymbol{\theta}}\left[(T(X_1, \ldots, X_n) - g(\boldsymbol{\theta}))^2\right].$$

A further desirable property of estimators is that of consistency.

Definition 2.6. An estimator $T(X_1, \ldots, X_n)$ of a parameter $g(\boldsymbol{\theta})$ is said to be *consistent* if

$$T(X_1, \ldots, X_n) \xrightarrow{P} g(\boldsymbol{\theta}), \text{ as } n \to \infty.$$

Consistency is the minimal property required of an estimator. In order to illustrate some of these concepts, we shall consider the exponential family of distributions.

Definition 2.7. The density of a real valued random variable X is said to be a member of the *exponential family* if its density is of the form

$$f(x;\theta) = h(x) \exp[\eta(\theta) t(x) - K(\theta)]. \tag{2.4}$$

where $h(x), \eta(\theta), t(x), K(\theta)$ are known functions. We will assume for simplicity that $\eta(\theta) = \theta$.

It follows that

$$\begin{aligned} E_\theta[t(X)] &= K'(\theta) \\ Var_\theta(t(X)) &= K''(\theta). \end{aligned}$$

The density $f(x;\theta)$ is sometimes called the exponential tilting of $h(x)$ with respect to the mean K' and variance $K.''$ The exponential family includes as special cases several of the commonly used distributions, such as the normal, exponential, binomial, and Poisson.

Suppose that we have a random sample from the exponential family. It follows that the joint density is given by

$$f(x_1, \ldots, x_n; \theta) = h(x_1, \ldots, x_n) \exp\left[\theta \sum_{i=1}^{n} t(x_i) - nK(\theta)\right]$$

which can again be identified as being a member of the exponential family. An important result in the context of estimation is the following.

Theorem 2.6 (Cramr-Rao Inequality). *Suppose that $\{X_1, \ldots, X_n\}$ is a random sample from a distribution having density $f(x;\theta)$. Under certain regularity conditions which permit the exchange of the order of differentiation and integration, the variance of any estimator $T(X_1, \ldots, X_n)$ is bounded below by*

$$Var_\theta(T(X_1, \ldots, X_n)) \geq \frac{[b'(\theta)]^2}{I(\theta)}, \tag{2.5}$$

where $b(\theta) = E_\theta[T(X_1, \ldots, X_n)]$ and

$$I(\theta) = -E_\theta\left[\frac{\partial^2 \log f(X_1, \ldots, X_n; \theta)}{\partial \theta^2}\right]. \tag{2.6}$$

The expression in (2.6) is known as the *Fisher information* and it plays a key role in estimation and hypothesis testing. The regularity conditions are satisfied by members of the exponential family. Under those conditions, it can be shown that

$$E_\theta \left[\frac{\partial f (X_1, \ldots, X_n; \theta)}{\partial \theta} \right] = 0$$

and

$$I(\theta) = E_\theta \left[\left(\frac{\partial \log f (X_1, \ldots, X_n; \theta)}{\partial \theta} \right)^2 \right] = -E_\theta \left[\frac{\partial^2 \log f (X_1, \ldots, X_n; \theta)}{\partial \theta^2} \right] < \infty.$$

Example 2.6. Suppose we have a random sample from the normal distribution with mean μ and variance σ^2. Then, $\bar{X}_n$ is a consistent and unbiased estimator of the population mean μ whose variance is σ^2/n. The Fisher information can be calculated to be $n\sigma^{-2}$ and hence the Cramr-Rao lower bound for the variance of any unbiased estimator is σ^2/n. Consequently, the sample mean has the smallest variance among all unbiased estimators.

A multi-parameter version of the Cramr-Rao inequality also exists. Suppose that $\boldsymbol{\theta}$ is a p-dimensional parameter. The $p \times p$ Fisher information matrix $I(\boldsymbol{\theta})$ has the (i, j) entry:

$$I_{ij}(\boldsymbol{\theta}) = -E_{\boldsymbol{\theta}} \left[\frac{\partial^2 \log f (X_1, \ldots, X_n; \boldsymbol{\theta})}{\partial \theta_i \partial \theta_j} \right].$$

provided the derivatives exist.

Let $T(X_1, \ldots, X_n)$ be a vector-valued estimator of $\boldsymbol{\theta}$ and let

$$b(\boldsymbol{\theta}) = E_{\boldsymbol{\theta}} [T(X_1, \ldots, X_n)].$$

Then under certain regularity conditions, the multi-parameter Cramr-Rao lower bound states that in matrix notation

$$Cov_{\boldsymbol{\theta}} (T(X_1, \ldots, X_n)) \geq \left[\frac{\partial b(\boldsymbol{\theta})}{\partial \boldsymbol{\theta}} \right] [I(\boldsymbol{\theta})]^{-1} \left[\frac{\partial b(\boldsymbol{\theta})}{\partial \boldsymbol{\theta}} \right]'. \tag{2.7}$$

The matrix inequality above of the form $\boldsymbol{A} \geq \boldsymbol{B}$ is interpreted to mean that the difference $\boldsymbol{A} - \boldsymbol{B}$ is positive semi-definite. In general the regularity conditions require the existence of the Fisher information and demand that either the density function has bounded support and the bounds do not depend on θ or the density has infinite support, is continuously differentiable and its support is independent of $\boldsymbol{\theta}$.

Example 2.7. Suppose that in Example 2.6 we would like to estimate $\boldsymbol{\theta} = (\mu, \sigma^2)'$ where both the mean μ and the variance σ^2 are unknown. Let

$$T(X_1, \ldots, X_n) = \left(\bar{X}_n, S_n^2 \right)'$$

be the vector of the sample mean and sample variance respectively. We have

$$I(\boldsymbol{\theta})^{-1} = \begin{pmatrix} \frac{\sigma^2}{n} & 0 \\ 0 & \frac{2\sigma^4}{n} \end{pmatrix}.$$

It can be shown that $\bar{X}_n$ and S_n are uncorrelated and

$$Var\left(\bar{X}_n\right) = \frac{\sigma^2}{n}, \quad Var\left(S_n^2\right) = \frac{2\sigma^4}{n-1}, n > 1.$$

Consequently, we conclude that $\bar{X}_n$ attains the Cramr-Rao lower bound whereas S_n^2 does not.

By far the most popular method for finding estimators is the method of maximum likelihood developed by R.A. Fisher.

Definition 2.8. An estimator of a parameter $\boldsymbol{\theta}$, denoted $\hat{\boldsymbol{\theta}}_n$, is a *maximum likelihood estimator* if it maximizes the likelihood function $L(\boldsymbol{\theta})$ in (2.3).

Provided the derivatives exist, the maximization of the likelihood may sometimes be done by maximizing instead the logarithm of the likelihood since

$$\frac{\partial L(\boldsymbol{\theta})}{\partial \boldsymbol{\theta}} = \mathbf{0} \Longrightarrow \frac{1}{L(\boldsymbol{\theta})} \frac{\partial L(\boldsymbol{\theta})}{\partial \boldsymbol{\theta}} = \frac{\partial \log L(\boldsymbol{\theta})}{\partial \boldsymbol{\theta}} = \mathbf{0}.$$

Example 2.8. Let $\{X_1, \ldots, X_n\}$ be a random sample from a normal distribution with mean μ and variance σ^2. The log likelihood function is given by

$$\log L\left(\mu, \sigma^2\right) = -\frac{n}{2}\log(2\pi) - \frac{n}{2}\log\left(\sigma^2\right) - \frac{\sum (x_i - \mu)^2}{2\sigma^2}.$$

The maximum likelihood equations are then

$$\begin{aligned} \frac{\partial \log L\left(\mu, \sigma^2\right)}{\partial \mu} &= \frac{\sum (x_i - \mu)}{\sigma^2} = 0 \\ \frac{\partial \log L\left(\mu, \sigma^2\right)}{\partial \sigma^2} &= -\frac{n}{2\sigma^2} + \frac{\sum (x_i - \mu)^2}{2\sigma^4} = 0 \end{aligned}$$

from which we obtain the maximum likelihood estimators $\bar{X}_n$ and $\frac{(n-1)S_n^2}{n}$ for the mean and variance respectively.

Example 2.9 (Multivariate Normal Distribution). Suppose that $\{\boldsymbol{Y}_1, \ldots, \boldsymbol{Y}_n\}$ is a random sample of size n from the multivariate normal distribution $\mathcal{N}_p(\boldsymbol{\mu}, \boldsymbol{\Sigma})$. The maximum likelihood estimators of $\boldsymbol{\mu}$ and $\boldsymbol{\Sigma}$ are given respectively by

$$\bar{\boldsymbol{Y}}_n = \frac{1}{n}\sum_{i=1}^{n}\boldsymbol{Y}_i$$

$$\hat{\boldsymbol{\Sigma}} = \frac{1}{n}\sum_{i=1}^{n}\left(\boldsymbol{Y_i}-\bar{\boldsymbol{Y}}_{\boldsymbol{n}}\right)\left(\boldsymbol{Y_i}-\bar{\boldsymbol{Y}}_{\boldsymbol{n}}\right)'.$$

Example 2.10. Suppose we have a random sample $\{X_1,\ldots,X_n\}$ from a uniform distribution on the interval $(0,\theta)$. Since the range of the density depends on θ we cannot take the derivative of the likelihood function. Instead we note that the likelihood function is given by

$$L(\theta) = \begin{cases} \frac{1}{\theta^n} & \max_{1\le i\le n} X_i < \theta \\ 0 & \text{elsewhere} \end{cases}$$

and hence, the maximum likelihood estimator of θ is $\max\limits_{1\le i\le n} X_i$.

Example 2.11 (Multinomial Distribution)**.** Consider a generalization of the binomial distribution which allows for trial one of M possible categorical outcomes. Suppose that at the kth trial we observe the jth outcome, $k = 1,\ldots,n, j = 1,\ldots,M$. Let $\boldsymbol{X}_k = (0,\ldots,0,1,0,\ldots,0)$ be the M-dimensional vector which records of 1 at the jth position if jth categorical outcome is observed and 0 otherwise. Let the frequency vector $\boldsymbol{N} = (N_1,\ldots,N_k)$ denote the number of occurrences of the categories in n repetitions of the same experiment conducted under independent and identical conditions (i.e., i.i.d.). It follows that

$$\boldsymbol{N} = \sum_{k=1}^{n}\boldsymbol{X}_k.$$

Denote the probability vector by $\boldsymbol{p} = (p_1,\ldots,p_M)$ where p_j is the probability of observing the jth outcome. The probability distribution associated with $\boldsymbol{N}$ is given by

$$P(N_j = n_j, j = 1,\ldots,M) = \frac{n!}{n_1\ldots,n_M}p_1^{n_1}\ldots p_M^{n_M},\quad \sum_{j=1}^{M}p_j = 1,\ \sum_{j=1}^{M}n_j = n,$$

and it is called the *multinomial distribution.* The $\{\boldsymbol{X}_k\}$ are i.i.d. with covariance matrix having (i,j) entry σ_{ij} given by

$$\sigma_{ij} = \begin{cases} p_i(1-p_i) & i = j \\ -p_ip_j & i \neq j \end{cases}$$

Also the covariance matrix of $\boldsymbol{N}$ is not of full rank and is given in matrix notation by

$$\Sigma = Cov\left(\boldsymbol{N}\right) = n\left[\left(diag\left(\mathbf{p}\right)\right) - \mathbf{p}\mathbf{p}'\right].$$

It can be shown that for large n,

$$\frac{1}{\sqrt{n}}\left(\boldsymbol{N} - n\boldsymbol{p}\right) \to \boldsymbol{\mathcal{N}}\left(0, \Sigma\right).$$

2.3.2. Hypothesis Testing

Suppose that $\{X_1, \ldots, X_n\}$ is a random sample from a distribution where the density of X is given by $f\left(x; \boldsymbol{\theta}\right), \boldsymbol{\theta} \in \mathbb{R}^p$. In hypothesis testing it is common to formulate two hypotheses about $\boldsymbol{\theta}$: the null hypothesis denoted H_0 represents the status quo (no change). The alternative, denoted H_1, represents the hypothesis that we are hoping to accept. To illustrate, suppose that we are interested in assessing whether or not a new drug represents an effective treatment compared to a placebo. The null hypothesis states that the new drug is not more effective than a placebo, whereas the alternative states that it is. Since the assessment is made on the basis of evidence contained in a random sample, there is the possibility of error. A Type I error results when one falsely rejects the null hypothesis in favor of the alternative. A Type II error occurs when one falsely accepts the null hypothesis when it is false. In our example, a Type I error means we would incorrectly adopt an ineffective drug whereas a Type II error means we incorrectly forgo an effective drug. Both types of error cannot simultaneously be controlled. It is customary to prescribe a bound on the probability of committing the Type I error and to then look for tests that minimize the Type II error. The Neyman-Pearson lemma makes concrete these ideas.

Formally we are interested in testing the null hypothesis

$$H_0 : \boldsymbol{\theta} \in \Theta_0$$

against the alternative hypothesis

$$H_1 : \boldsymbol{\theta} \in \Theta_1,$$

where Θ_0 and Θ_1 are subsets of the parameter space. In the situation where both Θ_0, Θ_1 consist of single points θ_0 and θ_1 respectively, the Neyman-Pearson lemma provides an optimal solution. Set $\boldsymbol{x} = (x_1, \ldots, x_n)$ and let $\phi\left(\boldsymbol{x}\right)$ be a critical function which represents the probability of rejecting the null hypothesis when $\boldsymbol{x}$ is observed. Also let α denote the prescribed probability of rejecting the null hypothesis when it is true (also known as the size of the test).

Lemma 2.4 (Neyman-Pearson Lemma). *Suppose that there exists some dominating measure μ with respect to which we have densities $f(x;\theta_0)$ and $f(x;\theta_1)$. Then the most powerful test of $H_0 : \theta = \theta_0$ against $H_1 : \theta = \theta_1$ is given by*

$$\phi(\boldsymbol{x}) = \begin{cases} 1 & \text{if } \frac{\prod_{i=1}^n f(x_i;\theta_1)}{\prod_{i=1}^n f(x_i;\theta_0)} > k \\ \gamma & \text{if } \frac{\prod_{i=1}^n f(x_i;\theta_1)}{\prod_{i=1}^n f(x_i;\theta_0)} = k \\ 0 & \text{if } \frac{\prod_{i=1}^n f(x_i;\theta_1)}{\prod_{i=1}^n f(x_i;\theta_0)} < k \end{cases}$$

where k is chosen such that $E_{\theta_0}[\phi(\boldsymbol{X})] = \alpha$. The power function of the test ϕ is defined to be

$$E_\theta[\phi(\boldsymbol{X})] = \int \phi(\boldsymbol{X}) \prod_{i=1}^n f(x_i;\theta)\, dx_1 \ldots dx_n$$

and it represents the probability of rejecting the null hypothesis for a given θ.

Example 2.12. Given a random sample $\{X_1, \ldots, X_n\}$ from a normal distribution with unknown mean μ and known variance σ^2, suppose that we are interested in testing the null hypothesis

$$H_0 : \mu = \mu_0$$

against the alternative hypothesis

$$H_1 : \mu = \mu_1 > \mu_0.$$

Then it can be shown that the uniformly most powerful test is given by

$$\phi(\boldsymbol{x}) = \begin{cases} 1 & \bar{X}_n > k \\ \gamma & \bar{X}_n = k \\ 0 & \bar{X}_n < k \end{cases}$$

where $k = \mu_0 + z_\alpha \frac{\sigma}{\sqrt{n}}$, and z_α is the upper α-point of the standard normal distribution.

Example 2.13. In the case of a random sample from an exponential distribution with mean θ, describing the lifetimes of light bulbs, the uniformly most powerful test of

$$H_0 : \theta = \theta_0$$

against the alternative hypothesis

$$H_1 : \theta = \theta_1 < \theta_0.$$

is given by

$$\phi(\boldsymbol{x}) = \begin{cases} 1 & \bar{X}_n > k \\ \gamma & \bar{X}_n = k \\ 0 & \bar{X}_n < k \end{cases}$$

where k is a solution of $\Gamma(n, nk/\theta_0) = \alpha(n-1)!$. Here, $\Gamma(a,b)$ is an upper incomplete gamma function defined as $\int_b^\infty u^{a-1}e^{-u}\,du$.

Uniformly most powerful (UMP) tests rarely exist in practice. An exception occurs when the family of densities possesses a monotone likelihood ratio.

Definition 2.9. We shall say that a family of densities $\{f(x;\theta), \theta \in \Theta\}$ has monotone likelihood ratio if the ratio $\frac{f(x;\theta_2)}{f(x;\theta_1)}$ is nondecreasing in some function $T(x)$ for all $\theta_1 < \theta_2$ in some interval Θ.

The exponential family (2.4) has likelihood ratio

$$\frac{f(x;\theta_2)}{f(x;\theta_1)} = \exp\{(\theta_2 - \theta_1)t(x) - (K(\theta_2) - K(\theta_1))\}$$

which is nondecreasing in x provided $t(x)$ is nondecreasing. In the situation where the family of densities has monotone likelihood ratio, it can be shown that there exists uniformly most powerful tests (Ferguson (1967), p. 210). Specifically, a test ϕ_0 is uniformly most powerful of size α if it has size α and if the power function satisfies

$$E_\theta[\phi_0(\boldsymbol{X})] \geq E_\theta[\phi(\boldsymbol{X})], \theta \in \Theta_1$$

As an application, we consider testing for the mean of a normal distribution the hypothesis

$$H_0 : \mu = \mu_0$$

against the alternative hypothesis

$$H_1 : \mu > \mu_0.$$

The family has monotone likelihood ratio in x. The uniformly most powerful test of

$$H_0 : \mu = \mu_0$$

against the alternative hypothesis

$$H_1 : \mu = \mu_1 > \mu_0$$

for μ_1 fixed is given by the Neyman-Pearson lemma becomes

$$\phi(\boldsymbol{x}) = \begin{cases} 1 & \bar{X}_n > k \\ \gamma & \bar{X}_n = k \\ 0 & \bar{X}_n < k \end{cases}$$

This test has nondecreasing power function

$$\Phi\left(-z_\alpha + \frac{\sqrt{n}(\mu - \mu_0)}{\sigma}\right) \geq \alpha, \text{ for } \mu > \mu_0.$$

and hence it is uniformly most powerful when the alternative is the composite $\mu > \mu_0$.

A generalization of the Neyman-Pearson lemma when both Θ_0 and Θ_1 are composite sets is based on the likelihood ratio

$$\Lambda_n = \frac{\sup_{\theta \in \Theta_0} \prod_{i=1}^n f(x_i; \theta)}{\sup_{\theta \in \Theta} \prod_{i=1}^n f(x_i; \theta)} \quad \text{with } \Theta = \Theta_0 \cup \Theta_1.$$

The likelihood ratio test rejects the null hypothesis whenever Λ_n is small enough. The factorization theorem shows that the likelihood ratio Λ_n is based on the sufficient statistic and moreover, it is invariant under transformations of the parameter space that leave the hypothesis and alternative hypotheses invariant. For a random sample of size n from the exponential family (2.4) we have

$$\log \Lambda_n = n \sup_{\theta_0 \in \Theta_0} \inf_{\theta \in \Theta} \left[(\theta_0 - \theta)\bar{t}_n - K(\theta_0) + K(\theta)\right],$$

where $\bar{t}_n = \sum_{i=1}^n t(x_i)/n$.

In certain situations, as in the case of a Cauchy location family, a uniformly powerful test may not exist. On the other hand, a *locally most powerful test* which maximizes the power function at θ_0 may exist. Provided we may differentiate under the integral sign the power function with respect to θ we see upon using the generalized Neyman-Pearson lemma (see Ferguson (1967), p. 204) that a test of the form

$$\phi(\boldsymbol{x}) = \begin{cases} 1 & \text{if } \frac{\partial}{\partial\theta} \prod_{i=1}^n f(x_i; \theta_0) > k \prod_{i=1}^n f(x_i; \theta_0) \\ \gamma & \text{if } \frac{\partial}{\partial\theta} \prod_{i=1}^n f(x_i; \theta_0) = k \prod_{i=1}^n f(x_i; \theta_0) \\ 0 & \text{if } \frac{\partial}{\partial\theta} \prod_{i=1}^n f(x_i; \theta_0) < k \prod_{i=1}^n f(x_i; \theta_0) \end{cases}$$

maximizes the power function at θ_0 among all tests of size α.

Definition 2.10. Let $\{X_1, \ldots, X_n\}$ be a random sample from some distribution having density $f(x;\boldsymbol{\theta}), \boldsymbol{\theta} \in \mathbb{R}^p$. Let $L(\boldsymbol{\theta};\boldsymbol{x})$ be the likelihood function where $\boldsymbol{x} = (x_1, \ldots, x_n)'$. Let

$$\ell(\boldsymbol{\theta};\boldsymbol{x}) = \log L(\boldsymbol{\theta};\boldsymbol{x})$$

The derivative

$$U(\boldsymbol{\theta};\boldsymbol{x}) = \frac{\partial \ell(\boldsymbol{\theta};\boldsymbol{x})}{\partial \boldsymbol{\theta}}$$

is called the score function.

The locally most powerful test can be seen to be equivalently based on the score function since

$$\begin{aligned} U(\boldsymbol{\theta};\boldsymbol{x}) &= \frac{\partial}{\partial \boldsymbol{\theta}}\left[\log \prod_{i=1}^{n} f(x_i;\boldsymbol{\theta}_0)\right] \\ &= \frac{\frac{\partial}{\partial \theta} \prod_{i=1}^{n} f(x_i;\boldsymbol{\theta}_0)}{\prod_{i=1}^{n} f(x_i;\boldsymbol{\theta}_0)}. \end{aligned}$$

2.3.2.1. The Three "Amigos"

Lemma 2.5. *Let $U(\boldsymbol{\theta};\boldsymbol{X}) = \frac{\partial}{\partial \theta}\ell(\boldsymbol{\theta};\boldsymbol{X})$ where $\boldsymbol{\theta}$ is possibly a vector. Denote by $f(\boldsymbol{x};\boldsymbol{\theta})$ the joint density of $X_1, \ldots, X_n$. Then, if we can differentiate under the integral signs below, the following properties hold:*

(i) $E_\theta[U(\boldsymbol{\theta};\boldsymbol{X})] = 0$ and

(ii) $Cov_\theta[U(\boldsymbol{\theta};\boldsymbol{X})] = I(\boldsymbol{\theta})$

Proof. (i) Note that

$$\begin{aligned} E_\theta[U(\boldsymbol{\theta};\boldsymbol{X})] &= \int \frac{\partial \ell(\boldsymbol{\theta};\boldsymbol{x})}{\partial \boldsymbol{\theta}} f(\boldsymbol{x};\boldsymbol{\theta})\, d\boldsymbol{x} \\ &= \int \frac{\partial f(\boldsymbol{x};\boldsymbol{\theta})}{\partial \boldsymbol{\theta}} d\boldsymbol{x} \\ &= \frac{\partial}{\partial \boldsymbol{\theta}} \int f(\boldsymbol{x};\boldsymbol{\theta})\, d\boldsymbol{x} = 0 \end{aligned}$$

(ii) The (i,j) term of the matrix $Cov_\theta[U(\boldsymbol{\theta};\boldsymbol{X})]$ is given by $E_\theta\left[\frac{\partial \ell(\boldsymbol{\theta};\boldsymbol{X})}{\partial \theta_i}\frac{\partial \ell(\boldsymbol{\theta};\boldsymbol{X})}{\partial \theta_j}\right]$. Since

$$\begin{aligned} 0 &= \frac{\partial}{\partial \theta_i} \int \frac{\partial \ell(\boldsymbol{\theta};\boldsymbol{x})}{\partial \theta_j} f(\boldsymbol{x};\boldsymbol{\theta})\, d\boldsymbol{x} \\ &= \int \frac{\partial^2 \ell(\boldsymbol{\theta};\boldsymbol{x})}{\partial \theta_i \partial \theta_j} f(\boldsymbol{x};\boldsymbol{\theta})\, d\boldsymbol{x} + \int \frac{\partial \ell(\boldsymbol{\theta};\boldsymbol{x})}{\partial \theta_j} \frac{\partial}{\partial \theta_i} f(\boldsymbol{x};\boldsymbol{\theta})\, d\boldsymbol{x} \end{aligned}$$

$$= \int \frac{\partial^2 \ell(\boldsymbol{\theta}; \boldsymbol{x})}{\partial \theta_i \partial \theta_j} f(\boldsymbol{x}; \boldsymbol{\theta}) \, d\boldsymbol{x} + \int \frac{\partial \ell(\boldsymbol{\theta}; \boldsymbol{x})}{\partial \theta_j} \frac{\partial \ell(\boldsymbol{\theta}; \boldsymbol{x})}{\partial \theta_j} f(\boldsymbol{x}; \boldsymbol{\theta}) \, d\boldsymbol{x}$$

the result follows.

□

For any hypothesis testing problem, there are three distinct possible test statistics: the likelihood ratio test, the Wald test, and the Rao score test, all of which are asymptotically equivalent as the sample size gets large. In the lemma below, we outline the proof for the asymptotic distribution of the Rao score test.

Lemma 2.6 (Score Test). *Let $\{X_1, \ldots, X_n\}$ be a random sample from a continuous distribution having density $f(x; \boldsymbol{\theta}), \boldsymbol{\theta} \in \mathbb{R}^p$ and suppose we wish to test*

$$H_0 : \boldsymbol{\theta} \in \Theta_0$$

against the alternative

$$H_1 : \boldsymbol{\theta} \in \Theta_1.$$

Let $\hat{\boldsymbol{\theta}}_0$ be the maximum likelihood estimate of $\boldsymbol{\theta}$ under H_0. The test statistic under H_0

$$U'\left(\hat{\boldsymbol{\theta}}_0; \boldsymbol{X}\right) I^{-1}\left(\hat{\boldsymbol{\theta}}_0\right) U\left(\hat{\boldsymbol{\theta}}_0; \boldsymbol{X}\right)$$

has, as the sample size gets large, *under the null hypothesis a χ^2_k* distribution where k *is the number of constraints imposed by the null hypothesis.*

Proof. The result follows from the multivariate central limit theorem since the score is a sum of independent identically distributed random vectors

$$\begin{aligned} U(\boldsymbol{\theta}; \boldsymbol{X}) &= \frac{\partial}{\partial \boldsymbol{\theta}} \left[\log \prod_{i=1}^{n} f(X_i; \boldsymbol{\theta}) \right] \\ &= \sum_{i=1}^{n} \left[\frac{\partial}{\partial \boldsymbol{\theta}} \log f(X_i; \boldsymbol{\theta}) \right] \end{aligned}$$

with mean $\mathbf{0}$ and covariance matrix $I(\boldsymbol{\theta})$. □

Theorem 2.7 (The Three Amigos). *Let $\boldsymbol{X} = \{X_1, \ldots, X_n\}$ be a random sample from a continuous distribution having density $f(x; \boldsymbol{\theta}), \boldsymbol{\theta} \in \Theta$, the parameter space. Suppose we are interested in testing the general null hypothesis $H_0 : \boldsymbol{\theta} \in \Theta_0$ against the alternative $H_1 : \boldsymbol{\theta} \in \Theta_1 = \Theta - \Theta_0$. Then, the likelihood test, the Wald test, and the score test all reject the null hypothesis for large values and are asymptotically distributed as central chi-square distributions with k degrees of freedom as $n \to \infty$ where k is the number of constraints imposed by the null hypothesis. Specifically,*

(a) The likelihood ratio test rejects whenever,

$$-2\log\left[\frac{L\left(\hat{\boldsymbol{\theta}}_0;\boldsymbol{X}\right)}{L\left(\hat{\boldsymbol{\theta}};\boldsymbol{X}\right)}\right] > \chi^2_k(\alpha);$$

(b) The Wald test rejects whenever

$$W = \left(\hat{\boldsymbol{\theta}}-\boldsymbol{\theta}_0\right)' I\left(\hat{\boldsymbol{\theta}}\right)\left(\hat{\boldsymbol{\theta}}-\boldsymbol{\theta}_0\right) > \chi^2_k(\alpha);$$

(c) The score test rejects whenever

$$U'\left(\hat{\boldsymbol{\theta}}_0;\boldsymbol{X}\right) I^{-1}\left(\hat{\boldsymbol{\theta}}_0\right) U\left(\hat{\boldsymbol{\theta}}_0;\boldsymbol{X}\right) > \chi^2_k(\alpha),$$

where $\hat{\boldsymbol{\theta}}_0$ and $\hat{\boldsymbol{\theta}}$ are the maximum likelihood estimates of $\boldsymbol{\theta}$ under H_0 and $H_0 \cup H_1$ respectively.

Proof. By expanding the likelihood $L\left(\hat{\boldsymbol{\theta}};\boldsymbol{X}\right)$ around $\boldsymbol{\theta} \in \Theta_0$ in a second order Taylor series, it can be shown that these three tests are equivalent (see van der Vaart (2007), p. 231 or Cox and Hinkley (1974), p. 315 & p. 324). □

We note that it is possible to substitute the matrix of second partials of the log likelihood evaluate at $\hat{\boldsymbol{\theta}}$ for the theoretical Fisher information $I\left(\hat{\boldsymbol{\theta}}\right)$. The score test is used since it may be easier to maximize the log likelihood subject to constraints as we shall see later on. The likelihood ratio test requires computation of both $\hat{\boldsymbol{\theta}}_0$ and $\hat{\boldsymbol{\theta}}$. The Wald test requires computation of $\hat{\boldsymbol{\theta}}$ whereas the score test requires computation of $\hat{\boldsymbol{\theta}}_0$. All of these tests can be inverted to provide confidence regions for $\boldsymbol{\theta}$. In that case, the Wald test will lead to ellipsoidal regions. Finally, we see that the Wald test can be used to test a single hypothesis on multiple parameters as for example $H_0 : \boldsymbol{\theta} = \boldsymbol{\theta}_0$ or $H_0 : \boldsymbol{A\theta} = \boldsymbol{b}$.

We illustrate the three tests in the following examples.

Example 2.14. Let $\{X_1, \ldots, X_n\}$ be a random sample from the Poisson distribution with mean $\theta > 0$ given by

$$f(x;\theta) = \frac{e^{-\theta}\theta^x}{x!}, x = 0, 1, \ldots, \theta > 0.$$

Suppose we wish to test

$$H_0 : \theta = \theta_0$$

against the alternative

$$H_1 : \theta \neq \theta_0.$$

Then the score function evaluated at θ_0 is

$$U\left(\theta_0; \bar{X}\right) = \frac{n}{\theta_0}\left(\bar{X} - \theta_0\right)$$

whereas the Fisher information is given by $I(\theta) = \frac{n}{\theta}$. Hence the score test rejects whenever

$$\frac{\theta_0}{n}\left[\frac{n}{\theta_0}\left(\bar{X} - \theta_0\right)\right]^2 = \frac{n}{\theta_0}\left(\bar{X} - \theta_0\right)^2 > \chi_1^2(\alpha)$$

The Wald test rejects whenever

$$\frac{n}{\bar{X}}\left(\bar{X} - \theta_0\right)^2 > \chi_1^2(\alpha).$$

The likelihood ratio test rejects the null hypothesis whenever

$$-2\log\frac{L(\theta_0; \boldsymbol{X})}{L\left(\hat{\theta}; \boldsymbol{X}\right)} > \chi_1^2(\alpha) \tag{2.8}$$

It can be seen that

$$\log\frac{L(\theta_0; \boldsymbol{x})}{L\left(\hat{\theta}; \boldsymbol{x}\right)} = n\bar{x}\log\left(\frac{\theta_0}{\bar{x}}\right) - n\left(\theta_0 - \bar{x}\right) \tag{2.9}$$

Example 2.15. Let $\{X_1, \ldots, X_n\}$ be a random sample from the normal distribution with mean μ and variance σ^2. Suppose we wish to test

$$H_0 : \mu = \mu_0$$

against the alternative

$$H_1 : \mu \neq \mu_0.$$

Then the score function evaluated at μ_0 is

$$U\left(\mu_0; \bar{X}\right) = \frac{n}{\sigma^2}\left(\bar{X} - \mu_0\right)$$

whereas the Fisher information is given by $I(\theta) = \frac{n}{\sigma^2}$. Hence the score test rejects whenever

$$\frac{\sigma^2}{n}\left[\frac{n}{\sigma^2}\left(\bar{X} - \mu_0\right)\right]^2 = \frac{n}{\sigma^2}\left(\bar{X} - \mu_0\right)^2 > \chi_1^2(\alpha).$$

Since σ^2 is unknown, we may replace it by its maximum likelihood estimator $\frac{n-1}{n}S_n^2$ and hence the score test rejects whenever

$$\frac{n^2}{n-1}\left(\frac{\bar{X}-\mu_0}{S_n}\right)^2 > \chi_1^2(\alpha).$$

Notes

1. The score test geometrically represents the slope of the log-likelihood function evaluated at $\hat{\boldsymbol{\theta}}_0$. It is locally most powerful for small deviations from $\boldsymbol{\theta}_0$. This can be seen from a Taylor series expansion of the likelihood function $L(\boldsymbol{\theta}_0 + \boldsymbol{h})$ around $\boldsymbol{\theta}_0$ for small values of the vector $\boldsymbol{h}$.

2. The Wald test is not invariant under re-parametrization unlike the score test which is (Stein, 1956). It is easy to construct confidence intervals using the Wald statistic. However, since the standard error is calculated under the alternative, it may lead to poor confidence intervals.

3. Both the likelihood ratio test and the Wald test are asymptotically equivalent to the score test under the null as well as under Pitman alternative hypotheses (Serfling (2009), p. 155). All the statistics have limiting distributions that are weighted chi-squared if the model is misspecified. See (Lindsay and Qu, 2003) for a more extensive discussion of the score test and related statistics.

4. An interesting application of the score test is given by (Jarque and Bera (1987)) who consider the Pearson family of distributions.

In the next section we study the concept of contiguity and its consequences.

2.3.2.2. Contiguity

In the previous section, we were concerned with determining the asymptotic distribution of the test statistics under the null hypothesis. It is of interest for situations related to the efficiency of tests, to determine the asymptotic distribution of the test statistic under the alternative hypothesis as well. Since most of the tests considered in practice are consistent, they tend to perform well for alternatives "far away" from the null hypothesis. Consequently, the interest tends to focus on alternatives which are "close" to the null hypothesis in a manner to be made more precisely here. The concept of contiguity was introduced by Le Cam in connection with the study of local asymptotic normality. Specifically, the concept enables one to obtain the limiting distribution of a sequence of statistics under the alternative distribution from knowledge of the limiting distribution under the null hypothesis distribution. We follow closely the development and notation in (Hájek and Sidak (1967); van der Vaart (2007)) and we begin with some definitions.

Definition 2.11. Suppose that P and Q are two probability measures defined on the same measurable space $(\Omega, \mathcal{A})$. We shall say that Q is absolutely continuous with respect to P, denoted $Q \ll P$ if $Q(A) = 0$ whenever $P(A) = 0$ for all measurable sets $A \in \mathcal{A}$.

If $Q \ll P$, then we may compute the expectation of a function $f(\boldsymbol{X})$ of a random vector $\boldsymbol{X}$ under Q from knowledge of the expectation under P of the product $\left[f(\boldsymbol{X}) \frac{dQ}{dP}\right]$ through the change of measure formula

$$E_Q[f(\boldsymbol{X})] = E_P\left[f(\boldsymbol{X}) \frac{dQ}{dP}\right]$$

The expression $\left[\frac{dQ}{dP}\right]$ is known as the Radon-Nikodym derivative. The asymptotic version of absolute continuity is the concept of contiguity.

Definition 2.12. Suppose that P_n and Q_n are two sequences of probability measures defined by the same measurable space $(\Omega_n, \mathcal{A}_n)$. We shall say that Q_n is contiguous with respect to P_n, denoted $Q_n \lhd P_n$ if $Q_n(A_n) \to 0$ whenever $P_n(A_n) \to 0$ for all measurable sets $A_n \in \mathcal{A}_n$. The sequences P_n, Q_n are mutually contiguous if both $Q_n \lhd P_n$ and $P_n \lhd Q_n$.

Example 2.16. Suppose that under P_n we have a standard normal distribution whereas under Q_n we have a normal distribution with mean $\mu_n \to \mu$, and variance $\sigma^2 > 0$. Then it follows that P_n and Q_n are mutually contiguous. Note however that if $\mu_n \to \infty$, then for $A_n = [\mu_n, \mu_n + 1]$, we have that $P_n(A_n) \to 0$, yet $Q_n(A_n) \to$constant.

Definition 2.13. Le Cam proposed three lemmas which enable us to verify contiguity and to obtain the limiting distribution under one measure given the limiting distribution under another. Suppose that P_n and Q_n admit densities p_n, q_n respectively and define the likelihood ratio L_n for typical points $\boldsymbol{x}$ in the sample space

$$L_n(\boldsymbol{x}) = \begin{cases} \frac{q_n(\boldsymbol{x})}{p_n(\boldsymbol{x})} & p_n(\boldsymbol{x}) > 0 \\ 1 & p_n(\boldsymbol{x}) = q_n(\boldsymbol{x}) = 0 \\ 0 & p_n(\boldsymbol{x}) = 0 < q_n(\boldsymbol{x}) \end{cases}$$

Lemma 2.7 (Le Cam's First Lemma). *Let F_n be the cdf of L_n which converges weakly under P_n at continuity points to a distribution F for which*

$$\int_0^\infty x dF(x) = 1.$$

Then the densities q_n are contiguous to the densities $p_n, n \geq 1$.

As a corollary, if $\log L_n$ is under P_n asymptotically normal with mean $-\frac{\sigma^2}{2}$ and variance σ^2, then the densities q_n are mutually contiguous to the densities p_n. This follows from the fact that the moment generating function of a normal (μ, σ^2) evaluated at $t = 1$ must equal 1; that is for large n,

$$
\begin{aligned}
E\left[e^{\log L_n}\right] &\rightarrow e^{\mu+\frac{\sigma^2}{2}} \\
&= e^{-\frac{\sigma^2}{2}+\frac{\sigma^2}{2}} = 1.
\end{aligned}
$$

Le Cam's second lemma provides conditions under which a log likelihood ratio is asymptotically normal under the hypothesis probability measure.

Let $\boldsymbol{x} = (x_1, \ldots, x_n)'$ and

$$p_n(\boldsymbol{x}) = \prod_{i=1}^{n} f_{ni}(x_i)$$

and

$$q_n(\boldsymbol{x}) = \prod_{i=1}^{n} g_{ni}(x_i)$$

Then

$$\log L_n(\boldsymbol{X}) = \sum_{i=1}^{n} \log\left[g_{ni}(X_i)/f_{ni}(X_i)\right]$$

Let

$$W_n = 2\sum_{i=1}^{n}\left\{\left[g_{ni}(X_i)/f_{ni}(X_i)\right]^{\frac{1}{2}} - 1\right\}$$

Lemma 2.8 (Le Cam's Second Lemma). *Suppose that the following uniform integrability condition holds*

$$\lim_{n\to\infty} \max_{1\le i\le n} P_n\left(\left|g_{ni}(X_i)/f_{ni}(X_i) - 1\right| > \varepsilon\right) = 0, \text{ for all } \varepsilon > 0$$

and W_n is asymptotically $\mathcal{N}\left(-\frac{\sigma^2}{4}, \sigma^2\right)$ under P_n. Then

$$\lim_{n\to\infty} P_n\left(\left|\log L_n(\boldsymbol{X}) - W_n + \frac{\sigma^2}{4}\right| > \varepsilon\right) = 0, \varepsilon > 0$$

and under P_n, $\log L_n(\boldsymbol{X})$ is asymptotically $\mathcal{N}\left(-\frac{\sigma^2}{2}, \sigma^2\right)$.

The third lemma which is most often utilized establishes the asymptotic distribution under the alternative hypothesis provided the measures are contiguous.

Lemma 2.9 (Le Cam's Third Lemma). *Let P_n, Q_n be sequences of probability measures on measurable spaces $(\Omega_n, \mathcal{A}_n)$ such that $Q_n \lhd P_n$. Let Y_n be a sequence of k-dimensional* *random vectors such that under* P_n

$$(Y_n, \log L_n(\boldsymbol{X})) \xrightarrow{\mathcal{L}} \mathcal{N}_{k+1}\left(\begin{pmatrix} \mu \\ -\frac{\sigma^2}{2} \end{pmatrix}, \begin{pmatrix} \Sigma & \tau \\ \tau' & \sigma^2 \end{pmatrix}\right).$$

Then, under Q_n

$$Y_n \xrightarrow{\mathcal{L}} \mathcal{N}_k(\mu + \tau, \Sigma)$$

We may now consider an important application of Le Cam's lemmas. Suppose that we have two sequences of probability measures defined by densities $p_\theta, p_{\theta+h/\sqrt{n}}$.

Let $\ell_{\boldsymbol{\theta}}(x) = \log p_{\boldsymbol{\theta}}(x)$. Then under suitable regularity conditions, a second order Taylor series expansion yields under $p_{\boldsymbol{\theta}}$

$$\log\left[\prod_i \left(\frac{p_{\boldsymbol{\theta}+\boldsymbol{h}/\sqrt{n}}(X_i)}{p_{\boldsymbol{\theta}}(X_i)}\right)\right] = \frac{1}{\sqrt{n}}\sum_{i=1}^{n} \boldsymbol{h}'\dot{\ell}_{\boldsymbol{\theta}}(X_i) - \frac{\boldsymbol{h}'\boldsymbol{I}_{\boldsymbol{\theta}}\boldsymbol{h}}{2} + o(1) \tag{2.10}$$

where $\dot{\ell}_{\boldsymbol{\theta}}(X_i)$ is the k-dimensional vector of partial derivatives. The expansion in (2.10) is known as the local asymptotic normality property (LAN). It follows from the central limit theorem that the right-hand side in (2.10) is asymptotically normal $\mathcal{N}\left(-\frac{\boldsymbol{h}'\boldsymbol{I}_{\boldsymbol{\theta}}\boldsymbol{h}}{2}, \boldsymbol{h}'\boldsymbol{I}_{\boldsymbol{\theta}}\boldsymbol{h}\right)$ and consequently, by Le Cam's first lemma, the measures $p_{\boldsymbol{\theta}}, p_{\boldsymbol{\theta}+\boldsymbol{h}/\sqrt{n}}$ on the left-hand side are contiguous. Next, suppose that we have a sequence of test statistics T_n which are to first order, sums of i.i.d. random variables:

$$\sqrt{n}(T_n - \mu) = \frac{1}{\sqrt{n}}\sum_{i=1}^{n} \psi(X_i) + o_{p_{\boldsymbol{\theta}}}(1)$$

for some function ψ. Consequently, the joint distribution of $\frac{1}{\sqrt{n}}\sum_i \left(\psi(X_i), \dot{\ell}_{\boldsymbol{\theta}}(X_i)\right)$ is under p_θ asymptotically multivariate normal

$$\left(\sqrt{n}(T_n - \mu), \log\left[\prod_i \left(\frac{p_{\boldsymbol{\theta}+\boldsymbol{h}/\sqrt{n}}(X_i)}{p_{\boldsymbol{\theta}}(X_i)}\right)\right]\right) \xrightarrow{\mathcal{L}} \mathcal{N}_{k+1}\left(\begin{pmatrix} \boldsymbol{0} \\ -\frac{\boldsymbol{h}'\boldsymbol{I}_{\boldsymbol{\theta}}\boldsymbol{h}}{2} \end{pmatrix}, \begin{pmatrix} \boldsymbol{\Sigma} & \boldsymbol{\tau} \\ \boldsymbol{\tau}' & \boldsymbol{h}'\boldsymbol{I}_{\boldsymbol{\theta}}\boldsymbol{h} \end{pmatrix}\right)$$

It then follows from Le Cam's third lemma that $\sqrt{n}(T_n - \mu)$ is asymptotically normal under $p_{\boldsymbol{\theta}+\boldsymbol{h}/\sqrt{n}}$.

We may demonstrate contiguity in the case of an exponential family of distributions.

Example 2.17. Let $\boldsymbol{X} = (X_1, \ldots, X_n)'$ be a random sample from the exponential family for a real valued parameter θ

$$f(x;\theta) = \exp[\theta t(x) - K(\theta)]\, g(x)$$

The log likelihood and its derivatives are respectively

$$\begin{aligned}
\ell(\theta; \boldsymbol{X}) &= \left[\theta \sum_{i=1}^{n} t(X_i) - nK(\theta)\right] \\
\ell'(\theta; \boldsymbol{X}) &= \left[\sum_{i=1}^{n} t(X_i) - nK'(\theta)\right] \\
\ell''(\theta; \boldsymbol{X}) &= -nK''(\theta)
\end{aligned}$$

It follows that

$$\begin{aligned}
\log\left[\prod_{i=1}^{n}\left(\frac{p_{\theta+h/\sqrt{n}}(X_i)}{p_\theta(X_i)}\right)\right] &= \frac{h}{\sqrt{n}}\sum_{i=1}^{n} t(X_i) - n\left[K\left(\theta + \frac{h}{\sqrt{n}}\right) - K(\theta)\right] \\
&= \frac{h}{\sqrt{n}}\sum_{i=1}^{n}\left[t(X_i) - K'(\theta)\right] - \frac{h^2}{2}K''(\theta) + o(1)
\end{aligned}$$

and consequently, the measures $p_\theta = f(\boldsymbol{x}; \theta), p_{\theta+h/\sqrt{n}} = f\left(\boldsymbol{x}; \theta + \frac{h}{\sqrt{n}}\right)$ are contiguous.

The concept of contiguity will enable us to obtain the non-null asymptotic distribution of various linear rank statistics to be defined in later chapters.

2.3.3. Composite Likelihood

Composite likelihood is an approach whereby one can combine together by multiplying several component likelihoods each of which is either a conditional or a marginal density. The components are not necessarily independent and consequently, the resulting product itself may or may not be a likelihood function in its own right. The advantage of using composite likelihood methods is that it facilitates the modeling of dependence in smaller dimensions. We review below some of the basic properties. For a more detailed review, we refer the reader to Varin et al. (2011).

Let $\boldsymbol{X}$ be an m-dimensional random vector with density $f_X(\boldsymbol{x}; \boldsymbol{\theta})$ where $\boldsymbol{\theta}$ is a k-dimensional vector parameter taking values in $\boldsymbol{\Theta}$. Let $\{\mathcal{A}_1, \ldots, \mathcal{A}_K\}$ be a set of marginal or conditional events (subsets of the sample space) such that the corresponding likelihood function is given by

$$L_k(\boldsymbol{\theta}; \boldsymbol{x}) = f(\boldsymbol{x} \in \mathcal{A}_k; \boldsymbol{\theta}).$$

A composite likelihood is then defined to be the weighed product

$$L_C = \prod_{k=1}^{K} (L_k(\boldsymbol{\theta}; \boldsymbol{x}))^{w_k}$$

where the $\{w_k\}$ are nonnegative weights selected to improve efficiency. The composite log-likelihood is

$$cl(\boldsymbol{\theta};\boldsymbol{x}) = \sum_{k=1}^{K} w_k \ell_k(\boldsymbol{\theta};\boldsymbol{x})$$

with

$$\ell_k(\boldsymbol{\theta};\boldsymbol{x}) = \log\, L_k(\boldsymbol{\theta};\boldsymbol{x}).$$

In the simplest case of independence, we have a genuine likelihood function

$$\mathrm{L}_{\mathrm{ind}}(\boldsymbol{\theta};\boldsymbol{x}) = \prod_{\mathrm{r}=1}^{\mathrm{m}} f_{\mathrm{X}}(\mathrm{x}_{\mathrm{r}};\boldsymbol{\theta}).$$

This composite likelihood leads to inference on the parameters $\boldsymbol{\theta}$ as detailed in the previous sections. In cases where our interest is in parameters involving dependence, we shall consider likelihoods which for a fixed $\boldsymbol{x}$ are based on the pairwise differences

$$\prod_{r=1}^{m}\prod_{s\neq r}^{m} f_X(x_r - x_s;\boldsymbol{\theta}).$$

Similarly, we may be interested in likelihoods of the form

$$L_{diff}(\boldsymbol{\theta};\boldsymbol{x}) = \prod_{r=1}^{m-1}\prod_{s=r+1}^{m} f_X(x_r - x_s;\boldsymbol{\theta}).$$

Some asymptotic theory is available when we have a random sample. Let $\hat{\boldsymbol{\theta}}_{CL}$ be the maximum with respect to $\boldsymbol{\theta}$ of the composite log likelihood $cl(\boldsymbol{\theta};\boldsymbol{x})$. Define the following quantities.

$$U(\boldsymbol{\theta};\boldsymbol{x}) = \frac{\partial}{\partial\theta} cl(\boldsymbol{\theta};\boldsymbol{x}) \tag{2.11}$$

$$H(\boldsymbol{\theta}) = E_{\boldsymbol{\theta}}\left\{-\frac{\partial}{\partial\boldsymbol{\theta}} U(\boldsymbol{\theta};\boldsymbol{X})\right\} \tag{2.12}$$

$$J(\boldsymbol{\theta}) = var_{\boldsymbol{\theta}}\{U(\boldsymbol{\theta};\boldsymbol{X})\} \tag{2.13}$$

$$G(\boldsymbol{\theta}) = H(\boldsymbol{\theta})\, J(\boldsymbol{\theta})^{-1} H(\boldsymbol{\theta}) \tag{2.14}$$

$$I(\boldsymbol{\theta}) = var_{\boldsymbol{\theta}}\left\{\frac{\partial}{\partial\boldsymbol{\theta}}\log f(\boldsymbol{X};\boldsymbol{\theta})\right\} \tag{2.15}$$

where $G(\boldsymbol{\theta})$ is the Godambe information matrix in a single observation. If $cl(\boldsymbol{\theta};\boldsymbol{x})$ is a true log-likelihood, then $G(\boldsymbol{\theta}) = I(\boldsymbol{\theta})$. Then under some regularity conditions on the log densities, we have that as $n \to \infty$,

$$\sqrt{n}\left(\hat{\boldsymbol{\theta}}_{CL} - \boldsymbol{\theta}\right) \xrightarrow{\mathcal{L}} \mathcal{N}_p\left(0, G^{-1}(\boldsymbol{\theta})\right),$$

where $\mathcal{N}_p\left(\boldsymbol{\mu}, G^{-1}(\boldsymbol{\theta})\right)$ is a p-dimensional normal distribution with vector mean $\boldsymbol{\mu}$ and variance-covariance matrix $G^{-1}(\boldsymbol{\theta})$.

2.3.4. Bayesian Methods

In traditional parametric statistical inference, the probability joint density function of a random sample $\boldsymbol{X} = (X_1, \ldots, X_n)'$, say $f(\boldsymbol{x}|\theta)$ is a function of a parameter θ assumed to be an unknown but fixed constant. A radical departure from this paradigm consists of treating θ as a random variable itself with its own prior distribution function given by $p(\theta; \alpha)$ where α is a prespecified hyper-parameter. The marginal density of $\boldsymbol{X}$ is calculated to be

$$\int f(\boldsymbol{x}|\theta)\, p(\theta; \alpha)\, d\theta$$

and the conditional density of θ given $\boldsymbol{X} = \boldsymbol{x}$, labeled the posterior density of θ is given by Bayes' theorem:

$$p(\theta|\boldsymbol{x}) = \frac{f(\boldsymbol{x}|\theta)\, p(\theta; \alpha)}{\int f(\boldsymbol{x}|\theta)\, p(\theta; \alpha)\, d\theta}$$

In view of the factorization theorem in Section 2.3.1, the posterior density is always a function of the sufficient statistic. The use of Bayesian methods enables us to update information about the prior. There have been countless applications of Bayesian inference in practice. We refer the reader to Box and Tiao (1973) for further details. We provide below some simple examples, whereas in Part III of this book, we consider a modern application of Bayesian methods to ranking data.

Example 2.18. Let $\boldsymbol{x} = \{x_1, \ldots, x_n\}$ be a set of observed data randomly drawn from a normal distribution with mean θ and variance σ^2. Assume that θ is itself a random variable having a normal prior distribution with unknown mean μ and known variance τ^2. Then the posterior distribution of θ given $\boldsymbol{x}$ is proportional to

$$\begin{aligned}
p(\theta|\boldsymbol{x}) &\approx \prod_{i=1}^{n} f(x_i|\theta)\, p(\theta; \alpha) \\
&\approx \exp\left\{-\frac{n\sigma^{-2}}{2}(\bar{x}_n - \theta)^2\right\} \exp\left\{-\frac{\tau^{-2}}{2}(\theta - \mu)^2\right\} \\
&\approx \exp\left\{-\frac{\tau_n^{-2}}{2}(\theta - \mu_n)^2\right\}
\end{aligned}$$

where

$$\begin{aligned} \mu_n &= \frac{n\sigma^{-2}\bar{x}_n + \tau^{-2}\mu}{n\sigma^{-2} + \tau^{-2}} \\ \tau_n^{-2} &= n\sigma^{-2} + \tau^{-2} \end{aligned}$$

We recognize therefore this posterior to be a normal density with mean μ_n and variance τ_n^2.

We provide another example below.

Example 2.19. Let $\{X_1, \ldots, X_n\}$ be a random sample from a Bernoulli distribution with parameter $0 < \theta < 1$:

$$f(x|\theta) = \theta^x (1-\theta)^{1-x}, x = 0, 1.$$

Suppose that the prior for θ is the Beta distribution with parameters $\alpha > 0, \beta > 0$

$$p(\theta; \alpha, \beta) = \frac{\Gamma(\alpha+\beta)}{\Gamma(\alpha)\Gamma(\beta)} \theta^{\alpha-1} (1-\theta)^{\beta-1}, 0 < \theta < 1.$$

Then the posterior distribution of θ given $\boldsymbol{x}$ is again a Beta distribution but with parameters

$$\alpha_n = \alpha + \sum x_i, \quad \beta_n = \beta + n - \sum x_i.$$

In any given problem involving Bayesian methods, the specification of the prior and the consequent computation of the posterior may pose some difficulties. In certain problems, this difficulty is overcome if the prior and the posterior come from the same family of distributions. The prior in that case is called a conjugate prior . In the previous example, both the prior and the posterior distributions were normal distributions but with different parameters. In many modern problems not involving conjugate priors, the Bayesian computation of the posterior distribution is a challenging task. The goal in many instances is to compute the expectation of some function $h(\theta)$ with respect to the posterior distributions:

$$E[h(\theta)|\boldsymbol{x}] = \int h(\theta) p(\theta|\boldsymbol{x}) d\theta = \frac{\int h(\theta) f(x|\theta) p(\theta; \alpha) d\theta}{\int f(x|\theta) p(\theta; \alpha) d\theta}$$

We encounter such integrals for example when one is interested in the posterior probability that $h(\theta)$ lies in an interval. If one can draw a random sample $\theta^{(1)}, \theta^{(2)}, \ldots, \theta^{(m)}$ from $p(\theta|\boldsymbol{x})$, the strong law of large numbers guarantees that $E[h(\theta)|\boldsymbol{x}]$ is well approximated by the sample mean of $h(\theta)$: $\frac{1}{m}\sum_{i=1}^m h(\theta^{(i)})$ if m is large enough. What happens if $p(\theta|\boldsymbol{x})$ is hard to sample? One way of proceeding is to approximate the posterior distribution by a multivariate normal density centered at the mode of θ obtained through an optimization method (see Albert (2008) p. 94). Another approach is to generate samples from the posterior distribution of θ indirectly via various simulation methods such as importance sampling, rejection sampling, and Markov Chain Monte Carlo (MCMC) methods. We describe these methods below.

1. **Importance Sampling.**

   When it is hard to sample from $p(\theta|\boldsymbol{x})$ directly, one can resort to importance sampling. Suppose we are able to sample from another distribution $q(\theta)$. Then

$$E_p[h(\theta)] = \int h(\theta)p(\theta|\boldsymbol{x})d\theta = \int h(\theta)\frac{p(\theta|\boldsymbol{x})}{q(\theta)}q(\theta)d\theta = E_q[h(\theta)w(\theta)],$$

   with a weight $w(\theta) = p(\theta|\boldsymbol{x})/q(\theta)$. One can draw a random sample $\theta^{(1)}, \theta^{(2)}, \ldots, \theta^{(m)}$ from $q(\theta)$ and $E[h(\theta)|\boldsymbol{x}]$ can be approximated by $\widehat{h(\theta)} = \frac{1}{m}\sum_{i=1}^{m} h(\theta^{(i)})w(\theta^{(i)})$. In practice, $q(\theta)$ should be chosen so that it is easy to be sampled and can achieve a small estimation error. For Monte Carlo error, one can choose $q(\theta)$ to minimize the variance of $\widehat{h(\theta)}$, see Robert and Casella (2004).

2. **Rejection Sampling.**

   Rejection sampling consists of identifying a proposal density say $q(\theta)$ which "resembles" the posterior density with respect to location and spread and which is easy to sample from. As well, we ask that for all θ and a constant c,

$$p(\theta|\boldsymbol{x}) \le cq(\theta)$$

   Rejection sampling then proceeds by repeatedly generating independently a random variable from both q and a uniformly distributed random variable U on the interval $(0, 1)$. We accept θ as coming from the posterior if and only if

$$U \le \frac{p(\theta|\boldsymbol{x})}{cq(\theta)}.$$

   We may justify the procedure as follows. Suppose in general that Y has density $f_Y(y)$ and V has density $f_V(v)$ with common support with

$$M = \sup_y \frac{f_Y(y)}{f_V(y)}$$

   To generate a value from f_Y, we consider the following rejection algorithm:

   (a) Generate instead a uniform variable U and the variable V from f_V independently.

   (b) If $U < \frac{1}{M}\frac{f_Y(V)}{f_V(V)}$, set $Y = V$. Otherwise, return to (a).

It is easy to see that the value of Y generated from this rejection algorithm is distributed as f_Y:

$$\begin{aligned} P(Y \leq y) &= P\left(V \leq y | U < \frac{1}{M}\frac{f_Y(V)}{f_V(V)}\right) \\ &= \frac{P\left(V \leq y, U < \frac{1}{M}\frac{f_Y(V)}{f_V(V)}\right)}{P\left(U < \frac{1}{M}\frac{f_Y(V)}{f_V(V)}\right)} \\ &= \frac{\int_{-\infty}^{y}\int_{0}^{\frac{1}{M}\frac{f_Y(v)}{f_V(v)}} f_V(v)\,dudv}{\int_{-\infty}^{\infty}\int_{0}^{\frac{1}{M}\frac{f_Y(v)}{f_V(v)}} f_V(v)\,dudv} \\ &= \int_{-\infty}^{y} f_Y(v)\,dv. \end{aligned}$$

3. **Markov Chain Monte Carlo Methods (MCMC).**

- The *Gibbs Sampling* algorithm was developed by Geman and Geman (1984). See Casella and George (1992) for an intuitive exposition of the algorithm. This algorithm helps generate random variables indirectly from the posterior distribution without relying on its density. The following describes the algorithm.

  Suppose we have to estimate m parameters $\theta_1, \theta_2, \ldots, \theta_{m-1}$ and θ_m. Let $p(\theta_i \mid \boldsymbol{x}, \theta_1, \theta_2, \ldots, \theta_{i-1}, \theta_{i+1}, \ldots, \theta_m)$ be the full conditional distribution of θ_i and $\theta_i^{(t)}$ be the random variate of θ_i simulated in the tth iteration. Then the procedure of Gibbs Sampling is

  (a) Set initial values for the m parameters $\{\theta_1^{(0)}, \theta_2^{(0)}, \ldots, \theta_m^{(0)}\}$.

  (b) Repeat the steps below. At the tth iteration,

  (1) draw $\theta_1^{(t)}$ from $p(\theta_1^{(t)} \mid \boldsymbol{x}, \theta_2^{(t-1)}, \theta_3^{(t-1)}, \theta_4^{(t-1)}, \ldots, \theta_m^{(t-1)})$

  (2) draw $\theta_2^{(t)}$ from $p(\theta_2^{(t)} \mid \boldsymbol{x}, \theta_1^{(t)}, \theta_3^{(t-1)}, \theta_4^{(t-1)}, \ldots, \theta_m^{(t-1)})$

  (3) draw $\theta_3^{(t)}$ from $p(\theta_3^{(t)} \mid \boldsymbol{x}, \theta_1^{(t)}, \theta_2^{(t)}, \theta_4^{(t-1)}, \ldots, \theta_m^{(t-1)})$

  ⋮

  (m) draw $\theta_m^{(t)}$ from $p(\theta_m^{(t)} \mid \boldsymbol{x}, \theta_1^{(t)}, \theta_2^{(t)}, \theta_3^{(t)}, \ldots, \theta_{m-1}^{(t)})$

  Tierney (1994) showed that if this process is repeated many times, the random variates drawn will converge to a single variable drawn from the joint posterior distribution of $\theta_1, \theta_2, \ldots, \theta_{m-1}$ and θ_m given $\boldsymbol{x}$. Suppose that the process is repeated $M + N$ times. In practice, one can discard the first M iterates as the burn-in period. In order to eliminate the autocorrelations of the iterates, one in every a observations is kept in the last N iterates. The choice of M can be determined by examining the plot of traces of the Gibbs iterates for stationary of the Gibbs iterates. The value of a could be selected by examining the autocorrelation plots of the Gibbs iterates.

- The *Metropolis-Hastings* (M-H) algorithm was developed by Metropolis et al. (1953) and subsequently generalized by Hastings (1970). The algorithm gives rise to the Gibbs sampler as a special case (Gelman, 1992). This algorithm is similar to that of the acceptance-rejection method which either accepts or rejects the candidate except that the value of the current state is retained when rejection occurred in the M-H algorithm. For more details, see Liang et al. (2010).

  When a random variable θ follows a *nonstandard* posterior distribution $p(\theta|\boldsymbol{x})$, the M-H algorithm helps to simulate from the posterior distribution indirectly. In this algorithm, we need to specify a proposed density $q(\theta|\theta^{(t-1)})$ from which we can generate the next θ given $\theta^{(t-1)}$. The procedure of the Metropolis-Hastings algorithm is as follows.

  (1) Set a starting value $\theta^{(0)}$ and $t = 1$.

  (2) Draw a candidate random variate θ^* from $q(\theta|\theta^{(t-1)})$.

  (3) Accept $\theta^{(t)} = \theta^*$ with the acceptable probability $\min(1, R)$, where

  $$R = \frac{p\left(\theta^*|\boldsymbol{x}\right) q\left(\theta^{(t-1)}|\theta^*\right)}{p\left(\theta^{(t)}|\boldsymbol{x}\right) q\left(\theta^*|\theta^{(t-1)}\right)}.$$

  Otherwise, set $\theta^{(t)} = \theta^{(t-1)}$. Update $t \leftarrow t + 1$.

  (4) If steps (2) and (3) are repeated for a large number of times, the random variates come from the posterior distribution $p(\theta|\boldsymbol{x})$. In practice, we repeat steps (2) and (3) for $M + N$ times, discard the first M iterates and then select one in every ath iterate to reduce the autocorrelations of the random variates.

2.3.5. Variational Inference

Variational inference can provide a deterministic approximation to an intractable posterior distribution. Suppose the posterior distribution $p(\theta|\boldsymbol{x})$ to be sampled is intractable. The basic idea is to pick an approximation $q(\theta)$ chosen from some tractable family, such as a Gaussian distribution, and then to try to make q as close to $p\left(\theta|\boldsymbol{x}\right)$ as possible. The variational Bayesian (VB) method defines the "closeness" in terms of the Kullback-Liebler (KL) divergence:

$$KL(q(\theta)|p(\theta|\boldsymbol{x})) = \int q(\theta) \log \frac{q(\theta)}{p(\theta|\boldsymbol{x})} d\theta.$$

However, KL may be hard to compute as evaluating $p(\theta|\boldsymbol{x})$ requires evaluating the intractable normalizing constant $p(\boldsymbol{x})$. It is easy to show that

$$KL(q(\theta)|p(\theta|\boldsymbol{x})) = -L(q) + \log p(\boldsymbol{x}), \quad L(q) = \int q(\theta) \log \frac{p(\boldsymbol{x}, \theta)}{q(\theta)} d\theta.$$

Since the KL divergence is always nonnegative, $L(q)$ is the lower bound for the model log-likelihood $\log p(x)$. So minimizing the KL divergence is equivalent to maximizing $L(q)$ as $\log p(\boldsymbol{x})$ is a constant.

One of the most popular forms of VB is the so-called mean-field approximation which factorizes q into independent factors of the form:

$$q(\theta) = \prod_i q_i(\theta_i).$$

To retain the possible dependency among the θ_i's, one may adopt a structural factorization of $q(\theta)$ as

$$q(\theta) = q_1(\theta_1) q_2(\theta_2|\theta_1) q_3(\theta_3|\theta_2, \theta_1).$$

Under either the mean-field approximation or the structural factorization, we can convert the problem of minimizing the KL divergence with respect to the joint distribution $q(\theta)$ to that of minimizing KL divergences with respect to individual univariate distribution q_i's. For an application to an angle-based ranking model, see Chapter 11.

2.4. Exercises

Exercise 2.1. Suppose that the conditional density of Y given μ is normal with mean μ and variance σ^2. Suppose that the distribution of μ is normal with mean m and variance τ^2. Show that $\frac{Y-E[Y]}{\sqrt{Var(Y)}}$ is also normally distributed.

Exercise 2.2. Let $\{X_1, X_2, \ldots, X_n\}$ be a random sample of size n from a cumulative distribution $F(x)$ with density $f(x)$.

(a) Find the joint density of $X_{(i)}, X_{(j)}$

(b) Find the distribution of the range $X_{(n)} - X_{(1)}$ when F is given by the uniform on the interval $(0, 1)$.

(c) Find the distribution of the sample median when n is an odd integer.

Exercise 2.3. Suppose that in Exercise 2.2, $F(x)$ is the exponential cumulative distribution with mean 1.

(a) Show that the pairwise differences $X_{(i)} - X_{(i-1)}$ for $i = 2, \ldots, n$ are independent.

(b) Find the distribution of the $X_{(i)} - X_{(i-1)}$.

Exercise 2.4. Consider the usual regression model

$$Y_i = \boldsymbol{x}_i'\boldsymbol{\beta} + e_i, i = 1, \ldots, n,$$

where $\boldsymbol{\beta}$ and $\boldsymbol{x}_i$ are $p \times 1$ vectors and $\{e_i\}$ are i.i.d. $\mathcal{N}(0, \sigma^2)$ random error terms. Suppose we wish to test

$$H_0 : \boldsymbol{A\beta} = 0$$

against

$$H_1 : \boldsymbol{A\beta} \neq 0,$$

where $\boldsymbol{A}$ is a $k \times p$ known matrix.

(a) Obtain the likelihood ratio test.

(b) Obtain the Wald test.

(c) Obtain the Rao score test.

Exercise 2.5. By using a Taylor series expansion of $\log(1+\delta)$ in (2.9) for $\delta = \left(\frac{\theta_0}{\bar{x}} - 1\right)$, show that the likelihood ratio test in Example 2.13 is asymptotically equivalent to the Wald test.

Exercise 2.6. Suppose that we have a random sample of size n from a Bernoulli distribution with mean θ. Suppose that we impose a Beta(α, β) prior distribution on θ. Find the posterior distribution of θ.

Exercise 2.7. Suppose that we have a random sample of size n from some distribution having density $f(x|\theta)$ conditional on θ. Suppose that there is a prior density $g(\theta)$ on θ. We would like to estimate θ using square error loss, $L(\theta, a) = (\theta - a)^2$.

(a) Show that the mean of the posterior distribution minimizes the Bayes risk $EL(\theta, a)$;

(b) Find the mean of the posterior distribution when

$$f(x|\theta) = \theta e^{-\theta x}, x > 0, \theta > 0$$

and

$$g(\theta) = \frac{\beta^\alpha}{\Gamma(\alpha)} x^{\alpha-1} e^{-\beta\theta}, \theta > 0, \alpha > 0, \beta > 0$$

is the Gamma density.

Exercise 2.8. Suppose that X has a uniform density on the interval $\left(\theta - \frac{1}{2}, \theta + \frac{1}{2}\right)$.

(a) Show that for a random sample of size n, the estimator $\frac{X_{(n)}+X_{(1)}}{2}$ has mean square error $\frac{1}{2(n+2)(n+1)}$ where $X_{(n)}, X_{(1)}$ are respectively the maximum and minimum of the sample.

(b) Show that the estimator $X_{(n)} - \frac{1}{2}$ is consistent and has mean square error $\frac{2}{(n+2)(n+1)}$.

Exercise 2.9. Suppose that under P_n we have a uniform distribution on the interval $[0, 1]$ whereas under Q_n the distribution is uniform on the interval $[a_n, b_n]$ with $a_n \to 0, b_n \to 1$. Show that P_n, Q_n are mutually contiguous.

3. Tools for Nonparametric Statistics

Nonparametric statistics is concerned with the development of distribution free methods to solve various statistical problems. Examples include tests for a monotonic trend, or tests of hypotheses that two samples come from the same distribution. One of the important tools in nonparametric statistics is the use of ranking data. When the data is transformed into ranks, one gains the simplicity that objects may be more easily compared. For example, if web pages are ranked in accordance to some criterion, one obtains a quick summary of choices. In this chapter, we will study linear rank statistics which are functions of the ranks. These consist of two components: regression coefficients and a score function. The latter allows one to choose the function of the ranks in order to obtain an "optimal score" while the former is tailor made to the problem at hand. Two central limit theorems, one due to Hajek and Sidak and the other to Hoeffding, play important roles in deriving asymptotic results for statistics which are linear functions of the ranks.

The subject of U statistics has played an important role in studying global properties of the underlying distributions. Many well-known test statistics including the mean and variance of a random sample are U statistics. Here as well, Hoeffding's central limit theorem for U statistics plays a dominant role in deriving asymptotic distributions. In the next sections we describe in more detail some of these key results.

3.1. Linear Rank Statistics

The subject of linear rank statistics was developed by (Hájek and Sidak (1967)). In this section we mention briefly some of the important properties. Let $\{X_1, \ldots, X_n\}$ constitute a random sample from some continuous distribution. The rank of X_i denoted R_i is defined to be the ith largest value among the $\{X_j\}$. A functional definition can be given as

$$R_i = 1 + \sum_{j=1}^{n} I\left[X_i - X_j\right],$$

M. Alvo, P. L. H. Yu, *A Parametric Approach to Nonparametric Statistics*,
Springer Series in the Data Sciences, https://doi.org/10.1007/978-3-319-94153-0_3

where

$$I[u] = \begin{cases} 1 & u > 0, \\ 0 & u < 0. \end{cases}$$

We begin with the definition of a linear rank statistic.

Definition 3.1. Let $X_1, \ldots, X_n$ constitute a random sample from some continuous distribution and let $R_1, \ldots, R_n$ represent the corresponding ranks. A statistic of the form

$$S_n = \sum_{i=1}^{n} a(i, R_i)$$

is called a linear rank statistic where the $\{a(i,j)\}$ represents a matrix of values. The statistic is called a simple linear rank statistic if

$$a(i, R_i) = c_i a(R_i)$$

where $\{c_1, \ldots, c_n\}$ and $\{a(1), \ldots, a(n)\}$ are given vectors of real numbers. The $\{c_i\}$ are called regression coefficients whereas the $\{a_i\}$ are labeled scores.

To illustrate why the $\{c_i\}$ are called regression coefficients, suppose that $X_1, \ldots, X_n$ are a set of independent random variables such that

$$X_i = \alpha + \Delta c_i + \varepsilon_i,$$

where $\{\varepsilon_i\}$ are i.i.d. continuous random variables with cdf F having median 0. We are interested in testing hypotheses about the slope Δ. Denoting the ranks of the differences $\{X_i - \Delta c_i\}$ by $R_1, \ldots, R_n$ and assuming that $c_1 \leq c_2 \leq \ldots \leq c_n$ we see that the usual sample correlation coefficient for the pairs $(c_i, a(R_i))$ is

$$\frac{\sum_i (c_i - \bar{c})(a(R_i) - \bar{a})}{\sqrt{\left[\sum_i (c_i - \bar{c})^2 \sum_i (a(R_i) - \bar{a})^2\right]}},$$

where $a(.)$ is an increasing function and $\bar{c}$ and $\bar{a}$ are respectively the means of the $\{c_i\}$and$\{a(R_i)\}$. Some simplification shows this is a simple linear rank statistic.

Many test statistics can be expressed as simple linear rank statistics. For example, suppose that we have a random sample of n observations with corresponding ranks $\{R_i\}$ and we would like to test for a linear trend. In that case, we may use the simple linear rank statistic

$$S_n = \sum_{i=1}^{n} i R_i$$

which correlates the ordered time scale with the observed ranks of the data.

Table 3.1.: Two-sample Problems of location

	a_i
Wilcoxon	$\left(i - \frac{n+1}{2}\right)$
Terry-Hoeffding	$E\left(V_{(i)}\right)$
Van der Warden	$\Phi^{-1}\left(\frac{i}{n+1}\right)$

Table 3.2.: Two-sample Problems of scale

	a_i
Mood	$\left(i - \frac{n+1}{2}\right)^2$
Freund-Ansari-Bradley	$\left\lvert i - \frac{n+1}{2}\right\rvert$
Siegel-Tukey	$\begin{cases} 2i & \text{i even, } 1 < i \leq \frac{n}{2} \\ 2i-1 & \text{i odd, } 1 \leq i < \frac{n}{2} \\ 2(n-i)+2 & \text{i even, } \frac{n}{2} < i \leq n \\ 2(n-i)+1 & \text{i odd, } \frac{n}{2} < i < n \end{cases}$
Klotz	$\left[\Phi^{-1}\left(\frac{i}{n+1}\right)\right]^2$
Normal	$\left[E\left(V_{(i)}\right)\right]^2$

In another example, suppose we have a random sample of size n from one population (X) and $N-n$ from another (Y). We are interested in testing the null hypothesis that the two populations are the same against the alternative that they differ in location. We may rank all the N observations together and retain only the ranks of the second population (Y) denoted $R_i, i = n+1, \ldots, N$, by choosing

$$c_i = \begin{cases} 0 & i = 1, \ldots, n \\ 1 & i = n+1, \ldots, N. \end{cases}$$

The test statistic takes the form $T_n = \sum_{i=n+1}^{N} c_i R_i$. Properties of T_n under the null and alternative hypotheses are given in Gibbons and Chakraborti (2011).

We may generalize this test statistic by defining functions $\{a(R_i), i = 1, \ldots, N\}$ of the ranks and choosing

$$T_n = \sum_{i=n+1}^{N} c_i a(R_i).$$

These functions may be chosen to reflect emphasis either on location or on scale. For both the problems of location and scale, Tables 3.1 and 3.2 list values of the constants for location and scale alternatives. Here, $V_{(i)}$ represents the ith order statistic from a standard normal with cumulative distribution function $\Phi(x)$. We shall see in later chapters that this results in the well-known two-sample Wilcoxon statistic.

Lemma 3.1. *Suppose that the set of rankings $(R_1, \ldots, R_n)$ are uniformly distributed over the set of $n!$ permutations of the integers 1,2,...,n. Let $S_n = \sum_{i=1}^{n} c_i a(R_i)$. Then,*

(a) for $i = 1, \ldots, n$,

$$E[R_i] = \frac{n+1}{2}, Var(R_i) = \frac{(n^2-1)}{12}$$

and for $i \neq j$,

$$Cov(R_i, R_j) = -\frac{n+1}{12}.$$

(b) $E[S_n] = n\bar{c}\bar{a}$, where $\bar{c} = \sum_{i=1}^{n} c_i/n$ and $\bar{a} = \sum_{i=1}^{n} a(i)/n$.

(c) $Var(S_n) = \frac{1}{n-1}\sum_{i=1}^{n}(c_i - \bar{c})^2 \sum_{i=1}^{n}(a(i) - \bar{a})^2$.

(d) Let $T_n = \sum_{i=1}^{n} d_i b(R_i)$, for regression coefficients $\{d_i\}$ and score function $b(.)$. Then

$$Cov(S_n, T_n) = \sigma_{ab} \sum_{i=1}^{n} (c_i - \bar{c})\left(d_i - \bar{d}\right),$$

where $\bar{d} = \sum_{i=1}^{n} d_i/n$ and $\sigma_{ab} = \frac{1}{n-1}\sum_{i=1}^{n}(a(i) - \bar{a})(b(i) - \bar{b})$.

Proof. The proof of this lemma is straightforward and is left as an exercise for the reader. □

Example 3.1. The simple linear rank statistic $S_n = \sum_{i=1}^{n} iR_i$ can be used for testing the hypothesis that two continuous random variables are independent. Suppose that we observe a random sample of pairs $(X_i, Y_i), i = 1, \ldots, n$. Replace the $X's$ by their ranks, say $R_1, \ldots, R_n$ and the $Y's$ by their ranks, say $Q_1, \ldots, Q_n$. A nonparametric measure of the correlation between the variables is the Spearman correlation

$$\begin{aligned}\rho &= \frac{\sum_i (R_i - \bar{R})(Q_1 - \bar{Q})}{\sqrt{\sum_i (R_i - \bar{R})^2 \sum_i (Q_1 - \bar{Q})^2}} \\ &= \frac{\sum_i \left(i - \frac{n+1}{2}\right)\left(R_i - \frac{n+1}{2}\right)}{\sum_i \left(i - \frac{n+1}{2}\right)^2}\end{aligned}$$

Under the hypothesis of independence we have from Lemma 3.1 (a),

$$E[\rho] = 0, Var(\rho) = 1.$$

Under certain conditions, linear rank statistics are asymptotically normally distributed. We shall consider square integrable functions ϕ defined on the interval $(0,1)$ which have the property that they can be written as the difference between two nondecreasing functions and satisfy

$$0 < \int_0^1 \left[\phi(u) - \bar{\phi}\right]^2 du < \infty$$

where $\bar{\phi} = \int_0^1 \phi(u)\, du$. An important result for proving limit theorems is the Projection Theorem.

Theorem 3.1 (Projection Theorem). *Let $T(R_1, \ldots, R_n)$ be a rank statistic and set*

$$\hat{a}(i,j) = E\left[T | R_i = j\right]$$

Then the projection of T into the family of linear rank statistics is given by

$$\hat{T} = \frac{n-1}{n} \sum_{i=1}^{n} \hat{a}(i, R_i) - (n-2)\, E[T].$$

Proof. The proof of this theorem is given in (Hájek and Sidak (1967), p. 59). □

Example 3.2 (Hájek and Sidak (1967), p. 60). Suppose that $R_1, \ldots, R_n$ are uniformly distributed over the set of $n!$ permutations of the integers $1, 2, \ldots, n$. The projection of the Kendall statistic defined by

$$\tau = \frac{\sum_{i \neq j} \operatorname{sgn}(i-j) \operatorname{sgn}(R_i - R_j)}{n(n-1)}$$

into the family of linear rank statistics is given by

$$\begin{aligned} \hat{\tau} &= \frac{8 \sum \left(i - \frac{n+1}{2}\right)\left(R_i - \frac{n+1}{2}\right)}{n^2(n-1)} \\ &= \frac{2}{3}\left(\frac{n+1}{n}\right)\rho. \end{aligned} \tag{3.1}$$

The proof proceeds by noting $E[\hat{\tau}] = 0$ and

$$E\left[\operatorname{sgn}(R_i - R_h) | R_k = j\right] = \begin{cases} 0 & k \neq i, k \neq h \\ \frac{2j-n-1}{n-1} & k = i, 1 \leq j \leq n \\ \frac{n+1-2j}{n-1} & k = h, 1 \leq j \leq n \end{cases}$$

and hence

$$\begin{aligned} E\left[\hat{\tau}|R_k = j\right] &= \frac{1}{n(n-1)}\left[\sum_h \operatorname{sgn}(k-h)\frac{2j-n-1}{n-1} + \sum_i \operatorname{sgn}(i-k)\frac{n+1-2j}{n-1}\right] \\ &= \frac{8\sum_i \left(i-\frac{n+1}{2}\right)\left(R_i - \frac{n+1}{2}\right)}{n(n-1)^2}. \end{aligned}$$

The result now follows from the projection theorem.

Theorem 3.2. *Suppose that $R_1, \ldots, R_n$ are uniformly distributed over the set of $n!$ permutations of the integers 1,2,...,n. Let the score function be given by any one of the following three functions:*

$$\begin{aligned} a(i) &= \phi\left(\frac{i}{n+1}\right) \\ a(i) &= n\int_{(i-1/n)}^{i/n} \phi(u)\, du \\ a(i) &= E\left[\phi\left(U_n^{(i)}\right)\right] \end{aligned}$$

where ϕ is a square integrable function on $(0,1)$ and $U_n^{(1)} < \ldots < U_n^{(n)}$ are the order statistics from a sample of n uniformly distributed random variables. Let

$$S_n = \sum_{i=1}^n c_i a(R_i).$$

Then, provided the following Noether condition holds

$$\frac{\sum_{i=1}^n (c_i - \bar{c})^2}{max_{i\leq n}(c_i - \bar{c})^2} \longrightarrow \infty,$$

as $n \longrightarrow \infty$,

$$\frac{S_n - n\bar{c}\bar{a}}{\sigma_n} \xrightarrow{\mathcal{L}} \mathcal{N}(0,1)$$

where

$$\begin{aligned} \sigma_n^2 &= \frac{1}{n-1}\sum_{i=1}^n (c_i - \bar{c})^2 \sum (a_i - \bar{a})^2 \\ &\approx \sum_{i=1}^n (c_i - \bar{c})^2 \int_0^1 \left[\phi(u) - \bar{\phi}\right]^2 du. \end{aligned}$$

Proof. The details of the proof of this theorem, given in (Hájek and Sidak (1967), p. 160), consist in showing that the $\left\{\frac{R_i}{n+1}\right\}$ behave asymptotically like a random sample of uniformly distributed random variables and then that S_n is equivalent to a sum of independent random variables to which we can apply the Lindeberg-Feller central limit theorem (Theorem 2.4). □

The asymptotic distribution of various linear rank statistics under the alternative was obtained using the concept of contiguity (Hájek and Sidak (1967) which we further describe in Chapter 9 when we discuss efficiency. See also (Hajek (1968)) for a more general result.

3.2. U Statistics

The theory of U statistics was initiated and developed by Hoeffding (1948) who used it to study global properties of a distribution function. Let $\{X_1, \ldots, X_n\}$ be a random sample from some distribution F. Let $h(x_1, \ldots, x_m)$ be a real valued measurable function symmetric in its arguments such that

$$E\left[h\left(X_1, \ldots, X_m\right)\right] = \theta_F.$$

The smallest integer m for which $h(X_1, \ldots, X_m)$ is an unbiased estimate of θ_F called the degree of θ_F.

Definition 3.2. A U statistic for a random sample of size $n \geq m$ is defined to be

$$U_n = \binom{n}{m}^{-1} \sum h\left(X_{i_1}, \ldots, X_{i_m}\right),$$

where the summation is taken over all combinations of m integers $(i_1 < i_2 < \ldots < i_m)$ chosen from $(1, 2, \ldots, n)$.

An important property of a U statistic is that it is a minimum variance unbiased estimator of θ_F. There are numerous examples of U statistics which include the moments of a distribution, the variance, and the serial correlation coefficient. We present some below.

Example 3.3. (i) The moments of a distribution are given by the choice $h(x) = x^r$.

(ii) The sample variance is obtained from the choice of $h(x_i, x_j) = \frac{(x_i - x_j)^2}{2}$ from which we get

$$S_n^2 = \binom{n}{2}^{-1} \sum_{i<j} \frac{(x_i - x_j)^2}{2}, n > 1.$$

(iii) Let (x_i, y_i) be a sequence of n pairs of real numbers and construct for each coordinate, the $n(n-1)$ signs of differences $\{\text{sgn}(x_i - x_j), i \neq j\}$ $\{\text{sgn}(y_i - y_j), i \neq j\}$. Define the kernel

$$h((x_i x_j), (y_i, y_j)) = \text{sgn}(x_i - x_j)\,\text{sgn}(y_i - y_j).$$

Then the U statistic

$$\frac{1}{n(n-1)} \sum_{i \neq j} [\text{sgn}(x_i - x_j)\,\text{sgn}(y_i - y_j)]$$

with the sum extending over all possible pairs of indices is the covariance between the signs of the differences between the two sets. This is the Kendall statistic often used for measuring correlation.

(iv) Gini's mean difference statistic for a set of n real numbers $\{x_i\}$ given by

$$\frac{1}{n(n-1)} \sum_{i \neq j} |x_i - x_j|$$

is a U statistic which has seen many applications in economics. It measures the spread of a distribution.

(v) The serial coefficient for a set of n real numbers $\{x_i\}$ given by

$$\sum_{i=1}^{n-1} x_i x_{i+1}$$

is a U statistic.

We may obtain a general expression for the variance of a U statistic. Denote for $c = 0, 1, \ldots, m$, the conditional expectation of $h(X_1, \ldots, X_m)$ given $(X_1 = x_1, ..X_c = x_c, X_{c+1}, \ldots, X_m)$ by

$$h_c(x_1, \ldots, x_c) = E[h(x_1, \ldots, x_c, X_{c+1}, \ldots, X_m)].$$

Theorem 3.3. *The variance of a U statistic is given by*

$$Var(U_n) = \binom{n}{m}^{-1} \sum_{c=1}^{m} \binom{m}{c} \binom{n-m}{m-c} \sigma_c^2,$$

where $\sigma_c^2 = Var[h_c(X_1, \ldots, X_c)]$. Moreover, the variances are nondecreasing:

$$\sigma_1^2 \leq \sigma_2^2 \leq \ldots \leq \sigma_m^2$$

and for large n, if $\sigma_m^2 < \infty$,

$$Var\,(U_n) \cong m^2\sigma_1^2/n$$

Proof. See Ferguson (1996). □

Definition 3.3. A U statistic has a degeneracy of order k if $\sigma_i^2 = 0$ for $i \le k$ and $\sigma_{k+1}^2 > 0$.

Example 3.4. Let $X_1, \ldots, X_n$ be a random sample for which $E\,[X] = 0$. Let the kernel function be defined as the product

$$h\,(x_1, x_2) = x_1 x_2.$$

Then,

$$E\,[U_n] = 0$$

and

$$h_1\,(x) = E\,[h\,(X_1, X_2)\,|X_1 = x] = 0$$

This implies $\sigma_1^2 = 0$ and hence we have a degeneracy of order 1.

An important property of U statistics is that under certain conditions, they have limiting distributions as the next theorem states.

Theorem 3.4. *Let* $\sigma_m^2 < \infty$.

(i) Suppose that $\sigma_1^2 > 0$. *Then for large n*

$$\sqrt{n}\,(U_n - \theta_F) \xrightarrow{\mathcal{L}} \mathcal{N}\left(0, m^2\sigma_1^2\right).$$

(ii) If the U statistic has a degeneracy of order 1, then

$$n\,(U_n - \theta_F) \xrightarrow{\mathcal{L}} \sum_{1}^{\infty} \lambda_j \left(Z_j^2 - 1\right),$$

where the $\{Z_j\}$ *are i.i.d.* $\mathcal{N}\,(0,1)$ *and the* $\{\lambda_j\}$ *are the eigenvalues satisfying*

$$h\,(x_1, x_2) - \theta_F = \sum \lambda_j \varphi_k\,(x_1)\,\varphi_k\,(x_2)$$

for orthonormal functions $\{\varphi_k\,(x)\}$ *for which*

$$E\,[\varphi_k\,(X)\,\varphi_j\,(X)] = \delta_{kj}.$$

Proof. See Ferguson (1996). □

It is also possible to define a two-sample version of a U statistic.

Definition 3.4. Consider two independent random samples $X_1, \ldots, X_{n_1}$ from F and $Y_1, \ldots, Y_{n_2}$ from G. Let $h(x_1, \ldots, x_{m_1}; y_1, \ldots, y_{m_2})$ be a kernel function symmetric in the $x's$ and separately symmetric in the $y's$ with finite expectation

$$E\left[h\left(X_1, \ldots, X_{m_1}; Y_1, \ldots, Y_{m_2}\right)\right] = \theta_{F,G}.$$

A two-sample U statistic is defined to be

$$U_{n_1,n_2} = \binom{n_1}{m_1}^{-1} \binom{n_2}{m_2}^{-1} \sum h\left(X_{i_1}, \ldots, X_{i_{m_1}}; Y, \ldots, Y_{j_{m_2}}\right),$$

where the sum is taken over all subscripts $1 \leq i_1 < \ldots < i_{m_1} \leq n_1$ chosen from $1, 2, \ldots, n_1$ and subscripts $1 \leq j < \ldots < j_{m_2} \leq n_2$ chosen from $1, 2, \ldots, n_2$ respectively.

In analogy with the one-sample case, define for i, j, the conditional expectation of $h(X_1, \ldots, X_{m_1}; Y_1, \ldots, Y_{m_2})$ given
$(X_1 = x_1, ..X_i = x_i, X_{i+1}, \ldots, X_{m_1})$ and $(Y_1 = y_1, ..Y_j = y_j, Y_{j=1}, \ldots, Y_{m_2})$ by

$$h_{ij}(x_1, \ldots, x_i; y_1, \ldots, y_j) = E\left[h\left(x_1, \ldots, x_i, X_{i+1}, \ldots, X_{m_1}; y_1, \ldots, y_j, Y_{j+1}, \ldots, Y_{m_2}\right)\right].$$

Theorem 3.5. *The variance of the U statistic U_{n_1,n_2} is given by*

$$Var\left(U_{n_1,n_2}\right) = \binom{n_1}{m_1}^{-1} \binom{n_2}{m_2}^{-1} \sum_{i=1}^{m_1} \sum_{j=1}^{m_2} \binom{m_i}{i} \binom{n_1 - m_1}{m_1 - i} \binom{m_2}{j} \binom{n_2 - m_2}{m_2 - j} \sigma_{ij}^2,$$

where $\sigma_{ij}^2 = Var\left[h_{ij}(X_1, \ldots, X_i; Y_1, \ldots, Y_j)\right]$.

If $\sigma_{m_1 m_2}^2 < \infty$, and $\frac{n_1}{n_1+n_2} \to \lambda$, $0 < \lambda < 1$ as $n_1, n_2 \to \infty$, then

$$\sqrt{n_1 + n_2}\left(U_{n_1,n_2} - \theta_{F,G}\right) \xrightarrow{\mathcal{L}} \mathcal{N}\left(0, \sigma^2\right),$$

where

$$\sigma^2 = \frac{m_1^2}{\lambda}\sigma_{10}^2 + \frac{m_2^2}{1-\lambda}\sigma_{01}^2.$$

Proof. See (Bhattacharya et al. (2016), or Lehmann (1975), p. 364) for a slight variation of this result. The proof as in the one-sample case is based on a projection argument. □

Example 3.5. Let $X_1, \ldots, X_{n_1}$ and $Y_1, \ldots, Y_{n_2}$ be two independent random samples from distributions $F(x)$ and $G(y)$ respectively. Let $h(X, Y)$ be a two-sample kernel function and let the corresponding U statistic be given by

$$U_{n_1,n_2} = \frac{1}{n_1 n_2} \sum_{i=1}^{n_1} \sum_{j=1}^{n_2} h\left(X_i, Y_j\right).$$

Set $m_1 = m_2 = 1$ in Theorem 3.5 and let $h_{10}(x) = E[h(x, Y)]$ and $h_{01}(y) = E[h(X, y)]$.
Then, as $n_1 + n_2 \to \infty$, with $\frac{n_1}{n_1+n_2} \to \lambda$,

$$\sqrt{n_1 + n_2}\left(U_{n_1,n_2} - \theta\right) \xrightarrow{\mathcal{L}} \mathcal{N}\left(0, \sigma^2\right),$$

where

$$\sigma^2 = \frac{\sigma_{10}^2}{\lambda} + \frac{\sigma_{01}^2}{1-\lambda}$$

and

$$\sigma_{10}^2 = Var\left[h_{10}(X)\right], \sigma_{01}^2 = Var\left[h_{01}(Y)\right].$$

An immediate application of Example 3.5 will be seen in Theorem 5.1 when we consider the Wilcoxon two-sample statistic.

U statistics may arise as we shall see when using composite likelihood. Consider a density defined for a kernel $T = h(X_1, \ldots, X_r)$

$$f(t; \theta) = \exp\left[\theta t - K(\theta)\right] g_T(t),$$

where $g_T(t)$ is the density of T under the null hypothesis $H_0 : \theta = 0$.

For a sample of size n, we may construct the composite log likelihood function which is proportional to

$$l(\theta) \sim \left[\sum \theta\, h(x_{i_1}, \ldots, x_{i_r}) - \tbinom{n}{r} K(\theta)\right], \tag{3.2}$$

where the summation is over all possible subsets of r indices chosen $(1, \ldots, n)$. It is then seen that

$$\ell(\theta) \sim \left[\theta\, U_n(x_1, \ldots, x_n) - K(\theta)\right].$$

The log likelihood obtained in (3.2) will be shown to lead to well-known test statistics.

Remark. A modern detailed account of multivariate U statistics may be found in Lee (1990) and Gotze (1987).

3.3. Hoeffding's Combinatorial Central Limit Theorem

Let $(R_{n1}, \ldots, R_{nn})$ be a random vector which takes the $n!$ permutations of $(1, \ldots, n)$ with equal probabilities $1/n!$. Set

$$S_n = \sum_{i=1}^{n} a_n\left(i, R_{ni}\right),$$

where $a_n(i,j), i,j = 1,\ldots,n$ are n^2 real numbers. Let

$$d_n(i,j) = a_n(i,j) - \frac{1}{n}\sum_{g=1}^{n} a_n(g,j) - \frac{1}{n}\sum_{h=1}^{n} a_n(i,h) + \frac{1}{n^2}\sum_{g=1}^{n}\sum_{h=1}^{n} a_n(g,h).$$

Theorem 3.6 (Hoeffding (1951a)). *Then the distribution of S_n is asymptotically normal with mean*

$$E[S_n] = \frac{1}{n}\sum_i\sum_j a_n(i,j)$$

and variance

$$Var(S_n) = \frac{1}{n-1}\sum_i\sum_j d_n^2(i,j)$$

provided

$$\lim_{n\to\infty} \frac{\frac{1}{n}\sum_i\sum_j d_n^r(i,j)}{\left[\frac{1}{n}\sum_i\sum_j d_n^2(i,j)\right]^{\frac{r}{2}}} = 0, r = 3,4,\ldots \tag{3.3}$$

Equation (3.3) is satisfied if

$$\lim_{n\to\infty} \frac{\max_{1\le i,j\le n} d_n^2(i,j)}{\left[\frac{1}{n}\sum_i\sum_j d_n^2(i,j)\right]} = 0.$$

In the special case, $a_n(i,j) = a_n(i)\, b_n(j)$, equation (3.3) is satisfied if

$$\lim_{n\to\infty} n \frac{\max\left(a_n(i) - \bar{a}_n\right)^2}{\sum\left(a_n(i) - \bar{a}_n\right)^2} \frac{\max\left(b_n(i) - \bar{b}_n\right)^2}{\sum\left(b_n(i) - \bar{b}_n\right)^2} = 0,$$

where

$$\bar{a}_n = \frac{1}{n}\sum_i a_n(i), \bar{b}_n = \frac{1}{n}\sum_i b(i).$$

3.4. Exercises

Exercise 3.1 (Hájek and Sidak (1967), p. 81). Show that $Var(\tau) = \frac{2(2n+5)}{9n(n-1)}$.
Hint: Define the Kendall statistics τ_n and τ_{n-1} based on $(X_1,\ldots,X_n)$ and $(X_1,\ldots,X_{n-1})$ respectively. Using the fact that

$$R_i = 1 + \frac{1}{2}\sum_{j\neq i} \operatorname{sgn}(X_i - X_j),$$

show that

$$\tau_n = \frac{n-2}{n}\tau_{n-1} + \frac{4}{n(n-1)}\left(R_n - \frac{n+1}{2}\right)$$

from which we obtain the telescopic recursion for $h_i = [i(i-1)]^2 Var(\tau_i)$ and

$$h_i - h_{i-1} = \frac{4}{3}\left(i^2 - 1\right).$$

Hence,

$$h_n = \sum_{i=1}^{n}\left(h_i - h_{i-1}\right).$$

Exercise 3.2. Suppose that $R_1, \ldots, R_n$ are uniformly distributed over the set of $n!$ permutations of the integers $1, 2, \ldots, n$. Show that the conditions of Theorem 3.5 are satisfied for the statistic $S_n = \sum_{i=1}^{n} iR_i$.

Exercise 3.3. Apply the projection method to the sample variance for a random sample of size n

$$\begin{aligned} S_n^2 &= \frac{1}{n-1}\sum_{i=1}^{n}\left(X_i - \bar{X}\right)^2 \\ &= \binom{n}{2}^{-1}\sum\sum_{i<j}\frac{\left(X_i - X_j\right)^2}{2} \end{aligned}$$

to show that, properly standardized, it is asymptotically normal provided $E[X^4] < \infty$. Hint: Show that

$$E\left[\left.\frac{\left(X_i - X_j\right)^2}{2}\right| X_i = x\right] = \begin{cases} E\left[\frac{(x - X_k)^2}{2}\right] & i = j, k \\ E\left[\frac{(X_j - X_k)^2}{2}\right] & \text{otherwise} \end{cases}$$

Exercise 3.4.

(a) Show that $Var(\hat{\tau}) = \frac{4}{9}\left(\frac{n+1}{n}\right)^2 \frac{1}{n-1}$. Hence, as $n \to \infty$,

$$\frac{Var(\hat{\tau})}{Var(\tau)} \to 1.$$

(b) Show that this implies that the Kendall and Spearman statistics are asymptotically equivalent.

Exercise 3.5. Let $X_1, \ldots, X_{n_1}$ and $Y_1, \ldots, Y_{n_2}$ be two independent random samples from distributions $F(x)$ and $G(y)$, respectively. Let M be the number of pairs (X_i, Y_j) whereby $X_i < Y_j$ and let W represent the sum of the ranks of the Y's in the combined samples. Show that

$$W = M + \frac{n_2(n_2+1)}{2}.$$

Hint: For any j,

$$\sum_i I(X_i < Y_j) + j = Rank(Y_j)$$

Exercise 3.6. In Example 3.5, find the projection of U_{n_1,n_2} onto the space of linear rank statistics for the $X's$ and for the $Y's$.

Exercise 3.7 (Randles and Wolfe (1979))**.** Find the mean and variance of the statistic

$$S_n = \sum_{i=1}^{n} i^2 \log\left(\frac{R_i}{n+1}\right)$$

and show that as $n \to \infty$, it has an asymptotic normal distribution.

Exercise 3.8. Consider the Spearman Footrule distance between two permutations μ, ν of length n defined as

$$d(\mu,\nu) = \sum_{i=1}^{n} |\mu(i) - \nu(i)|$$

It was shown in Diaconis and Graham (1977) that when μ, ν are independent and are uniformly distributed over the integers $1, 2, \ldots, n$

$$\begin{aligned} E[d(\mu,\nu)] &= \frac{n^2}{3} + O(n) \\ Var[d(\mu,\nu)] &= \frac{2n^3}{45} + O(n^2). \end{aligned}$$

Use Hoeffding's combinatorial central limit theorem with $a_n(i,j) = |i-j|$ to show that $d(\mu,\nu)$ is asymptotically normal.

Exercise 3.9. An alternate form of the projection Theorem 3.1 (see van der Vaart (2007), p. 176) is as follows.

Let $\boldsymbol{R} = (R_1, \ldots, R_N)'$ be the ranks of an i.i.d. sample from a uniform distribution $U_1, \ldots < U_N$ on $(0,1)$. Let

$$a(i) = E\left[\phi\left(U_N^{(i)}\right)\right]$$

where ϕ is a square integrable function on $(0,1)$. Let

$$\tilde{S}_N = N\bar{a}_N\bar{c}_N + \sum_{i=1}^{N}(c_i - \bar{c}_N)\,\phi(F(X_i))$$

and

$$S_N = \sum_{i=1}^{N} c_i E[\phi(U_i)\,|\boldsymbol{R}].$$

Then the projection of $\tilde{S}_N$ onto the space of linear rank statistics is S_N in the sense that

$$E[S_N] = E\left[\tilde{S}_N\right]$$

and

$$\frac{Var\left[S_N - \tilde{S}_N\right]}{Var\left[\tilde{S}_N\right]} \to 1, \text{as } N \to \infty.$$

Use the above result to show that the Wilcoxon two-sample statistic defined in Section 3.1

$$T_N = \frac{1}{N}\sum_{i=m+1}^{N} R_i$$

is asymptotically equivalent to

$$\frac{1}{N}\left(-n\sum_{i=1}^{m} F(X_i) + m\sum_{j=1}^{n} F(Y_i)\right).$$

Part II.

Nonparametric Statistical Methods

4. Smooth Goodness of Fit Tests

Goodness of fit problems have had a long history dating back to Pearson (1900). Such problems are concerned with testing whether or not a set of observed data emanate from a specified distribution. For example, suppose we would like to test the hypothesis that a set of n observations come from a standard normal distribution. Pearson proposed to first divide the interval $(-\infty, \infty)$ into d subintervals and then calculate the statistic

$$X^2 = \sum_{i=1}^{d} \frac{(o_i - e_i)^2}{e_i}$$

where o_i and e_i represent the number of observed and expected observations appearing in the ith interval. The expected values $\{e_i\}$ are calculated as

$$e_i = np_i,$$

where p_i is the probability that a standard normal random variable falls in the ith interval. The test rejects the null hypothesis that the data come from a standard normal distribution whenever $X^2 \geq \chi^2_{d-1}(\alpha)$ where $\chi^2_{d-1}(\alpha)$ represents the upper $100(1-\alpha)\%$ point of a chi-square distribution with $(d-1)$ degrees of freedom.

Apart from having to specify the number and the choice of subintervals, one of the drawbacks of the Pearson test is that the alternative hypothesis is vague leaving the researcher in a quandary if in fact the test rejects the null hypothesis. Similarly, the usual tests for goodness of fit proposed by Kolmogorov-Smirnov and Cramér von Mises are omnibus tests that lack power when the alternatives specify departures in location, scale, skewness, or kurtosis. Neyman (1937) in an attempt to deal with those issues, reconsidered the problem by embedding it into a more general framework. This is perhaps the first application of what has become known as exponential tilting whereby the density specified by the null hypothesis is exponentially tilted to provide a density under the alternative. Moreover, that transition from the null to the alternative occurs in a smooth manner.

M. Alvo, P. L. H. Yu, *A Parametric Approach to Nonparametric Statistics*,
Springer Series in the Data Sciences, https://doi.org/10.1007/978-3-319-94153-0_4

4.1. Motivation for the Smooth Model

The distribution for the smooth model to be defined in connection with the goodness of fit problem can be motivated as follows. Suppose that one would like to estimate the mean μ of a population with a confidence interval. Suppose that we observe a set of L values $x_1, \ldots, x_L$ of a random variable X. We may estimate μ by a weighted estimate using a set of weights $\{w_i\}$ satisfying $w_i \geq 0, \sum w_i = 1$ and on which there is a distribution

$$\hat{\mu} = \sum w_i x_i.$$

The Kullback-Leibler information number for choosing between two distributions of weights $\boldsymbol{w}$ and $\boldsymbol{w}_0$ is defined to be

$$D(\boldsymbol{w} \| \boldsymbol{w_o}) = \sum_{i=1}^{L} w_i \log(w_i / w_{0i}),$$

where $\boldsymbol{w}$ represents the true distribution. The Kullback-Leibler measure is not a metric since it does not satisfy the metric properties of symmetry and triangle inequality. However, it satisfies the Gibbs inequality $D(\boldsymbol{w} \| \boldsymbol{w}_0) \geq 0$. This follows from the fact that $-\log x$ is a strictly convex function and hence

$$-\sum w_i \log(w_{0i}/w_i) \geq -\log \sum w_i (w_{0i}/w_i) = 0.$$

Let $\boldsymbol{w}_0 = \left(\frac{1}{L}, \ldots, \frac{1}{L}\right)$. Minimizing with respect to the $\{w_i\}$ the Lagrange multiplier expression

$$D(\boldsymbol{w} \| \boldsymbol{w}_0) + \theta\left(\mu - \sum w_i x_i\right) + \lambda\left(1 - \sum w_i\right)$$

leads to the choice

$$\begin{aligned} w_i &= \frac{e^{\theta x_i}}{\sum e^{\theta x_i}} \\ &= \frac{e^{\theta(x_i - \hat{\mu})}}{\sum e^{\theta(x_i - \hat{\mu})}}. \end{aligned}$$

There is an interpretation for the parameter θ (Efron, 1981) as follows. Since the $\{w_i\}$ determine an estimate $\hat{\mu}$, we may interpret $\hat{\mu}$ as indexed by θ as a contender for being in a confidence interval for μ. Hence, varying θ leads to different estimates as one would find in a confidence interval.

Suppose now that we fix the $\{x_i\}$ and resample them n times using weights $\{w_i\}$ so that

$$P(X = x_i) = w_i.$$

Let n_i be the number of occurrences of x_i with $\sum_{i=1}^{L} n_i = n$. Then the bootstrap distribution corresponds to the multinomial distribution

$$\begin{aligned} P(n_1, \ldots, n_k) &= \frac{n!}{n_1! \ldots n_L!} \prod w_i^{n_i} \\ &= \frac{n!}{n_1! \ldots n_L!} \prod \frac{e^{\theta n_i (x_i - \hat{\mu})}}{\left(\sum e^{\theta(x_i - \hat{\mu})}\right)^{n_i}} \\ &= \frac{n!}{n_1! \ldots n_L!} \frac{e^{\theta \sum n_i (x_i - \hat{\mu})}}{\left(\sum e^{\theta(x_i - \hat{\mu})}\right)^{n}} \\ &= \frac{n!}{n_1! \ldots n_L!} e^{\theta \sum n_i (x_i - \hat{\mu}) - nK^*(\theta)} \end{aligned}$$

where

$$K^*(\theta) = \log \left(\sum e^{\theta(x_i - \hat{\mu})} \right).$$

The non-bootstrapped distribution is given by

$$\begin{aligned} P_0(n_1, \ldots, n_k) &= \frac{n!}{n_1! \ldots n_L!} \prod \left(\frac{1}{L} \right)^{n_i} \\ &= \frac{n!}{n_1! \ldots n_L!} \left(\frac{1}{L} \right)^{n} \end{aligned}$$

so that the bootstrapped distribution appears as an exponentially tilted distribution

$$\begin{aligned} \frac{P(n_1, \ldots, n_L)}{P_0(n_1, \ldots, n_L)} &= e^{\theta \sum n_i (x_i - \hat{\mu}) - nK(\theta)} \\ &= e^{\theta n (\mu^* - \hat{\mu}) - nK(\theta)} \end{aligned}$$

where

$$K(\theta) = \log \left(\frac{1}{L} \sum e^{\theta(x_i - \hat{\mu})} \right)$$

and

$$\mu^* = \sum \left(\frac{n_i}{n} \right) x_i$$

is the bootstrapped value. Consequently, the bootstrapped distribution of μ^* is centered at the observed mean. These considerations have shown that an exponentially tilted distribution arises in a natural way in an estimation context.

4.2. Neyman's Smooth Tests

Suppose that we are presented with a random sample $X_1, \ldots, X_n$ from a continuous strictly increasing cdf $F(x)$ with continuous density $f(x)$ and that we would like to test the null hypothesis

$$H_0 : F(x) = F_0(x), \text{ for all } x$$

for known F_0 against the alternative

$$H_1 : F(x) \neq F_0(x), \text{ for some } x.$$

Using the probability transformation $Y = F_0(X)$, we see that under H_0, $Y_1, \ldots, Y_n$ is a random sample from the uniform distribution on the interval $[0, 1]$ and we may instead test for uniformity for the sample $F(X_1), \ldots, F(X_n)$. If the alternative were specified more precisely, one could use the Neyman-Pearson lemma to arrive at a uniformly most powerful test. More often than not, however, the alternative is vague. Various tests such as the Kolmogorov-Smirnov and Anderson-Darling test statistics have been proposed for this problem (Serfling, 2009). In most cases, however, such tests have little power and moreover, if the null hypothesis of uniformity is rejected, it is not clear what the alternative hypothesis should be.

Neyman (1937) considered the problem of testing for goodness of fit by embedding the uniform distribution into a larger class of alternatives defined by

$$\begin{aligned} \pi(x; \boldsymbol{\theta}) &= \exp\left(\sum_{j=1}^{d} \theta_j h_j(x) - K(\boldsymbol{\theta})\right), 0 < x < 1 \qquad (4.1)\\ &= exp(\boldsymbol{\theta}' \boldsymbol{h}(x)) - K(\boldsymbol{\theta}), \end{aligned}$$

where

$$\boldsymbol{h}(x) = (h_1(x), \ldots, h_d(x))'$$

and $K(\boldsymbol{\theta})$ is the normalizing constant depending on $\boldsymbol{\theta}$, chosen to make $\pi(x, \boldsymbol{\theta})$ a proper density and the $\{h_j(x)\}$ consist of orthonormal polynomials that satisfy

$$\int_0^1 h_i(x)\, dx = 0 \qquad (4.2)$$

$$\int_0^1 h_i(x) h_j(x)\, dx = \delta_{ij} = \begin{cases} 0 & i \neq j \\ 1 & i = j. \end{cases} \qquad (4.3)$$

The orthonormal polynomials may be selected in order to detect different alternatives and these lead to different tests. Corresponding to various underlying densities, some common choices of orthogonal functions (denoted with an upper asterisk) are listed in Table 4.1.

Table 4.1.: Orthogonal functions for various densities

Density	Name	Orthogonal function
Normal: $\frac{1}{\sqrt{2\pi}}\exp\left(-\frac{x^2}{2}\right)$	Hermite polynomials	$h_j^*(x) = \sum_{t=0}^{[j/2]} \frac{(-1/2)^t j!}{t!(j-2t)!} x^{j-2t}$
Poisson: $\exp^{-\lambda}\lambda^x/x!$	Poisson-Charlier	$h_j^*(x;\lambda) = \sqrt{\lambda^j/j!}\sum_{t=0}^{j}(-1)^{j-t}\binom{x}{t}!\lambda^{-t}\binom{j}{t}$
Exponential: e^{-x}		$h_j^*(x) = \sum_{t=0}^{j}\binom{j}{t}(-x)^t/t!$
Gamma: $x^{\alpha-1}\frac{exp(-x/\beta)}{\Gamma(\alpha)\beta^\alpha}$	Laguerre polynomials	$h_j^*(x) = \frac{e^x}{j!}\frac{d^j}{dx^j}(e^{-x}x^j)$
Hypergeometric $\frac{\binom{M}{x}\binom{N-M}{n-x}}{\binom{N}{n}}$	Chebyshev polynomials	$h_0(x,n) = 1, h_1(x,n) = 1 - \frac{2x}{n}, x = 0,1,\ldots,n$ $(i+1)(n-i)h_{i+1}(x,n) =$ $(2i+1)(n-2x)h_i(x,n) - i(n+i+1)h_{i-1}(x,n).$

Under the model in (4.1), the test for uniformity then becomes a test of

$$H_0 : \boldsymbol{\theta} = \mathbf{0} \text{ vs } H_1 : \boldsymbol{\theta} \neq \mathbf{0}. \tag{4.4}$$

For small values of θ_j close to 0, we see that $\pi(x; \boldsymbol{\theta})$ is close to the uniform density. Hence, the model in (4.1) provides a "smooth" transition from the null hypothesis and as we shall later see, leads to tests with optimal properties. We note that under this formulation, the original nonparametric problem has been cast as a parametric one. The next theorem specifies the test statistic.

Theorem 4.1. *Let $X_1, \ldots, X_n$ be a random sample from (4.1). The score test for testing (4.4) rejects whenever*

$$\sum_{j=1}^{d} U_j^2 > c_\alpha,$$

where

$$U_j = \frac{1}{\sqrt{n}} \sum_{i=1}^{n} h_j(X_i)$$

and c_α satisfies $P(\chi_d^2 > c_\alpha) = \alpha$.

Proof. The log likelihood for the density in (4.1) is

$$l(\boldsymbol{\theta}; \boldsymbol{x}) = \sum_{j=1}^{d} \theta_j \sum_{i=1}^{n} h_j(x_i) - nK(\boldsymbol{\theta})$$

which yields the score vector

$$U(\boldsymbol{\theta}; \boldsymbol{X}) = \frac{\partial l(\boldsymbol{\theta}; \boldsymbol{X})}{\partial \boldsymbol{\theta}}$$

with jth component

$$\frac{\partial l(\boldsymbol{\theta}; \boldsymbol{x})}{\partial \theta_j} = \sqrt{n} u_j - n \frac{\partial K(\boldsymbol{\theta})}{\partial \theta_j}$$

Differentiating with respect to θ_j and evaluating the derivative at $\boldsymbol{\theta} = \mathbf{0}$:

$$\frac{\partial}{\partial \theta_j} \int_0^1 \pi(x; \boldsymbol{\theta})\, dx = 1$$

leads to

$$\frac{\partial K(\boldsymbol{\theta})}{\partial \theta_j} = 0$$

in view of (4.2) and consequently

$$\frac{\partial l(\boldsymbol{\theta};\boldsymbol{x})}{\partial \theta_j} = \sqrt{n}u_j.$$

Also, the (i,j) component of the information matrix $I_n(\boldsymbol{\theta})$ in view of (4.3) evaluated at $\boldsymbol{\theta} = \mathbf{0}$ is

$$\frac{\partial^2 K(\boldsymbol{\theta})}{\partial \theta_i \partial \theta_j} = n\delta_{ij}.$$

It follows that the score test statistic at $\boldsymbol{\theta} = \mathbf{0}$ is

$$U'(\boldsymbol{\theta};\mathbf{X}) I_n^{-1}(\boldsymbol{\theta}) U(\boldsymbol{\theta};\mathbf{X}) = \sum_{j=1}^{d} U_j^2.$$

Lemma 2.5 then yields the asymptotic distribution of the score statistic. An alternative direct proof makes use of the fact that the $U_i's$ are sums of i.i.d. random variables. □

The change of measure or exponential tilting model introduced in (4.1) has been used in rare event simulation (Asmussen et al., 2016) as well as in rejection sampling and importance sampling. We illustrate the latter use in the following example.

Example 4.1 (Importance Sampling). Suppose that X is a random variable having a normal distribution with mean 0 and variance 1 and we would like to estimate the p-value

$$\begin{aligned} p &= P(X > c) = \int_c^{\infty} \frac{1}{\sqrt{2\pi}\sigma} \exp\left(-\frac{x^2}{2\sigma^2}\right) dx \\ &\equiv E_1[I(X > c)] \end{aligned}$$

where $c > 0$ and $I(.)$ is the indicator function. This can be done in one of two ways. In the first case, we may take a random sample of size n and calculate the unbiased estimator

$$\hat{p} = \frac{1}{n}\sum_{i=1}^{n} I(X_i > c)$$

whose variance is equal to

$$\frac{p(1-p)}{n} = \frac{1}{n}\left\{E[I(X > c)] - p^2\right\}. \tag{4.5}$$

Alternatively, we note that

$$\begin{aligned} p &= \int_c^{\infty} \Lambda(x) \frac{1}{\sqrt{2\pi}\sigma} \exp\left(-\frac{(x-c)^2}{2\sigma^2}\right) dx \\ &\equiv E_2[\Lambda(X) I(X > c)] \end{aligned}$$

where

$$\Lambda(x) = \frac{\frac{1}{\sqrt{2\pi}\sigma}\exp\left(-\frac{x^2}{2\sigma^2}\right)}{\frac{1}{\sqrt{2\pi}\sigma}\exp\left(-\frac{(x-c)^2}{2\sigma^2}\right)} = \exp\left(-\frac{1}{2\sigma^2}c(2x-c)\right)$$

is the likelihood ratio. We may now take a random sample of size n and calculate the unbiased estimator with respect to the change of measure $\frac{1}{\sqrt{2\pi}\sigma}\exp\left(-\frac{(x-c)^2}{2\sigma^2}\right)$:

$$\hat{p}_c = \frac{1}{n}\sum_{i=1}^{n}\Lambda(X_i)\, I(X_i > c)$$

whose variance is

$$\frac{1}{n}\left\{E_2\left[\Lambda(X)^2 I(X > c)\right] - p^2\right\}. \tag{4.6}$$

Since on the set $I(X > c)$

$$\Lambda(X) \leq 1$$

then the inequality

$$E_2\left[\Lambda(X)^2 I(X > c)\right] \leq E_2\left[\Lambda(X) I(X > c)\right] = E_1\left[I(X > c)\right]$$

shows that the variance (4.6) of the second estimator will not exceed that of the first in (4.5). See Siegmund (1976) for further discussion on importance sampling and its application to the calculation of error probabilities connected to the sequential probability ratio test.

In this book, we shall make use of exponential tilting to describe many common nonparametric statistics. Setting

$$\pi(x;\theta) = \exp(\theta x - K(\theta))\, g_X(x)$$

we record in Table 4.2 various examples of the change of measure for different densities $g_X(x)$ where the latter is determined from $\pi(x;\theta)$ under $\theta = 0$. In all cases, the normalizing constant $K(\theta)$ is the cumulant or log of the moment generating function computed under $g_X(x)$

$$K(\theta) = \log(E[\exp(\theta X)]).$$

We note that $K(\theta)$ is a strictly convex function such that

$$E_\theta[X] = K'(\theta), Var_\theta(X) = K''(\theta).$$

Table 4.2.: Examples of change of measure distributions

Density $g_X(x)$	$\pi(x;\theta)$	$K(\theta)$
$\mathcal{N}(\mu,\sigma^2)$	$\mathcal{N}(\mu+\theta\sigma^2,\sigma^2)$	$\frac{\theta^2\sigma^2}{2}+\theta\mu$
$\beta\exp(-\beta x), x>0, \beta>0$	$(\beta-\theta)\exp(-(\beta-\theta)x), \theta<\beta$	$-\log\left(1-\frac{\theta}{\beta}\right)$
$\binom{n}{p} p^x (1-p)^{n-x}, x=0,\dots,n$	$\binom{n}{p} p_\theta^x (1-p_\theta)^{n-x}, x=0,\dots,n, 0<p_\theta<1, p_\theta=\frac{pe^\theta}{1-p+pe^\theta}$	$-\log(1-p_\theta)$
$\frac{e^{-\mu}\mu^x}{x!}, \mu>0, x=0,1,\dots$	$\frac{e^{-\mu_\theta}\mu_\theta^x}{x!}, \mu_\theta=\mu e^\theta, x=0,1,\dots$	$\mu\left(e^\theta-1\right)$
$\frac{\beta^\alpha}{\Gamma(\alpha)}x^{\alpha-1}e^{-\beta x}, \alpha,\beta>0$	$\frac{(\beta-\theta)^\alpha}{\Gamma(\alpha)}x^{\alpha-1}e^{-(\beta-\theta)x}, \alpha, \beta-\theta>0$	$-\alpha\log\left(1-\frac{\theta}{\beta}\right)$
$\mathcal{N}(\mu,\Sigma)$	$\mathcal{N}(\mu+\Sigma\theta,\Sigma)$	$\frac{\theta^2\Sigma}{2}+\theta\mu$
$\frac{1}{2^{r/2}\Gamma(r/2)}x^{r/2-1}e^{-x/2}, x, r>0$	$\frac{(1/2-\theta)^{r/2}}{\Gamma(r/2)}x^{r/2-1}e^{-(1/2-\theta)x}, r>0, \theta<1/2$	$(-r/2)\log(1-2\theta), \theta<\frac{1}{2}$
$(1-p)^{x-1}p, x=1,2,\dots; p>0$	$(1-p_\theta)^{x-1}p_\theta, x=1,2,\dots; p_\theta=(1-\theta-\log(1-p))>0$	$\log\left[\frac{pe^\theta}{1-(1-p)e\theta}\right]$

Example 4.2. Suppose that we are given a random sample of size n from a normal distribution with mean μ and variance σ^2. We would like to test a null hypothesis on the mean μ. Under the smooth alternative formulation,

$$\pi\left(x;\theta\right)=\frac{1}{\sqrt{2\pi}\sigma}\exp\left[-\frac{\left(x-\mu-\theta\sigma^{2}\right)^{2}}{2\sigma^{2}}\right].$$

Based on a random sample of size n, the log of the likelihood function as a function of θ is proportional to

$$\left[n\theta\left(\bar{x}_n-\mu\right)-\frac{n}{2}\sigma^2\theta^2\right],$$

where $\bar{x}_n$ is the sample mean. Setting the derivative of the log with respect to θ equal to 0 shows that the maximum likelihood estimate of θ is

$$\hat{\theta}=\frac{\left(\bar{x}_n-\mu\right)}{\sigma^2}.$$

Consequently, $\hat{\theta}$ represents the shift of the sample mean from the null hypothesis mean μ_0 as specified by the density $g_X(x)$. The score statistic is given by the derivative of the log likelihood evaluated at $\theta=0$:

$$n\left[\bar{x}_n-\mu\right].$$

Moreover, from the above equation, the information function is given by $K''(0)=n\sigma^2$. Hence the Rao score statistic from Theorem 2.7 becomes

$$\begin{aligned} W &= n^2\left[\bar{X}_n-\mu\right]\left(K''\left(0\right)\right)^{-1}\left[\bar{X}_n-\mu\right] && (4.7)\\ &= \frac{n}{\sigma^2}\left[\bar{X}_n-\mu\right]^2. && (4.8)\end{aligned}$$

The null hypothesis is rejected for large values of W which has asymptotically as $n\to\infty$, a chi-square distribution with one degree of freedom. The advantage of the Rao score test statistic (4.7) is that all the derivatives are computed under $\theta=0$. Asymptotically, it is equivalent to both the generalized likelihood ratio test and the Wald test as indicated in Theorem 2.7. It is seen then that (4.8) is the usual two-sided test statistic for the mean of a normal. We may generalize the previous sections to the case of composite hypotheses.

4.2.1. Smooth Models for Discrete Distributions

In this section, we describe some orthonormal expansions for two discrete distributions: the hypergeometric and the binomial. It is somewhat analogous to the smooth alternative model described in (4.1) where we invoke the use of Chebyshev polynomials. The Chebyshev polynomials defined over the integers $x=0,1,\ldots,n$ can be expressed in descending factorial form as

$$h_i(x,n) = \sum_{m=0}^{i}(-1)^m \binom{i}{m}\binom{i+m}{m}\binom{x}{m}\binom{n}{m}^{-1}, \tag{4.9}$$

where $0 \leq i \leq n$ (Ralston (1965), p. 238). There is also an ascending factorial form. For these polynomials we have the recursion

$$\begin{aligned} h_0(x,n) &= 1, h_1(x,n) = 1 - \frac{2x}{n}, \\ (i+1)(n-i)h_{i+1}(x,n) &= (2i+1)(n-2x)h_i(x,n) - i(n+i+1)h_{i-1}(x,n). \end{aligned}$$

For every n, the vectors

$$\boldsymbol{\varepsilon}_i = (h_i(0,n), \ldots, h_i(n,n))' \text{ for } i = 0, 1, \ldots, n.$$

form a basis in $(n+1)$-dimensional space. Hence, any function $\{g(x)\}$ defined over the integers $x = 0, \ldots, n$ can be expressed in vector notation

$$\boldsymbol{g} = \sum_{i=0}^{n} g_i \boldsymbol{\varepsilon}_i$$

where the vector $\boldsymbol{g} = (g(0), g(1), \ldots, g(n))'$ and the g_i are obtained from the relation

$$g_i = \frac{\boldsymbol{g}'\boldsymbol{\varepsilon}_i}{||\boldsymbol{\varepsilon}_i||^2}.$$

Alternatively, for each x

$$g(x) = \sum_{i=0}^{n} g_i h_i(x,n) \text{ for } x = 0, 1, \ldots, n.$$

Example 4.3 (Hypergeometric Distribution). The hypergeometric distribution is given by

$$p(x; M, N) = \frac{\binom{M}{x}\binom{N-M}{n-x}}{\binom{N}{n}} \text{ for } max(0, n-N+M) \leq x \leq min\,(n, M)\,.$$

The connection between the Chebyshev polynomials and the hypergeometric distribution is given by the following theorem proven in Alvo and Cabilio (2000).

Theorem 4.2. *The expected value of a function of a hypergeometric variable with parameters (M, N, n) is equal to a linear combination of the first n Chebyshev polynomials* *in parameters* M, N. *as*

$$E\left[g(X)\right] = \sum_{i=0}^{n} g_i h_i(M, N).$$

Proof. The proof uses the representation in (4.9) and proceeds by showing that

$$\sum_{x=0}^{n} h_i(x, n) p(x; M, N) = h_i(M, N) \text{ for all } i = 0, 1, \ldots, n, \text{and } M = 0, 1, \ldots, N.$$

□

A consequence of Theorem 4.2 is that if $g\left(x\right) = I\left(X = x\right)$, then

$$p(x; M, N) = \sum_{i=0}^{n} g_i h_i(M, N). \tag{4.10}$$

The R package `hypersampleplan` with subroutine hypergeotable (N, n) can be used to compute tables of hypergeometric probabilities simultaneously for all values of

$$x = 0, 1, \ldots, n \text{ and } M = 0, 1, \ldots, N.$$

Alvo and Cabilio (2000) discuss similar results to the binomial distribution.

Example 4.4 (Binomial Distribution)**.** The distribution of a binomial random variable X may be expressed in terms of (4.1) as

$$\begin{aligned} b(x; n, \rho) &= \tbinom{n}{x} \rho^x (1-\rho)^{n-x}, x = 0, 1, \ldots, n;\ 0 < \rho < 1. \\ &= \tbinom{n}{x} \exp\left(\theta x - K\left(\theta\right)\right) \end{aligned}$$

where

$$\theta = \log\left(\frac{\rho}{1-\rho}\right), K\left(\theta\right) = n \log\left(1 + e^{\theta}\right)$$

Here our interest would be in testing the null hypothesis that $\rho = 0.5$ corresponding to $\theta = 0$.

4.2.2. Smooth Models for Composite Hypotheses

Suppose that we wish to test the hypothesis that a random sample of size n arises from a given probability density $g\left(x; \boldsymbol{\beta}\right)$ where $\boldsymbol{\beta}$ is a q-dimensional vector of nuisance parameters. In the continuous case, we may embed this in the following parametric model

$$f(x;\boldsymbol{\theta},\boldsymbol{\beta}) = \exp\left[\sum_{i=1}^{k}\theta_i h_i(x;\boldsymbol{\beta}) - K(\boldsymbol{\theta},\boldsymbol{\beta})\right] g(x;\boldsymbol{\beta}),$$

where $K(\theta,\boldsymbol{\beta})$ is a normalizing constant, $\boldsymbol{\theta}$ is a k-dimensional vector of parameters, and $\{h_i(x;\boldsymbol{\beta})\}$ are a set of orthonormal functions with respect to $g(x;\boldsymbol{\beta})$. That is to say, they must satisfy

$$\begin{aligned}\int_{-\infty}^{\infty} h_i(x;\boldsymbol{\beta})\, g(x;\boldsymbol{\beta})\, dx &= 0\\ \int_{-\infty}^{\infty} h_i(x;\boldsymbol{\beta})\, h_j(x;\boldsymbol{\beta})\, g(x;\boldsymbol{\beta})\, dx &= \delta_{ij}.\end{aligned}$$

It is seen that since $f(x;\boldsymbol{\theta},\boldsymbol{\beta})$ is a proper density, $K(\boldsymbol{\theta},\boldsymbol{\beta})$ must satisfy

$$\frac{\partial K(\boldsymbol{\theta},\boldsymbol{\beta})}{\partial\theta_i} = E_{\boldsymbol{\theta}}\left[h_i(X;\boldsymbol{\beta})\right]$$

and

$$\frac{\partial K^2(\boldsymbol{\theta},\boldsymbol{\beta})}{\partial\theta_i\partial\theta_j} = Cov_{\boldsymbol{\theta}}\left[h_i(X;\boldsymbol{\beta}), h_j(X;\boldsymbol{\beta})\right].$$

The score test statistic is then given by

$$S\left(\hat{\boldsymbol{\beta}}\right) = \boldsymbol{U}'\left(\hat{\boldsymbol{\beta}}\right)\hat{\boldsymbol{\Sigma}}^{-1}\boldsymbol{U}\left(\hat{\boldsymbol{\beta}}\right),$$

where $\hat{\boldsymbol{\beta}}$ is the maximum likelihood estimate of $\boldsymbol{\beta}$ and $\boldsymbol{U}\left(\hat{\boldsymbol{\beta}}\right)$ has rth element $\sum_{i=1}^{n} h_j\left(X_i;\hat{\boldsymbol{\beta}}\right)$. Here $\hat{\boldsymbol{\Sigma}}$ is the estimated asymptotic covariance matrix of $\boldsymbol{U}\left(\hat{\boldsymbol{\beta}}\right)$. We refer the reader to (Rayner et al. (2009b), p. 100) for further details.

Example 4.5. Suppose that we have a random sample $X_1,\ldots,X_n$ from a normal distribution with mean μ and variance σ^2; set $\boldsymbol{\beta} = (\mu,\sigma^2)'$. A smooth test for the hypothesis

$$H_0 : \mu = \mu_0$$

against

$$H_1 : \mu \neq \mu_0$$

can be constructed using the normalized Hermite polynomials from Table 4.1 which satisfy

$$\begin{aligned}\int_{-\infty}^{\infty} h_i(x)\, e^{-\frac{x^2}{2}} dx &= 0,\\ \int_{-\infty}^{\infty} h_i(x)\, h_j(x)\, e^{-\frac{x^2}{2}} dx &= \sqrt{2\pi}\delta_{rs}.\end{aligned}$$

The Hermite polynomials with respect to the distribution of X are given by $\left\{h_i\left(\frac{x-\mu}{\sigma}\right)\right\}$. The maximum likelihood estimates of $\boldsymbol{\beta}$ are

$$\hat{\boldsymbol{\beta}} = \left(\bar{X}_n, \frac{\sum\left(X_i - \bar{X}_n\right)^2}{n}\right).$$

We note that

$$h_1\left(\frac{x-\mu}{\sigma}\right) = \left(\frac{x-\mu}{\sigma}\right), h_2\left(\frac{x-\mu}{\sigma}\right) = \frac{1}{\sqrt{2}}\left[\left(\frac{x-\mu}{\sigma}\right)^2 - 1\right]$$

and consequently for $j = 1, 2$,

$$\sum_{i=1}^{n} h_j\left(\frac{X_i - \bar{X}_n}{\sqrt{\frac{\sum\left(X_i-\bar{X}_n\right)^2}{n}}}\right) = 0.$$

It can be shown that the score statistic for testing $H_0 : \mu = \mu_0$ is given by

$$\sum_{j=3}^{k} \hat{V}_j^2,$$

where

$$\hat{V}_j = \frac{1}{\sqrt{n}} \sum_{i=1}^{n} h_j\left(\frac{X_i - \bar{X}_n}{\sqrt{\frac{\sum\left(X_i-\bar{X}_n\right)^2}{n}}}\right).$$

4.3. Smooth Models for Categorical Data

The model in (4.1) can be adapted to apply to categorical data. Suppose we have m categories and that

$$\pi\left(\boldsymbol{x}_j; \boldsymbol{\theta}\right) = \exp\left(\boldsymbol{\theta}'\boldsymbol{x}_j - K(\boldsymbol{\theta})\right) p_j,\ j = 1, \ldots, m \tag{4.11}$$

where $\boldsymbol{x}_j$ is the jth value of a k-dimensional random vector $\boldsymbol{X}$ and $\boldsymbol{p} = (p_j)'$ denotes the vector of probabilities under the null distribution when $\boldsymbol{\theta} = \boldsymbol{\theta}_0$. Here $K(\boldsymbol{\theta})$ is a normalizing constant for which

$$\sum_j \pi\left(\boldsymbol{x}_j; \boldsymbol{\theta}\right) = 1.$$

Let $\boldsymbol{T} = [\boldsymbol{x}_i, \ldots, \boldsymbol{x}_m]$ be the $k \times m$ matrix of possible vector values of $\boldsymbol{X}$. Then under the distribution specified by $\boldsymbol{p}$,

$$\begin{aligned}\boldsymbol{\Sigma} \equiv Cov_{\boldsymbol{p}}(\boldsymbol{X}) &= E_{\boldsymbol{p}}\left[(\boldsymbol{X} - E[\boldsymbol{X}])(\boldsymbol{X} - E[\boldsymbol{X}])'\right] &(4.12)\\ &= \boldsymbol{T}(diag(\boldsymbol{p}))\boldsymbol{T}' - (\boldsymbol{T}\boldsymbol{p})(\boldsymbol{T}\boldsymbol{p})' &(4.13)\end{aligned}$$

where the expectation is with respect to the model (4.11). As we shall see in Chapter 5, this particular situation arises often when dealing with the nonparametric randomized block design. Define

$$\boldsymbol{\pi}(\boldsymbol{\theta}) = (\pi(\boldsymbol{x}_1; \boldsymbol{\theta}), \ldots, \pi(\boldsymbol{x}_m; \boldsymbol{\theta}))'.$$

Example 4.6. Suppose that we would like to test

$$H_0 : \boldsymbol{\theta} = \boldsymbol{0} \; vs \; H_1 : \boldsymbol{\theta} \neq \boldsymbol{0}.$$

Letting $\boldsymbol{N}$ denote a multinomial random variable with parameters $(n, \boldsymbol{\pi}(\boldsymbol{\theta}))$, we see that the log likelihood as a function of $\boldsymbol{\theta}$ is, apart from a constant, proportional to

$$\begin{aligned}\sum_{j=1}^{m} n_j \log(\pi(\boldsymbol{x}_j; \boldsymbol{\theta})) &= \sum_{j=1}^{m} n_j (\boldsymbol{\theta}'\boldsymbol{x}_j - K(\boldsymbol{\theta}))\\ &= \boldsymbol{\theta}'\left(\sum_{j=1}^{m} n_j \boldsymbol{x}_j\right) - nK(\boldsymbol{\theta})\end{aligned}$$

The score vector under the null hypothesis is then given by

$$\begin{aligned}U(\boldsymbol{\theta}; \boldsymbol{X}) &= \sum_{j=1}^{m} N_j \left(\frac{1}{\pi_j(\boldsymbol{\theta})} \frac{\partial \pi_j(\boldsymbol{\theta})}{\partial \boldsymbol{\theta}}\right)\\ &= \boldsymbol{T}(\boldsymbol{N} - n\boldsymbol{p})\end{aligned}$$

Under the null hypothesis,

$$E[U(\boldsymbol{\theta}; \boldsymbol{X})] = 0$$

and the score statistic is given by

$$\frac{1}{n}[\boldsymbol{T}(\boldsymbol{N} - n\boldsymbol{p})]' \boldsymbol{\Sigma}^{-1} [\boldsymbol{T}(\boldsymbol{N} - n\boldsymbol{p})] = \frac{1}{n}(\boldsymbol{N} - n\boldsymbol{p})' \left(\boldsymbol{T}'\boldsymbol{\Sigma}^{-1}\boldsymbol{T}\right)(\boldsymbol{N} - n\boldsymbol{p}) \xrightarrow{\mathcal{L}} \chi_r^2 \quad (4.14)$$

where $r = rank\left(\boldsymbol{T}'\boldsymbol{\Sigma}^{-1}\boldsymbol{T}\right)$.

We shall return to this example in Section 7.2 when we consider applications to the randomized block design. In the following example, we consider the multinomial distribution in connection with the Pearson goodness of fit statistic.

Example 4.7 (Pearson's Goodness of Fit Statistic (Rayner et al. (2009b), p. 68)). We shall show that the Pearson goodness of fit statistic is given by

$$\sum_{j=1}^{m} \frac{(N_j - np_j)^2}{np_j} \xrightarrow{\mathcal{L}} \chi^2_{m-1},$$

where $\sum p_j = 1$ may be obtained using the smooth model formulation.

Define the random vector $\boldsymbol{x}^*$ as

$$\boldsymbol{x}^* = \begin{bmatrix} \boldsymbol{x} \\ 1 \end{bmatrix}$$

where $\boldsymbol{x}$ is of dimension $(m-1)$. Consider the smooth model

$$\pi\left(\boldsymbol{x}_j^*; \boldsymbol{\theta}\right) = \exp\left(\boldsymbol{\theta}'\boldsymbol{x}_j^* - K(\boldsymbol{\theta})\right) p_j, j = 1, \ldots, m$$

The matrix of possible values of $\boldsymbol{x}^*$ is an $m \times m$ matrix

$$\boldsymbol{T}^* = \begin{bmatrix} \boldsymbol{T} \\ \boldsymbol{1}_m' \end{bmatrix}$$

where $\boldsymbol{1}_m$ is an $m \times 1$ vector of ones. We note that

$$K'(0) = \begin{bmatrix} \sum \boldsymbol{x}_j p_j \\ \sum p_j \end{bmatrix}$$

and we may prescribe the $\{\boldsymbol{x}_j\}$ by imposing the orthonormality conditions

$$\sum \boldsymbol{x}_j p_j = 0$$

and

$$\begin{aligned} K''(0) &= \sum \boldsymbol{x}_j^* \boldsymbol{x}_j^{*'} p_j \\ &= \boldsymbol{T}^* \left(diag\left(p_j\right)\right) \boldsymbol{T}^{*'} \\ &= \boldsymbol{I}_m \end{aligned} \tag{4.15}$$

where $\boldsymbol{I}_m$ is the identity matrix of order m. It follows that since

$$\boldsymbol{I}_m = \begin{bmatrix} \boldsymbol{T}\left(diag\left(p_j\right)\right)\boldsymbol{T}' & \boldsymbol{T}\left(diag\left(p_j\right)\right)\boldsymbol{1}_m \\ \boldsymbol{1}_m'\left(diag\left(p_j\right)\right)\boldsymbol{T}' & \boldsymbol{1}_m'\left(diag\left(p_j\right)\right)\boldsymbol{1}_m \end{bmatrix}$$

and

$$\begin{aligned} \mathbf{1}_m' \left(diag\left(p_j\right)\right) \mathbf{1}_m &= \sum p_j = 1 \\ \mathbf{1}_m' \left(diag\left(p_j\right)\right) \boldsymbol{T}' &= \left(\sum x_j p_j\right)' = 0, \end{aligned}$$

then

$$\boldsymbol{T} \left(diag\left(p_j\right)\right) \boldsymbol{T}' = \boldsymbol{I}_{m-1}.$$

Now from (4.15)

$$\begin{aligned} diag\left(p_j^{-1}\right) &= \boldsymbol{T}^{*\prime} \boldsymbol{T}^{*} \\ &= \boldsymbol{T}'\boldsymbol{T} + \mathbf{1}_m \mathbf{1}_m'. \end{aligned}$$

Hence, the score statistic

$$\begin{aligned} \frac{1}{n} \left[(\boldsymbol{N} - n\boldsymbol{p}) \right]' \boldsymbol{T}'\boldsymbol{T} \left[(\boldsymbol{N} - n\boldsymbol{p}) \right] &= \frac{1}{n} (\boldsymbol{N} - n\boldsymbol{p})' \left[diag\left(p_j^{-1}\right) - \mathbf{1}_m \mathbf{1}_m' \right] (\boldsymbol{N} - n\boldsymbol{p}) \\ &= \sum_{j=1}^{m} \frac{\left(N_j - np_j\right)^2}{np_j} - \frac{1}{n} \left[\sum_{j=1}^{m} \left(N_j - np_j\right) \right]^2 \\ &= \sum_{j=1}^{m} \frac{\left(N_j - np_j\right)^2}{np_j}. \end{aligned}$$

4.4. Smooth Models for Ranking Data

In this section, we will study distance-based models within the context of smooth alternative distributions. We begin by describing the important Mallows models and their generalization before proceeding to the study of cyclic models.

4.4.1. Distance-Based Models

A ranking represents the order of preference one has with respect to a set of t objects. If we label the objects by the integers 1 to t, a ranking can then be thought of as a permutation of the integers $(1, 2, \ldots, t)$. We may denote such a permutation by

$$\boldsymbol{\mu} = (\boldsymbol{\mu}(1), \boldsymbol{\mu}(2), \ldots, \boldsymbol{\mu}(t))'$$

which may also be conceptualized as a point in t-dimensional space. It is natural to measure the spread or discrepancy between two individual rankings $\boldsymbol{\mu}, \boldsymbol{\nu}$ by means of a distance function.

The usual properties of a distance function between two rankings $\boldsymbol{\mu}$ and $\boldsymbol{\nu}$ are: (1) reflexivity: $d(\boldsymbol{\mu},\boldsymbol{\mu}) = 0$; (2) positivity: $d(\boldsymbol{\mu},\boldsymbol{\nu}) > 0$ if $\boldsymbol{\mu} \neq \boldsymbol{\nu}$; and (3) symmetry: $d(\boldsymbol{\mu},\boldsymbol{\nu}) = d(\boldsymbol{\nu},\boldsymbol{\mu})$. For rankings, we often require that the distance, apart from having these usual properties, must be right invariant,

$$d(\boldsymbol{\mu},\boldsymbol{\nu}) = d(\boldsymbol{\mu}\circ\boldsymbol{\tau},\boldsymbol{\nu}\circ\boldsymbol{\tau}), \text{ where } \boldsymbol{\mu}\circ\boldsymbol{\tau}(i) \equiv \boldsymbol{\mu}(\boldsymbol{\tau}(i)).$$

The requirement right invariance ensures that a relabeling of the objects has no effect on the distance. If a distance function satisfies the triangle inequality $d(\boldsymbol{\mu},\boldsymbol{\nu}) \leq d(\boldsymbol{\mu},\boldsymbol{\sigma}) + d(\boldsymbol{\sigma},\boldsymbol{\nu})$, the distance is said to be a *metric*. There are several examples of distance functions that have been proposed in the literature. Here are a few:
Spearman:

$$d_S(\boldsymbol{\mu},\boldsymbol{\nu}) = \frac{1}{2}\sum_{i=1}^{t}\left(\boldsymbol{\mu}(i) - \boldsymbol{\nu}(i)\right)^2 \tag{4.16}$$

Kendall:

$$d_K(\boldsymbol{\mu},\boldsymbol{\nu}) = \sum_{i<j}\left\{1 - \operatorname{sgn}\left(\boldsymbol{\mu}(j) - \boldsymbol{\mu}(i)\right)\operatorname{sgn}\left(\boldsymbol{\nu}(j) - \boldsymbol{\nu}(i)\right)\right\} \tag{4.17}$$

where $\operatorname{sgn}(x)$ is either 1 or -1 depending on whether $x > 0$ or $x < 0$.
Hamming:

$$d_H(\boldsymbol{\mu},\boldsymbol{\nu}) = t - \sum_{i=1}^{t}\sum_{j=1}^{t} I\left(\boldsymbol{\mu}(i) = j\right) I\left(\boldsymbol{\nu}(i) = j\right) \tag{4.18}$$

where $I(.)$ is the indicator function taking values 1 or 0 depending on whether the statement in brackets holds or not.
Spearman Footrule:

$$d_F(\boldsymbol{\mu},\boldsymbol{\nu}) = \sum_{i=1}^{t}\left|\boldsymbol{\mu}(i) - \boldsymbol{\nu}(i)\right| \tag{4.19}$$

Cayley:

$$d_C(\boldsymbol{\mu},\boldsymbol{\nu}) = n - \#\text{cycles in } \boldsymbol{\nu}\circ\boldsymbol{\mu}^{-1}$$

or equivalently, it is the minimum of transpositions needed to transform $\boldsymbol{\mu}$ into $\boldsymbol{\nu}$. Here, $\boldsymbol{\mu}^{-1} = \langle \boldsymbol{\mu}^{-1}(1), \ldots, \boldsymbol{\mu}^{-1}(t)\rangle$ denotes the inverse permutation that displays the objects receiving a specific rank.

Note that the Spearman Footrule, Kendall, Hamming, and Cayley distances are metrics but the Spearman distance, like the squared Euclidean distance, is not since it does not satisfy the triangular inequality property. We shall nonetheless for convenience refer to it as a distance function in this book. The Kendall distance counts the number of "discordant" pairs whereas the Hamming distance counts the number of "mismatches." The Hamming distance has found uses in coding theory.

Let $M = \left(d\left(\boldsymbol{\mu}_i, \boldsymbol{\mu}_j\right)\right)$ denote the matrix of all pairwise distances. If d is right invariant, then it follows that there exists a constant $c > 0$ for which

$$M\mathbf{1} = (ct!)\mathbf{1}$$

where $\mathbf{1} = (1, 1, \ldots, 1)'$ is of dimension $t!$. Hence, c is equal to the average distance. It is straightforward to show that for the Spearman and Kendall distances

$$c_S = \frac{t(t^2-1)}{12}, c_K = \frac{t(t-1)}{2}.$$

Turning attention to the Hamming distance, we note that if $e = (1, 2, \ldots, t)'$, then

$$\begin{aligned} \Sigma_{\boldsymbol{\mu}} d_H\left(\boldsymbol{\mu}, e\right) &= \Sigma_{\boldsymbol{\mu}} t - \Sigma_{\boldsymbol{\mu}} \Sigma_i \Sigma_j I\left(\boldsymbol{\mu}\left(i\right) = j\right) I\left(e\left(i\right) = j.\right) \\ &= t\left(t!\right) - t! \end{aligned}$$

and hence $c_H = t - 1$.

Example 4.8. Suppose that $t = 3$, and that the complete rankings are denoted by

$$\boldsymbol{\mu}_1 = (1,2,3)', \boldsymbol{\mu}_2 = (1,3,2)', \boldsymbol{\mu}_3 = (2,1,3)', \boldsymbol{\mu}_4 = (2,3,1)', \boldsymbol{\mu}_5 = (3,1,2)', \boldsymbol{\mu}_6 = (3,2,1)'$$

Using the above order of the permutations, we may write the matrix M of pairwise Spearman, Kendall, Hamming, and Footrule distances respectively as

$$M_S = \begin{pmatrix} 0 & 1 & 1 & 3 & 3 & 4 \\ 1 & 0 & 3 & 1 & 4 & 3 \\ 1 & 3 & 0 & 4 & 1 & 3 \\ 3 & 1 & 4 & 0 & 3 & 1 \\ 3 & 4 & 1 & 3 & 0 & 1 \\ 4 & 3 & 3 & 1 & 1 & 0 \end{pmatrix}$$

$$M_K = \begin{pmatrix} 0 & 2 & 2 & 4 & 4 & 6 \\ 2 & 0 & 4 & 2 & 6 & 4 \\ 2 & 4 & 0 & 6 & 2 & 4 \\ 4 & 2 & 6 & 0 & 4 & 2 \\ 4 & 6 & 2 & 4 & 0 & 2 \\ 6 & 4 & 4 & 2 & 2 & 0 \end{pmatrix}$$

$$M_H = \begin{pmatrix} 0 & 2 & 2 & 3 & 3 & 2 \\ 2 & 0 & 3 & 2 & 2 & 3 \\ 2 & 3 & 0 & 2 & 2 & 3 \\ 3 & 2 & 2 & 0 & 3 & 2 \\ 3 & 2 & 2 & 3 & 0 & 2 \\ 2 & 3 & 3 & 2 & 2 & 0 \end{pmatrix}$$

$$M_F = \begin{pmatrix} 0 & 2 & 2 & 4 & 4 & 4 \\ 2 & 0 & 4 & 2 & 4 & 4 \\ 2 & 4 & 0 & 4 & 2 & 4 \\ 4 & 2 & 4 & 0 & 4 & 2 \\ 4 & 4 & 2 & 4 & 0 & 2 \\ 4 & 4 & 4 & 2 & 2 & 0 \end{pmatrix}$$

These distances may alternatively be expressed in terms of a similarity function $\mathcal{A}$ in the form

$$d(\boldsymbol{\mu}, \boldsymbol{\nu}) = c - \mathcal{A}(\boldsymbol{\mu}, \boldsymbol{\nu}), \tag{4.20}$$

Spearman:

$$\mathcal{A}_S = \mathcal{A}_S(\boldsymbol{\mu}, \boldsymbol{\nu}) = \sum_{i=1}^{t} \left(\boldsymbol{\mu}(i) - \frac{t+1}{2}\right)\left(\boldsymbol{\nu}(i) - \frac{t+1}{2}\right) \tag{4.21}$$

Kendall:

$$\mathcal{A}_K = \mathcal{A}_K(\boldsymbol{\mu}, \boldsymbol{\nu}) = \sum_{i<j} \operatorname{sgn}\left(\boldsymbol{\mu}(j) - \boldsymbol{\mu}(i)\right) \operatorname{sgn}\left(\boldsymbol{\nu}(j) - \boldsymbol{\nu}(i)\right). \tag{4.22}$$

Hamming:

$$\mathcal{A}_H\left(\boldsymbol{\mu}, \boldsymbol{\nu}\right) = \sum_{i=1}^{t}\sum_{j=1}^{t} I\left([\boldsymbol{\mu}(i) = j] - \frac{1}{t}\right) I\left([\boldsymbol{\nu}(i) = j] - \frac{1}{t}\right) \tag{4.23}$$

Footrule:

$$\mathcal{A}_F\left(\boldsymbol{\mu}, \boldsymbol{\nu}\right) = \sum_{i=1}^{t}\sum_{j=1}^{t} I\left([\boldsymbol{\mu}(i) \leq j] - \frac{j}{t}\right) I\left([\boldsymbol{\nu}(i) \leq j] - \frac{j}{t}\right) \tag{4.24}$$

The similarity measures may also be interpreted geometrically as inner products which sets the groundwork for defining correlation (Alvo and Yu (2014)).

It is reasonable to assume that in a homogeneous population of judges, most of the judges will have rankings close to a modal ranking $\boldsymbol{\mu}_0$. According to this framework, Diaconis (1988a) developed a class of distance-based models over the set of all $t!$ rankings $\mathcal{P}$:

$$\boldsymbol{\pi}(\boldsymbol{\mu}|\lambda, \boldsymbol{\mu}_0) = \frac{e^{-\lambda d(\boldsymbol{\mu}, \boldsymbol{\mu}_0)}}{C(\lambda)}, \quad \boldsymbol{\mu} \in \mathcal{P}, \tag{4.25}$$

where $\lambda \geq 0$ is the dispersion parameter, $C(\lambda)$ is a normalizing constant and $d(\boldsymbol{\mu}, \boldsymbol{\sigma})$ is an arbitrary right invariant distance. In the particular case where we use Kendall as the distance function, the model is called the Mallows' ϕ-model (Mallows, 1957). Note that Mallows' ϕ-models also belong to the class of paired comparison models (Critchlow et al., 1991). Critchlow and Verducci (1992) and Feigin (1993) provided more details about the relationship between distance-based models and paired comparison models.

Motivated from Neyman's smooth alternative model (4.1), the distance-based model in (4.25) can be used to test for the uniform null distribution $\frac{1}{t!}$, i.e., $\lambda = 0$.

In distance-based models, the ranking probability is largest at the modal ranking $\boldsymbol{\mu}_0$ and the probability of a ranking will decay the further it is away from the modal ranking $\boldsymbol{\mu}_0$. The rate of the decay is governed by the parameter λ. For a small value of λ, the distribution of rankings will be more concentrated around $\boldsymbol{\mu}_0$. When λ becomes very large, the distribution of rankings will look more uniform. The closed form for the normalizing constant $C(\lambda)$ only exists for some distances. In principle, it can be solved numerically by summing the value $e^{-\lambda d(\boldsymbol{\mu}, \boldsymbol{\mu}_0)}$ over all possible $\boldsymbol{\mu}$ in $\mathcal{P}$. This numerical calculation could be time-consuming, as the computational time increases exponentially with the number of objects. In fact, various methods to avoid the problem of normalizing constant estimation have been proposed in the literature. For example, it suffices to estimate the ratio of two normalizing constants evaluating at two consecutive iterates of a simulated annealing algorithm, see Yu and Xu (2018). If the objective is to estimate $\boldsymbol{\mu}_0$ only, its estimation does not depend on λ and hence the computation of $C(\lambda)$ is not needed.

Given a ranking data set $\{\boldsymbol{\mu}_k, k = 1, \ldots, n\}$ and a known modal ranking $\boldsymbol{\mu}_0$, the maximum likelihood estimator (MLE) $\hat{\lambda}$ of the distance-based model can be found by solving the following equation:

$$\frac{1}{n}\sum_{k=1}^{n} d(\boldsymbol{\mu}_k, \boldsymbol{\mu}_0) = E_{\hat{\lambda}, \boldsymbol{\sigma}}[d(\boldsymbol{\mu}, \boldsymbol{\mu}_0)], \tag{4.26}$$

which equates the observed mean distance with the expected distance calculated under the distance-based model in (4.24).

The MLE can be found numerically because the observed mean distance is a constant and the expected distance is a strictly decreasing function of $\hat{\lambda}$. For the ease of solving, we re-parametrize λ with ϕ where $\phi = e^{-\lambda}$. The range of ϕ lies in $(0, 1]$ and the value of $\hat{\phi}$ can be obtained using the method of bisection. Critchlow (1985) suggested applying the method with 15 iterations, which yields an error of less than 2^{-15}. Also, the central limit theorem holds for the MLE $\hat{\lambda}$, which is shown in Marden (1995).

If the modal ranking $\boldsymbol{\mu}_0$ is unknown, it can be estimated by the MLE $\hat{\boldsymbol{\mu}}_0$ which minimizes the sum of the distances over $\mathcal{P}$, that is:

$$\hat{\boldsymbol{\mu}}_0 = \underset{\boldsymbol{\mu}_0 \in \mathcal{P}}{\operatorname{argmin}} \sum_{k=1}^{n} d(\boldsymbol{\mu}_k, \boldsymbol{\mu}_0). \tag{4.27}$$

For large values of t, a global search algorithm for the MLE $\hat{\boldsymbol{\mu}}_0$ is not practical because the number of possible rankings is too large. Instead, as suggested in Busse et al. (2007), a local search algorithm should be used. They suggested iteratively searching for the optimal modal ranking with the smallest sum of distances $\sum_{k=1}^{n} d(\boldsymbol{\mu}_k, \boldsymbol{\mu}_0)$ over $\boldsymbol{\mu}_0 \in \Pi^{(m)}$, where $\Pi^{(m)}$ is the set of all rankings having a Cayley distance of 0 or 1 to the optimal modal ranking found in the mth iteration:

$$\hat{\boldsymbol{\mu}}_0^{(m+1)} = \underset{\boldsymbol{\mu}_0 \epsilon \Pi^{(m)}}{\operatorname{argmin}} \sum_{k=1}^{n} d(\boldsymbol{\mu}_k, \boldsymbol{\mu}_0).$$

A reasonable choice of the initial ranking $\hat{\boldsymbol{\mu}}_0^{(0)}$ can be formed by ordering the mean ranks. Recently, Yu and Xu (2018) found in their simulation that this method may cause the $\hat{\boldsymbol{\mu}}_0^{(m+1)}$ stuck at a local minimum and cannot reach the global minimum. Yu and Xu (2018) proposed to use simulated annealing, a faster algorithm to find the global solution of the minimization problem in (4.27). Their simulation results revealed that simulated annealing algorithm always performs better than the local search algorithm even when the number of objects t becomes large. The local search algorithm generally performs satisfactory for small t but its performance deteriorates heavily when t gets large, say $t \geq 50$.

Distance-based models can handle partially ranked data in several ways, with some modifications in the distance measures. Beckett (1993) estimated the model parameters using the EM algorithm. On the other hand, Adkins and Fligner (1998) offered a non-iterative maximum likelihood estimation procedure for Mallows' ϕ-model without using the EM algorithm. Critchlow (1985) suggested replacing the distance metric d by the Hausdorff metric d^*. The Hausdorff metric between two partial rankings $\boldsymbol{\mu}^*$ and $\boldsymbol{\sigma}^*$ equals

$$d^*(\boldsymbol{\mu}^*, \boldsymbol{\sigma}^*) = \max[\max_{\boldsymbol{\mu} \in \boldsymbol{\mu}^*} \min_{\boldsymbol{\sigma} \in \boldsymbol{\sigma}^*} d(\boldsymbol{\mu}, \boldsymbol{\sigma}), \max_{\boldsymbol{\sigma} \in \boldsymbol{\sigma}^*} \min_{\boldsymbol{\mu} \in \boldsymbol{\mu}^*} d(\boldsymbol{\mu}, \boldsymbol{\sigma})]. \tag{4.28}$$

4.4.2. ϕ-Component Models

Fligner and Verducci (1986) extended the distance-based models by decomposing the distance $d(\boldsymbol{\mu}, \boldsymbol{\sigma})$ into $(t-1)$ distances,

$$d(\boldsymbol{\mu}, \boldsymbol{\sigma}) = \sum_{i=1}^{t-1} d_i(\boldsymbol{\mu}, \boldsymbol{\sigma}), \tag{4.29}$$

where the $d_i(,\boldsymbol{\sigma})$'s are independent. Note that Kendall distance and Cayley distance can be decomposed in this form.Fligner and Verducci (1986) developed two new classes of ranking models, called ϕ-component models and cyclic structure models, for the decomposition.

Fligner and Verducci (1986) showed that the Kendall distance satisfies (4.29):

$$d_K(\boldsymbol{\mu}, \boldsymbol{\mu}_0) = \sum_{i=1}^{t-1} V_i, \tag{4.30}$$

where

$$V_i = \sum_{j=i+1}^{t} I\{[\boldsymbol{\mu}(\boldsymbol{\mu}_0^{-1}(i)) - \boldsymbol{\mu}(\boldsymbol{\mu}_0^{-1}(j))] > 0\}. \tag{4.31}$$

We note that V_j has the uniform distribution on the integers $0, 1, \ldots, t-j$ (see Feller (1968), p. 257), and V_1 represents the number of adjacent transpositions required to place the best object in $\boldsymbol{\mu}_0$ in the first position, then remove this item in both $\boldsymbol{\mu}$ and $\boldsymbol{\mu}_0$, and V_2 is the number of adjacent transpositions required to place the best remaining object in $\boldsymbol{\mu}_0$ in the first position of the remaining items, and so on. Therefore, the ranking can be described as $t-1$ stages, V_1 to V_{t-1}, where $V_i = m$ can be interpreted as m mistakes made in stage i.

By applying the dispersion parameter λ_i at stage V_i, the Mallow's ϕ-model can be extended to:

$$\boldsymbol{\pi}(\boldsymbol{\mu}|\lambda, \boldsymbol{\mu}_0) = \frac{e^{-\sum_{i=1}^{t-1}\lambda_i V_i}}{C_K(\boldsymbol{\lambda})}, \tag{4.32}$$

where $\boldsymbol{\lambda} = \{\lambda_i, i = 1, \ldots, t-1\}$ and the normalizing constant $C_K(\boldsymbol{\lambda})$ is equal to

$$\prod_{i=1}^{t-1} \frac{1 - e^{-(t-i+1)\lambda_i}}{1 - e^{-\lambda_i}}. \tag{4.33}$$

Once again, the model in (4.31) can be expressed as

$$\boldsymbol{\pi}(\boldsymbol{\mu}|\lambda, \boldsymbol{\mu}_0) = \frac{e^{-\boldsymbol{\theta}' V}}{C_K(\boldsymbol{\lambda})},$$

where $\boldsymbol{\theta} = (\lambda_1, \ldots, \lambda_{t-1})'$ and $V = (V_1, \ldots, V_{t-1})'$.

These models were named $t-1$ parameter models in Fligner and Verducci (1986), but were also named ϕ-component models in other papers (e.g., Critchlow et al., 1991). Mallow's ϕ-models are special cases of ϕ-component models when $\lambda_1 = \ldots = \lambda_{t-1}$.

Based on a ranking data set $\{\boldsymbol{\mu}_k, k = 1, \ldots, n\}$ and a given modal ranking $\boldsymbol{\mu}_0$, the maximum likelihood estimates $\hat{\lambda}_i$, i = 1,2, ... ,$t-1$ can be found by solving the equation

$$\frac{1}{n}\sum_{k=1}^{n} V_{k,i} = \frac{e^{-\hat{\lambda}_i}}{1-e^{-\hat{\lambda}_i}} - \frac{(t-i+1)e^{-(t-i+1)\hat{\lambda}_i}}{1-e^{-(t-i+1)\hat{\lambda}_i}}, \tag{4.34}$$

where

$$V_{k,i} = \sum_{j=i+1}^{t} I\{[\boldsymbol{\mu}_k(\boldsymbol{\mu}_0^{-1}(i)) - \boldsymbol{\mu}_k(\boldsymbol{\mu}_0^{-1}(j))] > 0\}. \tag{4.35}$$

The left- and right-hand sides of (4.33) can be interpreted as the observed mean and theoretical mean of V_i respectively.

The extension of distance-based models to $t-1$ parameters allows more flexibility in the model, but unfortunately, the symmetric property of distance is lost. Notice here that the so-called "distance" in ϕ-component models can be expressed as

$$\sum_{i<j} \lambda_i I\{[\boldsymbol{\mu}(\boldsymbol{\mu}_0^{-1}(i)) - \boldsymbol{\mu}(\boldsymbol{\mu}_0^{-1}(j))] > 0\}, \tag{4.36}$$

which is obviously not symmetric, and hence it is not a proper distance measure. For example, in ϕ-component model, let $\boldsymbol{\mu} = (2,3,4,1)$, $\boldsymbol{\mu}_0 = (4,3,1,2)$.

$$\begin{aligned} d(\boldsymbol{\mu}, \boldsymbol{\mu}_0) &= \lambda_1 V_1 + \lambda_2 V_2 + \lambda_3 V_3 = 3\lambda_1 + 0\lambda_2 + 1\lambda_3 \neq 1\lambda_1 + 2\lambda_2 + 1\lambda_3 \\ &= d(\boldsymbol{\mu}_0, \boldsymbol{\mu}). \end{aligned}$$

The symmetric property of distance is not satisfied. Lee and Yu (2012) and Qian and Yu (2018) introduced new weighted distance measures which can retain the properties of a distance and also allow different weights for different ranks.

4.4.3. Cyclic Structure Models

Cayley's distance can also be decomposed into $t-1$ independent metrics. Fligner and Verducci (1986) showed that $d_C(\boldsymbol{\mu}, \boldsymbol{\mu}_0)$ can be decomposed as

$$d_C(\boldsymbol{\mu}, \boldsymbol{\mu}_0) = \sum_{i=1}^{t-1} X_i(\boldsymbol{\mu}, \boldsymbol{\mu}_0), \tag{4.37}$$

where $X_i(\boldsymbol{\mu}, \boldsymbol{\mu}_0) = I(i \neq \max\{\boldsymbol{\sigma}(i), \boldsymbol{\sigma}(\boldsymbol{\sigma}(i)), \ldots\})$, and $\boldsymbol{\sigma}(i) = \boldsymbol{\mu}(\boldsymbol{\mu}_0^{-1}(i))$.

This generalization can be illustrated by an example found in (Fligner and Verducci (1986)). Suppose there are t lockers, and each locker has one key that can open it. The $\boldsymbol{\mu}(\boldsymbol{\mu}_0^{-1}(i))$th key will be placed inside the ith locker. Without loss of generality, let the cost of breaking a locker be one. The minimum possible total cost of opening all lockers will then be $d_C(\boldsymbol{\mu}, \boldsymbol{\mu}_0)$, and it can be decomposed as the sum of costs of opening locker i, $i = 1,2, \ldots t-1$, which equals $X_i(\boldsymbol{\mu}, \boldsymbol{\mu}_0)$.

If we relax the assumption that the costs of breaking every locker are equal, the total cost will become

$$\sum_{i=1}^{t-1} \theta_i X_i(\boldsymbol{\mu}, \boldsymbol{\mu}_0), \tag{4.38}$$

where θ_i is the cost of opening locker i. This "total cost" can be interpreted as a weighted version of Cayley's distance. Similar to the extension of Mallow's ϕ models to ϕ-component models, Fligner and Verducci (1986) developed the cyclic structure models using the weighted Cayley's distance. Under this model assumption, the probability of observing a ranking $\boldsymbol{\mu}$ is

$$\boldsymbol{\pi}(\boldsymbol{\mu}|\boldsymbol{\theta}, \boldsymbol{\mu}_0) = \frac{e^{-\sum_{i=1}^{t-1} \theta_i X_i(\boldsymbol{\mu}, \boldsymbol{\mu}_0)}}{C_C(\boldsymbol{\theta})}, \tag{4.39}$$

where $\boldsymbol{\theta} = \{\theta_i, i = 1, \ldots, t-1\}$ and $C_C(\boldsymbol{\theta})$ is the normalizing constant, which equals

$$\prod_{i=1}^{t-1} \{1 + (t-i)e^{-\theta_i}\}. \tag{4.40}$$

For a ranking data set $\{\boldsymbol{\mu}_k, k = 1, \ldots, n\}$ with a given modal ranking $\boldsymbol{\mu}_0$, the MLEs $\hat{\theta}_i$, $i = 1,2, \ldots, t-1$ can be found from the equation

$$\hat{\theta}_i = \log(t-i) - \log \frac{\bar{X}_i}{1 - \bar{X}_i}, \tag{4.41}$$

where

$$\bar{X}_i = \frac{\sum_{k=1}^{n} X_i(\boldsymbol{\mu}_k, \boldsymbol{\mu}_0)}{n}. \tag{4.42}$$

4.5. Goodness of Fit Tests for Two-Way Contingency Tables

We may also consider doubly ordered two-way $r \times c$ contingency tables of counts which we denote by

$$\{N_{ij}\}, i = 1, \ldots, r, j = 1, \ldots, c$$

We are interested in testing for independence and consequently define the cell probabilities as

$$\pi_{ij}(\boldsymbol{\theta}) = exp\left\{\sum_{u=1}^{k_1} \sum_{v=1}^{k_2} \theta_{uv} g_u(i) h_v(j)\right\} p_{i.} p_{.j} \tag{4.43}$$

where

$$\sum_{i=1}^{r}\sum_{j=1}^{c}p_{ij}=\sum_{i=1}^{r}p_{i.}=\sum_{j=1}^{c}p_{.j}=1.$$

The $\{g_u(i)\}$ are orthonormal functions on the marginal row probabilities and the $\{h_v(j)\}$ are orthonormal functions on the marginal column probabilities. The test for independence is then a test of

$$H_0:\theta=0 \; vs \; H_1:\theta\neq 0$$

where $\boldsymbol{\theta}=(\theta_{11},\ldots,\theta_{1k_2},\ldots,\theta_{k_1 1},\ldots,\theta_{k_1 k_2})'$.

Set

$$\hat{V}_{uv}=\sum_{i}^{r}\sum_{j}^{c}N_{ij}\hat{g}_u(i)\,\hat{h}_v(j)\,/\sqrt{n}$$

where $\sum_i^r\sum_j^c N_{ij}=n$. Here, $\{\hat{g}_u(i)\}$ are the set of polynomials orthogonal to $\{\hat{p}_{i.}\}$ where $\hat{p}_{i.}=\sum_j N_{ij}/n$. Similarly, $\left\{\hat{h}_v(j)\right\}$ are the set of polynomials orthogonal to $\{\hat{p}_{.j}\}$ where $\hat{p}_{,j}=\sum_i N_{ij}/n$. The following theorem, proven in (Rayner et al. (2009b)), shows that we may obtain the usual test statistic as a consequence of the smooth model (4.43).

Theorem 4.3. *The score statistic for testing H_0 vs H_1 is given by $\sum_{u=1}^{k_1}\sum_{v=1}^{k_2}\hat{V}_{uv}^2$ where under the null hypothesis, the components $\hat{V}_{uv}$ are asymptotically i.i.d. standard normal variables.*

When $k_1=(r-1)\,,k_2=(c-1)\,,$ the test statistic is the usual Pearson statistic

$$\mathcal{X}_P^2=\sum_i\sum_j\frac{\left(N_{ij}-n\hat{p}_{i.}\hat{p}_{.j}\right)^2}{n\hat{p}_{i.}\hat{p}_{.j}}\to\chi^2_{k_1k_2}$$

Chapter Notes

1. Smooth tests for goodness of fit were introduced by Neyman (1937) when no nuisance parameters were involved. Since a probability integral transformation can transform a distribution to a uniform, the orthogonal polynomials were taken to be those of Legendre. The tests developed were locally uniformly most powerful, symmetric, and unbiased. A good introduction along with several references are given in Rayner et al. (2009b). See also Rayner et al. (2009a) and Rayner et al. (2009b) for generalizations and extensions of smooth tests of fit including smooth tests in randomized blocks with ordered alternatives.

2. The polynomials may be taken to be the discrete or Hermite polynomials whose first two components are

$$g_1(j) = \left(j-\frac{k+1}{2}\right)\sqrt{\frac{12}{k^2-1}},$$

$$g_2(j) = \left\{ \left(j - \frac{k+1}{2} \right)^2 - \left(\frac{k^2-1}{12} \right) \right\} \sqrt{\frac{180}{(k^2-1)(k^2-4)}}.$$

Higher order components may be obtained using the recurrence equations as described in (Rayner et al. (2009b), p. 243). Additional polynomials may be obtained from the usual three term recursion formulas (Kendall and Stuart, 1979).

3. We note that the smooth testing approach here leads to a study of global properties of the data. This is to be contrasted with the approach in Chapter 7 whereby we will consider smooth tests that incorporate specific score functions such as those of Spearman and Kendall. In those instances, the vector parameter $\boldsymbol{\theta}$ places a weighting on the components of the score function. The approach enables us to study more precisely local properties.

4. Lancaster (1953) considered a decomposition of the chi-square statistic in connection with testing for goodness of fit in contingency tables which helps to assess the individual contributions of the components.

5. A simple application of Neyman's smooth tests is to the problem of combining p-values which under the null hypothesis are uniformly distributed (Rayner et al. (2009a), p. 63).

4.6. Exercises

Exercise 4.1. Prove Theorem 4.2.

Exercise 4.2. Suppose that we are given the smooth binomial distribution given in Table 4.2. Find the score statistic to test the hypothesis that $\theta = 0$.

Exercise 4.3. Repeat Exercise 4.2 using a sample of size n from the smooth Poisson distribution given in Table 4.2 and test the hypothesis that $\theta = 0$.

5. One-Sample and Two-Sample Problems

In this chapter we consider several one- and two-sample problems in nonparametric statistics. Our approach will have a common thread. We begin by embedding the nonparametric problem into a parametric paradigm. This is then followed by deriving the score test statistic and finding its asymptotic distribution. The construction of the parametric paradigm often involves the use of composite likelihood. It will then be necessary to rely on the use of either linear rank statistics or U-statistics in order to determine the asymptotic distribution of the test statistic. We shall see that the parametric paradigm provides new insights into well-known problems. Starting with the sign test, we show that the parametric paradigm deals easily with the case of ties. We then proceed with the Wilcoxon signed rank statistic and the Wilcoxon rank sum statistic for the two-sample problem.

5.1. Sign Test

Suppose that we have a random sample $X_1, \ldots, X_n$ from a population having a (not necessarily continuous) distribution $F_X(x)$ with a unique median M (i.e., $F_X(M) = 0.5$). We would like to test the hypotheses

$$H_0 : M = M_0 \text{ versus } H_1 : M > M_0. \tag{5.1}$$

Under the parametric paradigm, we first define a score function, sensitive to changes in the median. Let $Y = \text{sgn}\,(X - M_0)$ and define the change of measure by

$$\pi\,(y;\theta) = \exp\,(\theta y - K(\theta))\,g(y), \quad y = -1, 0, 1, \tag{5.2}$$

where $g(y)$ represents the null probability distribution of Y. Let $g(1) = p_+, g(0) = p_0$ and $g(-1) = p_-$ such that

$$p_+ + p_0 + p_- = 1.$$

M. Alvo, P. L. H. Yu, *A Parametric Approach to Nonparametric Statistics*,
Springer Series in the Data Sciences, https://doi.org/10.1007/978-3-319-94153-0_5

It is natural to assume that $p_+ = p_-$. Here $K(\theta)$ satisfies

$$p_+ e^{\theta} + p_- e^{-\theta} + p_0 = e^{K(\theta)}.$$

Since $p_+ + p_- + p_0 = 1$ and $p_+ = p_-$, it follows that $K(0) = K'(0) = 0$. It is easy to see that testing the hypotheses in (5.1) is the same as testing

$$H_0 : \pi(1; \theta) = \pi(-1; \theta)$$

versus

$$H_1 : \pi(1; \theta) > \pi(-1; \theta)$$

or equivalently

$$H_0 : \theta = 0 \text{ versus } H_1 : \theta \neq 0$$

In a sample of size n, let n_+, n_-, n_0 denote the observed number of cases where $y = 1, -1, 0$, respectively. Note that under H_0, the score function is given by

$$U(\theta; \boldsymbol{X}) = n_+ - n_-$$

and the Fisher information is $I = nK''(0) = n(1-\hat{p}_0) = n-n_0$. The score test statistic is

$$S_{sign} = \frac{[U(\theta; \boldsymbol{X})]^2}{I} = \frac{(n_+ - n_-)^2}{n - n_0} = \frac{4\left(n_+ + \frac{n_0}{2} - \frac{n}{2}\right)^2}{n - n_0} = \frac{4\left(n_+ - \frac{n-n_0}{2}\right)^2}{n - n_0}. \tag{5.3}$$

Consequently large values of S_{sign} lead to rejection of the null hypothesis H_0. We see then that under H_0,

$$S_{sign} \xrightarrow{\mathcal{L}} \chi_1^2$$

as $n \to \infty$.

Remark 5.1. We note that the sign test takes into account situations where ties (i.e., $X_i = M_0$) are possible and the score function leads naturally to the statistic $n_+ + \frac{n_0}{2}$ often suggested without justification in the literature. Note that the last expression of the score test statistic S_{sign} in (5.3) seems to recommend the usual treatment of ties, namely to reduce the sample size by discarding the tied observations.

Remark 5.2. In the case of no ties (i.e., $n_0 = p_0 = 0$), the score test statistic S_{sign} only depends on n_+ which has a binomial distribution with probability p_+, i.e., $n_+ \sim Bin(n, p_+)$. Under H_0, we have $p_+ = 0.5$ and hence $n_+ \sim Bin(n, 0.5)$. For $H_1 : M > M_0$, we have $p_+ > 0.5$ and hence large values of n_+ will lead to rejection of H_0. The p-value of the sign test is then $\Pr(B \geq n_+^*)$ where $B \sim Bin(n, 0.5)$ and n_+^* is the observed value n_+. Similarly, for $H_1 : M < M_0$, the p-value is $\Pr(B \leq n_+^*)$ and for $H_1 : M \neq M_0$, the p-value is $2 \times \min\{\Pr(B \geq n_+^*), \Pr(B \leq n_+^*)\}$.

Example 5.1. The weekly sales of new mobile phones at a mobile phone shop in a mall are collected over the past 12 weeks. The number of phones sold is recorded for each of 12 weeks and are given below:

45 32 39 29 64 55 38 212 187 124 320 188

Last year, the *median* weekly sales was 50 units. Is there sufficient evidence to conclude that median sales this year are higher than last year? Test the hypothesis at a significance level $\alpha = 0.05$.

Solution. Let M be population median weekly sales this year. The hypotheses are

$$H_0 : M = 50 \text{ versus } H_1 : M > 50,$$

$n = 12$ and $n_+ = 7$. Set $B \sim Bin(12, 0.5)$. Then, the p-value for the test is given by

$$p\text{-value} = \Pr(B \geq 7) = \sum_{i=7}^{12} \binom{12}{i} 0.5^{12} = 0.3872.$$

Since the p-value is greater than 0.05, we do not have enough evidence to reject H_0 at the 5% level of significance. Thus, there are no grounds upon which to conclude that the median sales are now higher than 50 units per week.

In R, the function `pbinom(x,n,p)` will calculate $\Pr(X \leq x)$ where $X \sim Bin(n, p)$.

Remark 5.3. In the presence of ties, it is easy to show that under H_0, $n_+ + \frac{n_0}{2}$ has mean $\frac{n}{2}$ and variance $\frac{n}{4}(1 - p_0)$ but it no longer has a binomial distribution. When the sample size n is large, we can apply the normal approximation to its null distribution.

Example 5.2. A food product is advertised to contain 75 mg of sodium per serving, while preliminary studies indicate that servings may contain more than that amount. We can formulate this problem as a test for the median M of the amount of sodium per serving:

$$H_0 : M = 75 \text{ versus } H_1 : M > 75.$$

Suppose that 40 packages of the company's food product are examined of which 26 packages were observed to contain sodium amounts per serving exceeding 75 mg. Then using the normal approximation to the binomial probability, the p-value of sign test is

$$\begin{aligned} \Pr(B \geq 26) &= \Pr(B > 25.5) && \text{(continuity correction)} \\ &= \Pr(Z > \frac{25.5 - 20}{\sqrt{0.25 \times 40}}) && (Z \sim \mathcal{N}(0,1) \text{ under } H_0) \\ &= \Pr(Z > 1.7393) = 0.041 \end{aligned}$$

which is close to the exact p-value 0.040. Since the p-value is less than 0.05, we conclude that the median amount of sodium per serving is greater than 75 mg at the 5% level of significance.

In R, the function `pnorm(x,μ,σ)` will calculate $\Pr(X \leq x)$ where $X \sim \mathcal{N}(\mu, \sigma^2)$.

Remark 5.4. Since only the signs of $\{X_i - M_0\}$ but not its magnitude are used, the sign test has the advantage that it can be utilized when only the signs are available. The test is also robust against outliers, as only the sign of the outlier is of interest no matter how far it is away from M_0. When either $p_0 = 0$ or $n_0 = 0$, we obtain the usual sign test (Hájek and Sidak (1967), p. 110).

Remark 5.5 (Paired Comparisons). There are several applications of the sign test for paired comparisons whereby the null hypothesis is that the distribution of the difference $Z = Y - X$ is symmetric about 0. As a special case, we can consider the shift model where $Z = (Y - \Delta) - X$ so that the treatment adds a shift of value Δ to the control. The asymptotic properties of the sign test are discussed in several textbooks. See, for example, Chapter 14 of van der Vaart (2007) and Chapter 4 of Lehmann (1975).

5.1.1. Confidence Interval for the Median

Definition 5.1. A confidence interval for a parameter ϕ is called *distribution-free* if

$$\Pr(L < \phi < U) = 1 - \alpha$$

is true no matter what the distribution F is where $L = f_1(X_1, \ldots, X_n)$ and $U = f_2(X_1, \ldots, X_n)$ are statistics.

Suppose the distribution of the data $F_X(x)$ is continuous. The procedure for constructing a symmetric two-sided distribution-free confidence interval for the median M with level of confidence $1 - \alpha$ is given below:

(a) Find c_α such that $\Pr(c_\alpha \leq B^* \leq n - c_\alpha) = 1 - \alpha$, with $B^* \sim Bin(n, 0.5)$.

(b) Let $X_{(1)} < X_{(2)} < \cdots < X_{(n)}$ be the **order statistics**.

(c) $(X_{(c_\alpha)}, X_{(n+1-c_\alpha)})$ is a $100(1 - \alpha)\%$ confidence interval for M as it satisfies $\Pr(X_{(c_\alpha)} < m < X_{(n+1-c_\alpha)}) = 1 - \alpha$.

We may show

$$\Pr(X_{(c_\alpha)} < M < X_{(n+1-c_\alpha)}) = 1 - \alpha.$$

By definition, c_α is chosen such that

$$\Pr(c_\alpha \leq B^* \leq n - c_\alpha) = 1 - \alpha.$$

Now

$$\begin{aligned} B^* \geq c_\alpha &\Leftrightarrow \text{ number of } (X_i > M) \geq c_\alpha \\ &\Leftrightarrow M < \text{ at least } c_\alpha \text{ of } X_i \Leftrightarrow M < X_{(n+1-c_\alpha)} < \cdots < X_{(n)} \\ &\Leftrightarrow M < X_{(n+1-c_\alpha)}. \end{aligned}$$

Similarly,

$$\begin{aligned} B^* \leq n - c_\alpha &\Leftrightarrow \text{ number of } (X_i > M) \leq n - c_\alpha \\ &\Leftrightarrow M < \text{ at most } n - c_\alpha \text{ of } X_i \\ &\Leftrightarrow M > \text{ at least } c_\alpha \text{ of } X_i \Leftrightarrow X_{(1)} < \cdots < X_{(c_\alpha)} < M \\ &\Leftrightarrow M > X_{(c_\alpha)}. \end{aligned}$$

Therefore, the confidence interval is $X_{(c_\alpha)} < M < X_{(n+1-c_\alpha)}$.

Example 5.3. Suppose we have a set of $n = 7$ observations

$$\{2, -9, 11, 40, 10, 18, 0\}.$$

Then, if $B^* \sim Bin(7, \frac{1}{2})$,

$$\Pr\{B^* \geq 6\} = 0.0625 \Leftrightarrow \Pr(2 \leq B^* \leq 5) = 1 - 2 \times 0.0625 = 0.875.$$

Thus $c_\alpha = 2$ and the 87.5 confidence interval for M is

$$(X_{(2)}, X_{(7+1-2)}) = (X_{(2)}, X_{(6)}) = (0, 18).$$

Remark 5.6. For a large sample, we may approximate $B^* \sim Bin(n, \frac{1}{2})$ by $B^* \sim \mathcal{N}(\frac{n}{2}, \frac{n}{4})$. Hence c_α can be approximated by

$$c_\alpha \approx \frac{n}{2} - z_{\frac{\alpha}{2}} \left(\frac{n}{4}\right)^{\frac{1}{2}}$$

since we have

$$\begin{aligned} &\Pr\left(-z_{\frac{\alpha}{2}} \leq \frac{B^* - \frac{n}{2}}{(\frac{n}{4})^{\frac{1}{2}}} \leq z_{\frac{\alpha}{2}}\right) = 1 - \alpha \\ \Leftrightarrow\ &\Pr\left[\frac{n}{2} - z_{\frac{\alpha}{2}} \left(\frac{n}{4}\right)^{\frac{1}{2}} \leq B^* \leq \frac{n}{2} + z_{\frac{\alpha}{2}} \left(\frac{n}{4}\right)^{\frac{1}{2}}\right] = 1 - \alpha. \end{aligned}$$

Remark 5.7. A similar procedure can be formulated to determine a confidence interval for the $(100p)^{th}$ percentile or cdf of a continuous distribution.

5.1.2. Power Comparison of Parametric and Nonparametric Tests

5.1.2.1. Parametric Test for Large Samples: CLT Test

Let $X_1, X_2, \ldots, X_n$ denote a random sample from a distribution with mean μ and variance σ^2. The central limit theorem (CLT) states that for large samples, $\bar{X} \xrightarrow{d} \mathcal{N}(\mu, \sigma^2)$, regardless of the form of the underlying distribution.

For a sample from a symmetric distribution, the population mean and the population median are the same, and thus any test for the mean is also a test for the median. The hypotheses

$$H_0 : M = M_0 \text{ vs } H_1 : M > M_0$$

for the sign test are therefore equivalent to:

$$H_0 : \mu = \mu_0 \text{ vs } H_1 : \mu > \mu_0$$

where $\mu_0 = M_0$ is the hypothesized value of the population mean of X. When the standard deviation σ is known, we use the test statistic:

$$Z = \frac{\bar{X} - \mu_0}{\sigma/\sqrt{n}}.$$

The null hypothesis H_0 is rejected in favor of H_1 at the significance level α if $Z > z_\alpha$, the upper $100\alpha\%$ point of the standard normal distribution.

Remark 5.8. The CLT test is not applicable for populations that come from a distribution with an infinite variance, such as the Cauchy distribution.

5.1.2.2. Comparison of Sign Test and CLT Test

To choose between the sign test and the CLT test, we pay attention to two statistical issues: the **Type I error** which consists of rejecting H_0 when H_0 is true and the power which is defined to be the probability of rejecting H_0.

For our discussion, we shall only consider the large sample case with σ known. The probability of committing a Type I error should be at least close to the significance level α. Though bias may occur when using an approximation, the normal approximation for both CLT and sign tests is considered to be quite good for large samples. Therefore, the stated probability of committing a Type I error will be essentially correct.

To compare the power of the two tests, we must consider the population distribution under H_1. Referring to Example 5.2, let us assume the true median of sodium content is 75.8 mg with $\sigma = 2.5$ and the sample is selected from a **normal** population. Then

$$\begin{aligned}
\text{power of CLT test} &= \Pr(\frac{\bar{X} - 75}{2.5/\sqrt{40}} \geq 1.645 \mid \mu = 75.8) \\
&= \Pr(\frac{\bar{X} - \mu}{2.5/\sqrt{40}} \geq 1.645 - \frac{\mu - 75}{2.5/\sqrt{40}} \mid \mu = 75.8) \\
&= 1 - \Phi(1.645 - \frac{75.8 - 75}{2.5/\sqrt{40}}) = 0.65
\end{aligned}$$

Consider the sign test, if $\mu = 75.8$, we have $p \equiv \Pr(X > 75) = 0.626$.

$$\begin{aligned}
\text{power of sign test} &= \Pr(\frac{B - 20}{\sqrt{0.25 \times 40}} \geq 1.645 \mid B \sim Bin(40, 0.626)) \\
&= \Pr(\frac{B - 40p}{\sqrt{40p(1-p)}} \geq 1.645\sqrt{\frac{0.25}{p(1-p)}} - \frac{40p - 20}{\sqrt{40p(1-p)}} \mid B \sim Bin(40, 0.626)) \\
&= 1 - \Phi(1.645\sqrt{\frac{0.25}{p(1-p)}} - \frac{40p - 20}{\sqrt{40p(1-p)}}) = 0.48
\end{aligned}$$

In this example, the CLT test is preferred since the power is greater.

Remark 5.9. When samples are taken from normal populations with a known variance, the CLT test has the greatest power among all tests (*uniformly most powerful test*). It's the test to use when sampling from a normal population. But for nonnormal populations, this is not the case. The sign test will have higher power than the CLT test for heavy-tailed distributions, including the Cauchy or Laplace distributions. For example, if the true distribution is Laplace, the power of the sign test is 0.76.

5.2. Wilcoxon Signed Rank Test

As we saw in Section 5.1, the sign test can be derived from the theory of U-statistics by defining the kernel function $h(Z_i)$ on the sign of the difference $Z_i = X_i - M$,

$$h(Z_i) = \text{sgn}(Z_i), \quad i = 1, 2, \ldots, n.$$

The Wilcoxon signed ranked test which takes into account both the rank and the sign is more powerful. It is used to test the hypothesis that the distribution of X is symmetric about M. Consider the two-variable kernel

$$Y_{ij} = h(Z_i, Z_j) = I(Z_i + Z_j > 0), \; i \leq j, \; i, j = 1, 2, \ldots, n$$

where $(Z_i + Z_j)$ are the Walsh sums and define the smooth model

$$\pi(y_{ij}; \theta) = \exp\left[\theta y_{ij} - K(\theta)\right] g_Y(y_{ij}),\ i \leq j,\ i, j = 1, 2, \ldots, n$$

where g_Y is the density of Y assumed to be symmetric around 0 under the null hypothesis,

$$H_0 : \theta = 0.$$

It is easy to see that

$$K(0) = 0, K'(0) = \frac{1}{2}.$$

The log of the composite likelihood function is proportional to

$$\ell(\theta) \sim \left[\theta \sum_{i \leq j} I(z_i + z_j > 0) - \binom{n+1}{2} K(\theta)\right].$$

The score vector is

$$W_n^+ = \sum_{i \leq j} I(z_i + z_j > 0). \tag{5.4}$$

The usual Wilcoxon signed-rank statistic, SR_+, is defined as

$$SR_+ = \sum_{i=1}^{n} R_i^+ I(Z_i > 0),$$

where R_i^+ be the rank of $|Z_i|$ among $\{|Z_j|, j = 1, \ldots, n\}$. The following lemma shows an equivalent form between W_n^+ and SR_+.

Lemma 5.1. $W_n^+ = SR_+$.

Proof. Suppose without loss of generality that the $Z_i's$ are ordered in absolute value:

$$|Z_1| < \ldots < |Z_n|.$$

Fix j and consider all the pairs $(Z_i, Z_j), i \leq j$. The sum $Z_i + Z_j > 0$ if and only if $Z_j > 0$. The number of indices for which this is true is equal to j, the rank of $|Z_j|$. □

Though itself not a U-statistic, W_n^+ is, in fact, the sum of two U-statistics:

$$\begin{aligned} W_n^+ &= \sum_{i \leq j} I(Z_i + Z_j > 0) \\ &= \sum_{i} I(Z_i > 0) + \sum_{i<j} I(Z_i + Z_j > 0) \\ &= nU_{1n} + \binom{n}{2} U_{2n} \end{aligned} \tag{5.5}$$

with respective kernels $I(Z_i > 0)$ and $I(Z_i + Z_j > 0)$. Under the null hypothesis, we have that

$$E\left(\sum_i I(Z_i > 0)\right) = \frac{n}{2}, \quad E\left(\sum_{i<j} I(Z_i + Z_j > 0)\right) = \frac{n(n-1)}{4}$$

and hence

$$E\left(W_n^+\right) = \frac{n(n+1)}{4}.$$

As well,

$$\begin{aligned} Var(W_n^+) &= Var\left(nU_{1n}\right) + Var\left(\binom{n}{2}U_{2n}\right) + 2Cov\left(nU_{1n}, \binom{n}{2}U_{2n}\right) \\ &= \frac{n}{4} + \frac{n(n^2-1)}{12} + \frac{n(n-1)}{8} \\ &= \frac{n(n+1)(2n+1)}{24} \end{aligned}$$

It can be shown that for large n, the asymptotic distribution of W_n^+ is determined by the distribution of the second term. The test rejects for large values of W_n^+. Specifically, as $n \to \infty$

$$\frac{W_n^+ - n(n+1)/4}{\sqrt{n(n+1)(2n+1)/24}} \xrightarrow{\mathcal{L}} \mathcal{N}(0,1).$$

An adjustment for the variance can be made in the case of ties. Specifically, we replace the variance above by

$$n(n+1)(2n+1)/24 - \frac{\sum t_l^3 - \sum t_l}{48},$$

where $\{t_l\}$ represent the number of absolute differences tied for a particular nonzero rank. See Lehmann (1975).

Remark 5.10. The extension to paired-sample problems is straightforward. Given a random sample of paired observations $(X_i, Y_i), i = 1, \ldots, n$, the basic idea of the Wilcoxon signed rank test for paired samples is to apply the test to the differences $Z_i = X_i - Y_i, i = 1, \ldots, n$, where the null hypothesis is that the population median of Z is 0. Similarly, we can also apply the sign test to paired samples in the same manner.

Example 5.4. A speed-typing course was given to 18 clerks in the hope of training them to be more efficient typists. Each clerk was tested on her typing speed (in w.p.m.) before and after the course. The results are given in the table below. With a 1% significance level use Wilcoxon signed rank test to see whether the course was effective.

After	53	33	54	61	55	57	40	59	58	53	59	62	51	43	64	68	48	60
Before	42	35	48	52	60	43	36	63	51	45	56	50	41	38	60	52	48	57

Solution. We are proposing to test these hypotheses:

H_0 : There is no difference between a clerk's typing speed before and after the course;

H_1 : A clerk's typing speed generally increases after completing the course.

If we do not assume any statistical distribution (say, a normal distribution) for a clerk's typing speed, a nonparametric Wilcoxon signed rank test is applicable. The test proceeds as follows:

Step 1. The speed differences (Z = After - Before) are first obtained. We are testing

$$H_0 : M = 0 \text{ against } H_1 : M > 0,$$

where M is the median of Z.

Step 2. Their signs being temporarily suspended, the magnitudes of the differences are ranked, starting from the smallest. Tied cases are given by the "average" of the ranks that would have been given if no ties were present.

Z_i	11	−2	6	9	−5	14	4	−4	7	8	3	12	10	5	4	16	−1	3
R_i^+	15	2	10	13	8.5	17	6	6	11	12	3.5	16	14	8.5	6	18	1	3.5

Step 3. The signs of the original differences are then restored to the ranks, and the sum of the positive ranks, $W_n^+ = 153.5$, is the value of the test statistic.

Step 4. Since H_1 is a one-sided hypothesis, a one-tailed test is appropriate. As a bigger SR_+ value produces a stronger support for H_1, one can refer to the table of critical values (see Table 20 of Lindley and Scott (1995)) with $n = 18$ and $\alpha = 0.01$ for decision making. This critical value is 139. We therefore reject H_0 at 1% significance level and conclude that the course was effective.

The Wilcoxon signed rank test can be implemented using R function `wilcox.test`. For example, `wilcox.test(x, mu=0, alternative = "greater")` produces an exact p-value of the test with alternative being the median greater than 0 if the sample contains less than 50 observations and has no ties. Otherwise, a normal approximation is used.

Remark 5.11. One can also apply the sign test in this example. For $n = 18$ the data exhibit 14 positive signs, which is insignificant at 1% level (critical value being 15). It is interesting to see that the sign test uses less information derived from data than the Wilcoxon signed rank test and the strength of evidence supporting H_1 is also lower.

5.3. Two-Sample Problems

Suppose that we are presented with two independent random samples, $X_1, \ldots, X_n$ and $Y_1, \ldots, Y_m$ with continuous cumulative distributions F and G, respectively. Such would be the case if one is interested in assessing the effect of a treatment on a group of patients.

One sample would serve as the control group and the other as the treatment group. We would like to test the null hypothesis

$$H_0 : F = G$$

against the alternative

$$H_1 : F \neq G.$$

It is typically assumed that in the shift model, the two distributions only differ by a location parameter Δ:

$$F(x) = G(x - \Delta).$$

That is, $X - \Delta$ and Y have the same distribution. If the treatment effect depends on x, then a more general alternative might be

$$H_1 : F(x) > G(x) \quad \text{or} \quad H_1 : \Delta < 0$$

which indicates that small values are more likely under F than under G. In that case, we say that Y is stochastically larger than X. We consider in the next section a test for this more general alternative.

Example 5.5. Consider the usual two-sample t-test involving independent samples whereby $X_1, X_2, \ldots, X_n$ are iid $\mathcal{N}(\mu_x, \sigma^2)$ and $Y_1, Y_2, \ldots, Y_m$ are iid $\mathcal{N}(\mu_y, \sigma^2)$. Then $F(x) = \Phi(\frac{x-\mu_x}{\sigma})$ and $G(y) = \Phi(\frac{y-\mu_y}{\sigma})$, where $\Phi(\cdot)$ is the cdf of $\mathcal{N}(0,1)$. Then it is easy to see that $\Delta = \mu_x - \mu_y$.

5.3.1. Permutation Test

We include for completeness a brief discussion of permutation tests. Permutation or randomization tests as they are sometimes called provide an effective though computationally intensive method for determining the sampling distribution of a test statistic under the null hypothesis of "no association." The basic idea which originated with R.A. Fisher (1935, Chapter 3) consists of permuting the labels of the observed data and computing the value of the test statistic. This operation is repeated a large number of times thereby creating a histogram of values that approximates the null distribution of the test statistic. A p-value may then be calculated by counting the number of cases which yield values at least as extreme as the one obtained from the observed data. Since the permutation test is conditional on the observed data, it is distribution-free. Permutation tests are particularly useful when the data sets are small and the underlying distribution of the test statistic is uncertain. Further results on permutation tests may be found in the book written by Mielke and Berry (2001).

Example 5.6. Let $X_1, \ldots, X_n$ be a random sample from a continuous cdf $F(x)$ and let $Y_1, \ldots, Y_m$ be an independent random sample from $F(x + \Delta)$, where Δ is finite. Let $N = n + m$. We would like to test the hypotheses

$$H_0 : \Delta = 0$$

against

$$H_1 : \Delta > 0.$$

Suppose that the test statistic to be used is given by the difference in the sample means

$$D = \bar{X} - \bar{Y}.$$

Let D_{obs} be the observed value of D. Under the null hypothesis, the combined sample of N observations is assumed to come from the same cdf F. Ignoring the labels which identify the distribution from which the observations came, there would be $\binom{N}{n}$ possible samples of n observations assigned to the first population and m observations assigned to the second. This is precisely the number of possible permutations of the labels and each is equally likely. For each of these permutations, we may compute a value of the statistic T thereby creating a reference distribution (which also includes the observed D_{obs}) and then count how many of these exceed the observed D_{obs} in order to calculate the p-value.

Example 5.7. Fungal infection was believed to have certain effect on the eating behavior of rodents. In an experiment, infected apples were offered to a group of eight randomly selected rodents, and sterile apples were offered to a group of four. The amounts consumed (grams of apple per kilogram of body weight) are listed in the table below. Test whether fungal infection significantly reduces the amounts of apples consumed by rodents.

Experimental Group (X_i)	11, 33, 48, 34, 112, 369, 64, 44
Control Group (Y_i)	177, 80, 141, 132

In this example, no assumption is made about the distribution of the data. Given that the assignment of rodents to either group is random, if there is no difference (H_0) between the two groups, partitions of the 12 scores into two groups of sizes 8 and 4 will be equally likely to occur. By permuting the 7 scores, we obtain a total of $\binom{12}{4} = 495$ partitions as follows:

Permuted Samples	Experimental	Control	Difference Between Means
1	11 33 48 34 112 64 44 80	369 177 141 132	−151.500
2	11 33 48 34 64 44 80 132	112 369 177 141	−144.000
3	11 33 48 34 64 44 80 141	112 369 177 132	−140.625
4	11 33 48 34 112 64 44 132	369 177 80 141	−132.000
⋮	⋮	⋮	⋮
135	11 33 112 64 44 177 141 132	48 34 369 80	−43.500
136*	11 33 48 34 112 369 64 44	177 80 141 132	−43.125
137	11 34 112 64 44 177 141 132	33 48 369 80	−43.125
138	11 33 48 112 64 177 141 132	34 369 44 80	−42.000
⋮	⋮	⋮	⋮
495	48 112 369 64 177 80 141 132	11 33 34 44	109.875

* refers to the observed data.

Using the difference between the sample means, i.e., $D = \overline{X} - \overline{Y}$, as our test statistic, we obtain the following permutation distribution of the difference:

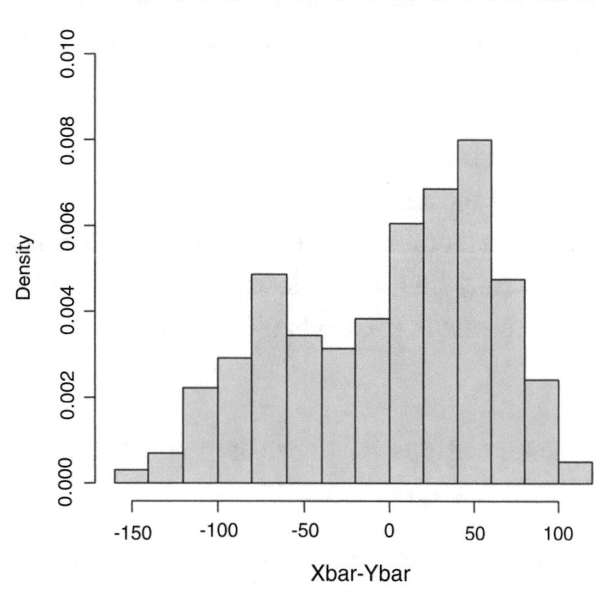

It can be seen from the above figure that under H_0 small differences tend to occur more frequently, which is reasonable as the difference will tend to fall around 0 if the two methods are the same.

In a lower-tailed test, to calculate the p-value of the observed difference is equivalent to find the probability of observing a difference of means of -43.125 or smaller under the assumption that the two groups do not differ. In this example, we have p-value $= \frac{137}{495} = 0.2768$.

Remark 5.12. There are other possible choices of the test statistic in the permutation test:

- Difference of means, $D = \overline{X} - \overline{Y}$
- Sum of the observations for one group, $T_1 = \sum_{i=1}^{n} X_i$ (or $T_2 = \sum_{j=1}^{m} Y_j$). Since $T = T_1 + T_2$ is fixed given the observed data, we have $D = \frac{T_1}{n} - \frac{T_2}{m} = T_1(\frac{1}{n} + \frac{1}{m}) - \frac{T}{m}$ and thus D and T_1 are equivalent test statistics.
- Two independent sample t-test statistic
- Difference of medians, $X_{0.5} - Y_{0.5}$. The median has the benefit of being robust to a few extreme observations (called outliers)
- Difference of trimmed means. In a trimmed mean, we remove extreme observations before taking the average. For example, if we have ten observations given by 2, 8, 9, 10, 11, 11, 12, 13, 14, 200, then to get a 20% trimmed mean we delete 2, 200 and take average, which is 11. The trimmed mean is useful because it is less sensitive to outliers than the mean.

Generally, deciding which statistic to use requires some advance knowledge of the population. The difference of the means is most commonly used, especially when the data come from an approximate normal distribution. But if the population has an asymmetric distribution, the median may be a more desirable indicator of the center of the data. The difference of trimmed means is used when the distribution is symmetric but likely to have outliers.

Here we summarize the general steps for a two-sample permutation test assuming that a large test statistic T tends to support $\Delta > 0$:

1. Based on the original data, compute the observed test statistic, T_{obs} (e.g., the difference between the two-sample means).
2. Permute the $n + m$ observations from the two treatments so that there are n observations for population 1 and m observations for population 2. Obtain all possible permutations, $\binom{n+m}{n}$ in total. Compute the value of the test statistic T for each permutation.

3. Calculate the p-value based on the test statistic T_{obs}:

$$P_{upper-tail} = \frac{\text{no. of } T's \geq T_{obs}}{\binom{m+n}{n}}$$

$$P_{lower-tail} = \frac{\text{no. of } T's \leq T_{obs}}{\binom{m+n}{n}}$$

$$P_{two-tail} = \frac{\text{no. of } |T|'s \geq |T_{obs}|}{\binom{m+n}{n}}$$

4. Declare the test to be significant if the p-value is less than or equal to the desired significance level.

Remark 5.13. The permutation test can be rather tedious as sample sizes m and n increase. For instance, $\binom{20}{10} = 184{,}756$ is already quite large. Fortunately, there is a simple way to obtain an approximate p-value in such cases. Rather than using all the possible permutations, we take a random sample of size say, 1,000 out of $\binom{m+n}{m}$ permutations and find the approximate p-value using the distribution formed by the 1000 statistics in the same manner as the exact permutation test.

5.3.2. Mann-Whitney-Wilcoxon Rank-Sum Test

Let X and Y be two independent random variables with cumulative continuous distributions F and G, respectively. Suppose that we are interested in testing the hypotheses,

$$H_0 : F(x) = G(x) \text{ vs } H_1 : F(x) \geq G(x) \tag{5.6}$$

for all x and strict inequality for some x.

Example 5.8. If $X \sim \mathcal{N}(\mu_X, \sigma^2)$ and $Y \sim \mathcal{N}(\mu_Y, \sigma^2)$, then the hypotheses become

$$H_0 : \mu_Y = \mu_X \text{ vs } H_1 : \mu_Y > \mu_X$$

so that the alternative points to a shift in the mean of the distribution of Y.

We may now embed the nonparametric problem given in (5.6) and we define the kernel function $h(x, y) = I(x < y)$ for the smooth model as

$$\pi(h(x, y); \theta) = \exp[\theta I(x < y) - K(\theta)]\, g(h(x, y)), \tag{5.7}$$

where $g(h(x, y)) = \frac{1}{2}, x \neq y$ is the null density of $h(X, Y)$ and $K(\theta)$ satisfies

$$e^{K(\theta)} = \frac{e^{\theta} + 1}{2}.$$

It follows that $K(0) = 0$. When $\theta = 0$, the model specified in (5.7) indicates that X and Y are independent and identically distributed with distribution F. Consequently, the hypotheses in (5.6) can be expressed in terms of θ as

$$H_0 : \theta = 0 \text{ vs } H_1 : \theta > 0.$$

For random samples of sizes n and m from F and G, respectively, the log of the composite likelihood function becomes proportional to

$$l\left(\theta; \boldsymbol{X}, \boldsymbol{Y}\right) \sim \theta \sum_{j=1}^{m} \sum_{i=1}^{n} I(X_i < Y_j) - nmK(\theta).$$

The score test statistic evaluated under H_0 is then given by the U-statistic

$$U\left(\boldsymbol{X}, \boldsymbol{Y}\right) = \sum_{j=1}^{m} \sum_{i=1}^{n} I(X_i < Y_j) = \#\left\{(X_i, Y_j) : X_i < Y_j\right\} \tag{5.8}$$

which rejects H_0 for large values. This is called the Mann-Whitney statistic and it is the counting form of the statistic. It is this version which is most often used for theoretical developments as will be seen in Chapter 8 when we consider notions of efficiency.

Lemma 5.2. *Let $Y_1, \ldots, Y_m$ and $X_1, \ldots, X_n$ be two independent random samples. Let $R(Y_j)$ be the rank of Y_j in the combined sample and let $S(Y_j)$ be the rank of Y_j among the $\{Y_j\}$. Denote the combined sample by $\{Z_j\}$. Then*

$$R(Y_j) = S(Y_j) + \sum_{i=1}^{n} I(Y_j > X_i).$$

Proof. It is easy to see that

$$\begin{aligned} R(Y_j) &= \sum_{i=1}^{n+m} I(Y_j > Z_i) + 1 \\ &= \sum_{i=1}^{n+m} I(Y_j > Z_i)\left[I(Z_i = Y_i) + I(Z_i = X_i)\right] + 1 \\ &= \left(\sum_{i=1}^{m} I(Y_j > Y_i) + 1\right) + \sum_{i=1}^{n} I(Y_j > X_i) \\ &= S(Y_j) + \sum_{i=1}^{n} I(Y_j > X_i). \end{aligned}$$

□

It can be seen from Lemma 5.2 that the score statistic is equal to

$$\begin{aligned} U(\boldsymbol{X},\boldsymbol{Y}) &= \sum_{j=1}^{m} R(Y_j) - \sum_{j=1}^{m} S(Y_j) \\ &= W - \frac{m(m+1)}{2}, \end{aligned} \tag{5.9}$$

where the sum

$$W = \sum_{j=1}^{m} R(Y_j) \tag{5.10}$$

is known as the Wilcoxon rank-sum statistic and it is equivalent to the Mann-Whitney statistic U. Both tests reject the null hypothesis for large values. Properties of the Wilcoxon test under the alternative can be derived from the properties of U-statistics.

Theorem 5.1. *Let $Y_1, \ldots, Y_m$ and $X_1, \ldots, X_n$ be two independent random samples with the X's having distribution F and the Y's having distribution G. Then for the statistic defined in (5.8)*

(a) $E(U(\boldsymbol{X},\boldsymbol{Y})) = mnq_1$, $Var(U(\boldsymbol{X},\boldsymbol{Y})) = mnq_1(1-q_1) + mn(n-1)(q_2 - q_1^2) + mn(m-1)(q_3 - q_1^2)$ *where*

$$\begin{aligned} q_1 &= P(X_1 < Y_1), \\ q_2 &= P(X_1 < Y_1, X_1 < Y_2), \\ q_3 &= P(X_1 < Y_1, X_2 < Y_1). \end{aligned}$$

(b) if $\min(n,m) \to \infty$, *with* $m/(n+m) \to \lambda > 0$, *then*

$$\frac{U(\boldsymbol{X},\boldsymbol{Y}) - E(U(\boldsymbol{X},\boldsymbol{Y}))}{\sqrt{Var(U(\boldsymbol{X},\boldsymbol{Y}))}} \xrightarrow{\mathcal{L}} \mathcal{N}(0,1).$$

(c) Under the null hypothesis whereby $F(x) = G(x)$ for all x,

$$\begin{aligned} q_1 &= \frac{1}{2} \\ q_2 = q_3 &= \frac{1}{3} \end{aligned}$$

and

$$\frac{U(\boldsymbol{X},\boldsymbol{Y}) - \frac{mn}{2}}{\sqrt{\frac{mn(m+n+1)}{12}}} \xrightarrow{\mathcal{L}} \mathcal{N}(0,1).$$

Proof. The proof is found in (Lehmann (1975), p. 335 and p. 364) and is a direct application of Example 3.5. □

Remark 5.14. The Mann-Whitney test is also based on the permutation test with U as the test statistic, and its critical value can also be tabulated accordingly. See, for example, Table 21 of Lindley and Scott (1995). To make inferences, we may either compute the p-value with the permutation method or compare the observed Mann-Whitney test statistic U_{obs} with the corresponding critical value in the table.

The Mann-Whitney test can be implemented using `R` function `wilcox.test`. For `wilcox.test(x, y, paired=FALSE)`: the test statistic is defined as "the number of pairs (X_i, Y_j) for which $Y_j \leq X_i$." Therefore, our Mann-Whitney U statistic can be obtained using `wilcox.test(y, x, paired=FALSE)`. By default, an exact p-value is computed if the samples contain less than 50 finite values and there are no ties. Otherwise, a normal approximation is used.

Remark 5.15. Under H_0, the m ranks associated with the Y-sample should be randomly selected from the finite population of $N = m + n$ ranks. From the theory of sampling from finite population, the expected value and variance of W under H_0 are given by

$$\begin{aligned} E(W) &= m\mu \\ Var(W) &= \frac{mn\sigma^2}{m+n-1}, \end{aligned}$$

where

$$\mu = \frac{\sum_{i=1}^{N} i}{N} = \frac{N+1}{2},$$

$$\sigma^2 = \frac{\sum_{i=1}^{N}(i-\mu)^2}{N} = \frac{1^2 + 2^2 + \cdots + N^2}{N} - \left(\frac{N+1}{2}\right)^2 = \frac{(N-1)(N+1)}{12}.$$

Hence, under H_0,

$$\begin{aligned} E(W) &= \frac{m(m+n+1)}{2} \\ Var(W) &= \frac{mn(m+n+1)}{12}. \end{aligned}$$

Note that from (5.9), $U(\boldsymbol{X}, \boldsymbol{Y}) = W - \frac{m(m+1)}{2}$. It can be shown that under H_0,

$$E(U(\boldsymbol{X}, \boldsymbol{Y})) = E(W) - \frac{m(m+1)}{2} = \frac{mn}{2}$$

and

$$Var(U(\boldsymbol{X}, \boldsymbol{Y})) = Var(W) = \frac{mn(m+n+1)}{12}.$$

Remark 5.16 (Wilcoxon Rank-Sum/Mann-Whitney Test Adjusted for Ties). The case of ties can be dealt with by counting each tied pair as $\frac{1}{2}$ in the Mann-Whitney statistic $U(\boldsymbol{X}, \boldsymbol{Y})$ or by assigning the average rank (also called mid-rank) to the tied observations in the Wilcoxon rank-sum statistic W. The mean of $U(\boldsymbol{X}, \boldsymbol{Y})$ or W does not change but the variance of $U(\boldsymbol{X}, \boldsymbol{Y})$ or W under H_0 should be adjusted downwards:

$$Var(U(\boldsymbol{X}, \boldsymbol{Y})) = Var(W) = \frac{mn(m+n+1)}{12} - \frac{mn\left(\sum_{i=1}^{k} d_i^3 - d_i\right)}{12(m+n)(m+n-1)}, \tag{5.11}$$

where k is the number of tied groups and d_i is the number of tied observations in the i th tied group, $i = 1, 2, \ldots, k$. See (Lehmann (1975) p. 355).

Example 5.9. A statistics course has two tutorial classes conducted by two tutors, Maya and Alyssa. Students of this course were given a mid-term test and some students were randomly drawn from each of two tutorial classes and their test scores are shown below:

Maya's class (X)	82	74	87	86	75			
Alyssa's class (Y)	88	77	91	88	94	93	83	94

We are testing the null hypothesis that there is no difference in statistics ability, as measured by this test, in the two classes.

Solution. $n = 5$ and $m = 8$

Combined Data	74	75	77	82	83	86	87	88	88	91	93	94	94
Ranks	1	2	3	4	5	6	7	8.5	8.5	10	11	12.5	12.5
Data from X	82	74	87	86	75								
Ranks	4	1	7	6	2								
Data from Y	88	77	91	88	94	93	83	94					
Ranks	8.5	3	10	8.5	12.5	11	5	12.5					

Then $W = 8.5 + 3 + 10 + \cdots + 12.5 = 71$. Using R, we obtain $U(\boldsymbol{X}, \boldsymbol{Y})$ = number of pairs (X_i, Y_j) for which $X_i < Y_j = 35$ and p-value = 0.0333 which is computed using normal approximation in R as there are ties in the data.

Given the above 13 adjusted ranks, the population mean and variance of these thirteen numbers are

$$\mu = \frac{1 + 2 + \cdots + 12.5 + 12.5}{13} = \frac{1 + 2 + 3 + \cdots + 13}{13} = 7$$

$$\sigma^2 = \frac{1^2 + 2^2 + \cdots + 12.5^2 + 12.5^2}{13} - 7^2 = 13.92.$$

The Wilcoxon rank-sum test statistic with adjusted ranks has, under H_0,

$$E(W) = m\mu = 8(7) = 56$$

$$Var(W) = \frac{mn\sigma^2}{m+n-1} = \frac{8(5)(13.92)}{12} = 46.41$$

Although one may apply the sampling formulas directly to the adjusted ranks for tied data, we can also the explicit formula (5.11) for $Var(W)$ under H_0. Note that there are 2 tied groups: {8.5, 8.5} and {12.5,12.5} and hence, $k = 2$, $d_1 = d_2 = 2$. Therefore,

$$\begin{aligned} Var(W) &= \frac{mn(m+n+1)}{12} - \frac{mn\left(\sum_{i=1}^{k} d_i^3 - d_i\right)}{12\,(m+n)\,(m+n-1)} \\ &= \frac{8(5)(8+5+1)}{12} - \frac{8(5)2(2^3-2)}{12\,(13)\,(12)} = 46.67 - 0.26 = 46.41. \end{aligned}$$

Using a normal approximation with continuity correction, letting $W \sim \mathcal{N}(56, 46.41)$, the p-value of the test is

$$2P(W \geq 71) = 2P(Z > \frac{70.5-56}{\sqrt{46.41}} | Z \sim \mathcal{N}(0,1)) = 0.0333.$$

5.3.3. Confidence Interval and Hodges-Lehmann Estimate for the Location Parameter Δ

Let $X_1, X_2, \ldots, X_n$ and $Y_1, Y_2, \ldots, Y_m$ be two independent random samples drawn from distributions $F(x)$ and $G(y)$, respectively, where $F(x)$ and $G(y)$ only differ by a location parameter Δ, i.e., $F(x) = G(x - \Delta)$, or equivalently, X and $Y + \Delta$ have the same distribution. We would like to construct a confidence interval for Δ.

First arrange all mn pairwise differences of the form $X_i - Y_j$ from the smallest to the largest. The median of these mn pairwise differences is called the *Hodges-Lehmann estimate* of Δ. It is a robust and nonparametric estimator of the difference of means $\overline{X} - \overline{Y}$, and is usually computed in conjunction with the confidence interval based on pairwise differences.

Let $D_{(i)}$ denote the ith smallest pairwise difference. To obtain the confidence interval for Δ, we look for integers a, b such that

$$\Pr(D_{(a)} < \Delta \leq D_{(b)}) = 1 - \alpha.$$

The inequality holds if and only if at least a and at most $b-1$ pairs of (X_i, Y_j) satisfy $X_i - Y_j < \triangle$ (or $X_i < Y_j + \triangle$). Since X_i and $Y_j' = Y_j + \triangle$ have the same distribution, we have

$$\Pr(D_{(a)} < \triangle \leq D_{(b)}) = \Pr(a \leq U \leq b-1) = 1-\alpha.$$

The values of a and b can be obtained according to the null distribution of $U(\boldsymbol{X}, \boldsymbol{Y})$:

$$a = l_{\frac{\alpha}{2}} + 1, \qquad b = u_{\frac{\alpha}{2}}$$

where $l_{\frac{\alpha}{2}}$ and $u_{\frac{\alpha}{2}}$ are the lower-$\frac{\alpha}{2}$ and upper-$\frac{\alpha}{2}$ percentile points of the $U(\boldsymbol{X}, \boldsymbol{Y})$ distribution.

Example 5.10. Refer to Example 5.9, we have $n = 5$, $m = 8$ and the Hodges-Lehmann estimate $\hat{\Delta} = -7.5$. From Table 21 of Lindley and Scott (1995), the lower-2.5th and upper-2.5th percentiles of U are $l_{0.025} = 6$ and $u_{0.025} = mn - 6 = 34$. So $a = 7$ and $b = 34$. Thus, a 95% confidence interval for Δ is (-17,-1].

Remark 5.17. To find a $100(1-\alpha)\%$ confidence interval for $\triangle$ for large samples, we can use a normal approximation with continuity correction.
Let $Z \sim \mathcal{N}(0,1)$). Then

$$\Pr(a \leq U(\boldsymbol{X},\boldsymbol{Y}) \leq b-1) = P\left(\frac{a-0.5-E(U(\boldsymbol{X},\boldsymbol{Y}))}{\sqrt{Var(U(\boldsymbol{X},\boldsymbol{Y}))}} < Z < \frac{b-0.5-E(U(\boldsymbol{X},\boldsymbol{Y}))}{\sqrt{Var(U(\boldsymbol{X},\boldsymbol{Y}))}}\right) = 1-\alpha.$$

For $\alpha = 0.05$, $\Pr(-1.96 < Z < 1.96) = 0.95$ and we can obtain a and b:

$$\begin{aligned} a &= 0.5 + E(U) - 1.96\sqrt{Var(U)}, \\ b &= 0.5 + E(U) + 1.96\sqrt{Var(U)}. \end{aligned}$$

Example 5.11. Referring to Example 5.10, we have $E(U(\boldsymbol{X},\boldsymbol{Y})) = \frac{mn}{2} = 20$ and $Var(U) = 46.41$. Then $a = \lfloor 7.15 \rfloor = 7$ and $b = \lceil 33.85 \rceil = 34$ so that the confidence coverage is at least 95%.

5.3.4. Test for Equality of Scale Parameters

The tests we have discussed so far are particularly designed to distinguish the difference between treatments when the observations from one treatment tend to be larger than those from the other. However, in some situations, the variability of the observations from the two treatments is important.

Suppose two machines for bottling coca cola are designed to fill the cans with 330 ml of the soft drink. It is expected that the observed data on the amount of coke in each can from the two machines are centered around 330 ml as they should be, but their variability may not be the same (see Figure 5.1).

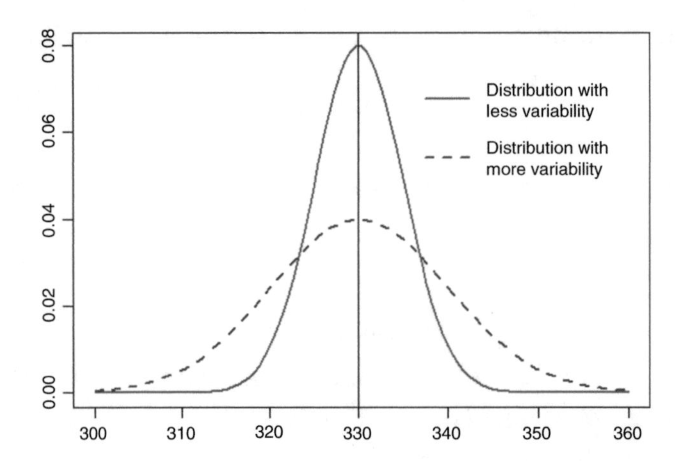

Figure 5.1.: Distributions with different scales

For two independent random samples: $X_1, \cdots, X_m$ from treatment 1 and $Y_1, \cdots, Y_n$ from treatment 2, assume

$$
\begin{aligned}
X_i &= \mu + \sigma_1 \varepsilon_{i1} \quad i = 1, \cdots, m \\
Y_j &= \mu + \sigma_2 \varepsilon_{j2} \quad j = 1, \cdots, n
\end{aligned}
$$

where all the ε's are *i.i.d* random variables with a median of 0. Note that both the X's and the Y's share the same location parameter μ. Here, we want to test $H_0 : \sigma_1 = \sigma_2$. A nonparametric test that makes use of Wilcoxon rank-sum test is the Siegel-Tukey test (Siegel and Tukey, 1960). The steps for carrying out the Siegel-Tukey test are as follows:

1. Arrange the combined data from smallest to largest.
2. Assign rank 1 to the smallest observation, rank 2 to the largest observation, rank 3 to the next largest observation, rank 4 to the next smallest observation, rank 5 to the next smallest observation, and so on.
3. Apply the Wilcoxon rank-sum test based on the rank sum of X.

In this test, we place lower ranks on the more extreme observations and higher ranks on the middle ones. A smaller rank sum indicates that X has more extreme observations than Y and hence X tends to have a larger variability than Y.

Remark. If X and Y do not have the same location parameter (median μ), we may apply the Siegel-Tukey test to $X - med_X$ and $Y - med_Y$, where med_X and med_Y are sample medians of X and Y, respectively.

Example 5.12. Consider the following two sets of data and calculate the Siegel-Tukey test statistic:

$$X: \; 1.4 \quad 7.8 \quad 6.3 \quad 8.9$$
$$Y: \; 3.9 \quad 2.5$$

Solution. We have $m = 4$, $n = 2$, $N = 6$, and $\binom{N}{m} = \binom{6}{4} = 15$.

Combined data:	1.4	2.5	3.9	6.3	7.8	8.9
Siegel-Tukey Ranks:	1	<u>4</u>	<u>5</u>	6	3	2

Hence, $W = 1 + 2 + 3 + 6 = 12$ and $W_2 = 4 + 5 = 9$. We may apply the permutation method to construct the null distribution of W:

Ranks of X	Ranks of Y	W	Ranks of X	Ranks of Y	W
3, 4, 5, 6	1, 2	18	1, 3, 4, 5	2, 6	13
2, 4, 5, 6	1, 3	17	1, 2, 5, 6	3, 4	14
2, 3, 5, 6	1, 4	16	1, 2, 4, 6	3, 5	13
2, 3, 4, 6	1, 5	15	1, 2, 4, 5	3, 6	12
2, 3, 4, 5	1, 6	14	1, 2, 3, 6	4, 5	12
1, 4, 5, 6	2, 3	16	1, 2, 3, 5	4, 6	11
1, 3, 5, 6	2, 4	15	1, 2, 3, 4	5, 6	10
1, 3, 4, 6	2, 5	14			

Under H_0, each occurs with probability $\frac{1}{15}$. For $H_1 : \sigma_1 > \sigma_2$, the p-value is $\Pr(W \leq 12) = \frac{4}{15} = 0.267$.

Remark. One of the difficulties of the Siegel-Tukey test is that if the ranking starts the alternative pattern with the largest observation instead of the smallest, the value of the Wilcoxon statistic will be different. The Ansari-Bradley test(Ansari and Bradley, 1960) helps overcome this problem. However, the corresponding rank-sum statistic will no longer follow the same distribution as the Wilcoxon rank-sum statistic. The critical values of the test can be found from a table in Ansari and Bradley (1960) or using the permutation method.

Chapter Notes

1. The power efficiency of the sign test relative to the Student's t test for the case of the normal distribution is 95% for small samples. The relative efficiency appears to decrease for increasing sample sizes and is about 75%. The relative efficiency decreases as well for increasing level of significance and for increasing alternative. See Wayne Daniel (1990, p. 33).

2. The asymptotic efficiency of the Wilcoxon signed test has been investigated by Noether (see p. 43 of Daniel (1990)). The Wilcoxon signed rank test has an asymptotic relative efficiency of 0.955 relative to the one-sample t test if the differences are normally distributed and an efficiency of 1 if the differences are uniformly distributed. The asymptotic efficiency of the sign test relative to the Wilcoxon signed rank test is 2/3 if the differences are normally distributed and an efficiency of 4/3 if the differences follow the double exponential distribution.

3. Tables for the Mann-Whitney test are found in Daniel (1990), p. 508.

4. The consistency of the Mann-Whitney test is discussed in Gibbons and Chakraborti (2011), p. 141. In fact $(U/mn) \to q_1$ as $\min(m,n) \to \infty$ since $Var(U/mn) \to 0$, thereby showing (U/mn) is a consistent estimator of q_1. Moreover, the test is consistent for the three subclasses of alternatives $q_1 < \frac{1}{2}$, $q_1 > \frac{1}{2}$, $q_1 \neq \frac{1}{2}$, since the power of the test when that alternative is true approaches 1 as the sample size tends to infinity. See p. 17 and p. 142 in Gibbons and Chakraborti (2011) and pp. 267–268 in Fraser (1957). By choosing $h(x_1, x_2, y_1, y_2) = I(x_1 < y_1, x_2 < y_1) + I(y_1 < x_1, y_2 < x_1)$, the test is consistent against all alternatives.

5.4. Exercises

Exercise 5.1. Refer to the data in Example 5.1. Last year, the average weekly sales was 100 units. Is there sufficient evidence to conclude that sales this year exceeds last year's sales? Test at $\alpha = 0.05$. The usual type of test to run for such a question is a one-sample t-test. For the given problem, the correct hypotheses are $H_0 : \mu = 100$ versus $H_1 : \mu > 100$ where μ is the true mean weekly sales of new mobile phones at the mobile phone shop this year.

(a) What is the result of the t-test? (Ans: $t = 0.4048$, $d.f. = 11$)

(b) What assumptions have you made in conducting the t-test?

(c) Are these assumptions valid? You may draw the histogram of the sales.

(d) Under what conditions do we feel safe to use the above t-test?

Exercise 5.2. Fifteen patients are trying a new diet. The differences between weights before and after the diet are (in the order from the smallest)

-7.8 -6.9 -4.0 3.7 6.5 8.7 9.1 10.1 10.8 13.6 14.4 15.6 20.2 22.4 23.5

(a) If the diet has no effect, what should the median weight loss be?

(b) Is it a one- or two-sided sign test? Hence, conduct the test at the 5% significance level.

(c) What assumptions do we need for the sign test in (b)?

(d) Find a 97.9% confidence interval for the population median.

Exercise 5.3. Denote $F(x)$ as the cumulative distribution function (cdf) of a continuous random variable X. Show that for a given x, an approximate $100(1-\alpha)\%$ confidence interval for $F(x)$:

$$\hat{F}(x) \pm z_{\frac{\alpha}{2}} \sqrt{\frac{\hat{F}(x)[1-\hat{F}(x)]}{n}},$$

where $\hat{F}(x)$ is the empirical cdf.

Exercise 5.4. Show that under the null hypothesis, the Wilcoxon signed rank statistic is equivalent in distribution to the sum

$$W^+ = \sum_j V_j$$

where V_j takes values $0, j$ with probability $1/2$ each.

Exercise 5.5.

(a) Two different groups of individuals were compared with respect to how quickly they respond to a changing traffic light. Test using the two-sided Wilcoxon statistic the hypothesis that there is no difference between the two groups on the basis of the following data

Group 1	19.0	14.4	18.2	15.6	14.5	11.2	13.9	11.6
Group 2	12.1	19.1	11.6	21.0	16.7	10.1	18.3	20.5

(b) Do the data suggest that the population variances differ? Carry out a Siegel-Tukey test at the 5% significance level.

(c) Suppose that those same groups are compared at a later time. Indicate how you would combine the two data sets using the smooth embedding approach in order to define a single test. (Hint: suppose in the embedding approach, the parameters for the individual groups are θ_1, θ_2, respectively. Define $\gamma_1 = \frac{\theta_1+\theta_2}{2}, \gamma_2 = \frac{\theta_1-\theta_2}{2}$. Then

$$\theta_1 = \gamma_1 + \gamma_2, \theta_2 = \gamma_1 - \gamma_2$$

and we may phrase the testing problem in terms of γ_1, γ_2.)

6. Multi-Sample Problems

In this chapter, we present a unified theory of hypothesis testing based on ranks. The theory consists of defining two sets of ranks, one consistent with the alternative and the other consistent with the data itself in a notion to be described. The test statistic is then constructed by measuring the distance between the two sets. Critchlow (1986, 1992) utilized a different definition for measuring the distance between sets. The problem can embedded into a smooth parametric alternative framework which then leads to a test statistic. It is seen that the locally most powerful tests can be obtained from this construction. We illustrate the approach in the cases of testing for ordered as well as unordered multi-sample location problems. In addition, we also consider dispersion alternatives. The tests are derived in the case of the Spearman and the Hamming distance functions. The latter were chosen to exemplify that different approaches may be needed to obtain the asymptotic distributions under the hypotheses.

6.1. A Unified Theory of Hypothesis Testing

In this section, we present a general approach for testing hypotheses. We begin by defining two sets of rankings, one set of rankings most in agreement with the data and the other set most in agreement with the alternative. Next, we define a distance function to measure the distance between a ranking from one set and a ranking from the other set. Finally we compute the test statistic defined to be the average distance over of all such pairwise rankings, one chosen from each set. The test then rejects the null hypothesis for small values of this average distance. After describing the specifics of this approach, we apply this general methodology to various testing problems. Our notation in this chapter differs from the notation of previous chapters. It will be convenient to denote observed ranks by μ since we will be constructing a set of permutations generated by it.

6.1.1. A General Approach

The following unified approach was described in Alvo and Pan (1997). Let H_0, H_1 be the null and alternative hypothesis respectively in a typical testing situation. Let $\mathcal{P}_n = \{\boldsymbol{\eta} : (\eta(1), \ldots, \eta(n))\}$ be the set of all permutations of the integers $1, \ldots, n$ and let $d(\boldsymbol{\mu}, \boldsymbol{\nu})$ be a measure of distance between permutations $\boldsymbol{\mu}, \boldsymbol{\nu}$ in $\mathcal{P}_n$.

M. Alvo, P. L. H. Yu, *A Parametric Approach to Nonparametric Statistics*,
Springer Series in the Data Sciences, https://doi.org/10.1007/978-3-319-94153-0_6

- Step 1: Let $X_1, \ldots, X_n$ be a collection of random variables from some continuous distribution and let $\mu(i)$ be the rank of X_i among the $X's$. The continuity assumption ensures that there are no ties in the permutation $\boldsymbol{\mu} = (\mu(1), \ldots, \mu(n))$ occur with probability one.

- Step 2: Define $\{\boldsymbol{\mu}\}$ to be the set of all rankings which are equivalent to the observed ranking $\boldsymbol{\mu}$ in the sense that ranks occupied by identically distributed random variables are exchangeable.

- Step 3: Define E to be the class of extremal rankings which are most in agreement with the alternative. The set E is not data dependent.

- Step 4: Define the distances between the subclass $\{\boldsymbol{\mu}\}$ and E

$$d(\{\boldsymbol{\mu}\}, E) = \sum_{\boldsymbol{\mu} \in \{\boldsymbol{\mu}\}} \sum_{\boldsymbol{\nu} \in E} d(\boldsymbol{\mu}, \boldsymbol{\nu}).$$

  Small values of $d(\{\boldsymbol{\mu}\}, E)$ are consistent with the alternative and lead to the rejection of the null hypothesis.

We may now integrate the unified theory in the context of the smooth alternative model. For each fixed $\boldsymbol{\mu} \in \{\boldsymbol{\mu}\}$, we may define the smooth alternative density as proportional to

$$\pi(\boldsymbol{\mu}; E, \theta) \sim \exp[-\theta\, d(\boldsymbol{\mu}, E) - K(\theta)], \tag{6.1}$$

where

$$d(\boldsymbol{\mu}, E) = \sum_{\boldsymbol{\nu} \in E} d(\boldsymbol{\mu}, \boldsymbol{\nu})$$

and $K(\theta)$ is a normalizing constant. Let $n_{\boldsymbol{\mu}}$ be the cardinality of $\{\boldsymbol{\mu}\}$. Under the null hypothesis, $H_0 : \theta = 0$, all the rankings $\boldsymbol{\mu} \in \{\boldsymbol{\mu}\}$ are equally likely. Under the alternative, $H_1 : \theta > 0$, rankings $\boldsymbol{\mu}$ which are closer to E are more likely than those which are further away. The distance model in (6.1) generalizes the distance-based models of Diaconis and Mallows described in (Alvo and Yu (2014)) as discussed earlier in Section 4.4.

The logarithm of the composite likelihood function constructed from (6.1) is then proportional to

$$\sum_{\boldsymbol{\mu}_i \in \{\boldsymbol{\mu}\}} [\theta d(\boldsymbol{\mu}_i, E) - K(\theta)] = \theta \sum_{\boldsymbol{\mu}_i \in \{\boldsymbol{\mu}\}} d(\boldsymbol{\mu}_i, E) - n_{\boldsymbol{\mu}} K(\theta)$$

and the score statistic obtained by calculating the derivative at $\theta = 0$ becomes

$$\boldsymbol{U}(\{\boldsymbol{\mu}\}) = \left(\sum_{\boldsymbol{\mu}_i \in \{\boldsymbol{\mu}\}} d(\boldsymbol{\mu}_i, E) - n_{\boldsymbol{\mu}} K'(0) \right).$$

The corresponding test statistic is given by

$$[\boldsymbol{U}(\{\boldsymbol{\mu}\})]' \boldsymbol{\Sigma}^{-1} [\boldsymbol{U}(\{\boldsymbol{\mu}\})], \tag{6.2}$$

where $\boldsymbol{\Sigma}$ is the variance-covariance matrix of $\boldsymbol{U}(\{\boldsymbol{\mu}\})$. In the next two sections, we consider direct applications of the unified theory. Specifically, we describe the multi-sample problem of location in both the ordered and unordered case. As well, we describe a test for umbrella alternatives and conclude the chapter with a discussion of tests for dispersion.

6.1.2. The Multi-Sample Problem in the Ordered Case

We may consider the general multi-sample location problem with ordered alternatives. Let $F_1(x), \ldots, F_r(x)$ be r continuous distributions and suppose we wish to test

$$H_0 : F_1(x) = \ldots = F_r(x), \text{ for all } x$$

against the alternative

$$H_1 : F_r(x) \leq \ldots \leq F_1(x)$$

with strict inequality for some x. Let $X_{N_{k-1}}, \ldots, X_{N_k}$ be a random sample of size n_k from $F_k(x)$ where $N_0 = 0$ and

$$N_k = n_1 + \ldots + n_k, k = 1, \ldots, r.$$

Rank all the N_r observations among themselves and write the permutation of observed ranks

$$\boldsymbol{\mu} = [\mu(1), \ldots, \mu(N_1) | \ldots | \mu(N_{r-1}+1), \ldots, \mu(N_r)]'$$

where ranks from the same distribution are placed together. Hence, $\mu(1), \ldots, \mu(N_1)$ represent the observed ranks of the n_1 observations from $F_1(x)$ the combined samples. The equivalent subclass $\{\boldsymbol{\mu}\}$ consists of all permutations of the integers $1, \ldots, N_r$ which assign the same set of ranks to the individual populations as $\boldsymbol{\mu}$ does. On the other hand, the extremal set E consists of all permutations which assign ranks $N_{k-1}+1, \ldots, N_k$ to population k. Alvo and Pan (1997) derived the test statistics corresponding to the Spearman, the Spearman Footrule, and the Hamming distances. They also obtained the asymptotic distributions under both the null and alternative hypotheses.

In order to illustrate the methodology, we consider the two-sample case. Suppose that we observe independent random variables X_1, X_2, X_3, X_4 with X_1, X_2 from F_1 and X_3, X_4 from F_2. The alternative hypothesis claims that $F_2 \leq F_1$. Among all rankings of the integers $(1, 2, 3, 4)$ one would expect that the small ranks would be most consistent with F_1 and larger ranks would be most consistent with F_2. Adopting the convention that the first two components of a ranking refer to population 1 and the next two to population 2, the exchangeable rankings compatible with the alternative hypothesis would be

$$(1,2|3,4)\,,(2,1|3,4)\,,(1,2|4,3)\,,(2,1|4,3)\,.$$

Returning now to the general situation, the Spearman and Hamming test statistics are as follows.

Spearman: In the case of the Spearman distance, the test statistic in the multi-sample case was shown to be

$$S=\sum_{i=1}^{N_r} c(i)\,\frac{\mu(i)}{N_r+1},$$

where for $1\le k\le r$,

$$c(i)=(N_{k-1}+N_k)\,,\ \text{for all } i \text{ such that } N_{k-1}<i\le N_k.$$

We note that

$$\begin{aligned}\bar{c} &= \frac{\sum_{k=1}^{r}\sum_{i=N_{k-1}+1}^{N_k}(N_{k-1}+N_k)}{N_r}\\ &= \frac{\sum_{k=1}^{r} n_k\,(N_{k-1}+N_k)}{N_r} &(6.3)\\ &= \frac{\sum_{k=1}^{r}(N_k-N_{k-1})\,(N_{k-1}+N_k)}{N_r}\\ &= \frac{\sum_{k=1}^{r}\left(N_k^2-N_{k-1}^2\right)}{N_r}=N_r. &(6.4)\end{aligned}$$

We recognize that S is a simple linear rank statistic and consequently, under the null hypothesis, S is asymptotically normal with mean

$$\bar{c}\frac{\sum_{i=1}^{N_r}\mu(i)}{N_r+1}=\frac{N_r\bar{c}}{2}$$

and variance

$$\sigma^2=\frac{N_r^3}{12}\sum_{k=1}^{r} w_k W_k W_{k-1},$$

where as $\min\{n_1,\ldots,n_r\}\to\infty$

$$\frac{n_k}{N_r}\to w_k,\,W_k\equiv\sum_{i=1}^{k} w_i.$$

The test rejects for large values of S. In the two-sample case, the statistic becomes:

Example 6.1. The statistic in the two-sample ordered location problem based on the Spearman distance is

$$\begin{aligned} S &= \frac{n_1}{n+1}\sum_{i=1}^{n_1}\mu(i) + \frac{(n_1+n)}{n+1}\sum_{i=n_1+1}^{n_1+n_2}\mu(i) \\ &= \frac{n_1 n}{2} + \frac{n}{n+1}W \end{aligned}$$

where W is the Wilcoxon statistic for which the exact mean and variance from Section 5.3.3 are

$$E[W] = \frac{n_2(n+1)}{2}, Var[W] = \frac{n_1 n_2 (n+1)}{12}.$$

Hamming: In the case of the Hamming distance, the test statistic in the multi-sample case was shown to be

$$H = \sum_{i=1}^{N_r} a_{i\mu(i)},$$

where

$$a_{ij} = \begin{cases} \frac{1}{n_k} & i.j \in \{N_{k-1}+1,\ldots,N_k\}, 1 \le k \le r \\ 0 & \text{otherwise} \end{cases}$$

Equivalently, we may express Hamming's statistic as

$$H = \frac{Y_1}{n_1} + \ldots + \frac{Y_r}{n_r},$$

where Y_k is the number of observed rankings in the set $\{N_{k-1}+1,\ldots,N_k\}$. The test rejects for large values of H. We may apply Hoeffding's theorem (Section 3.3) to obtain a central limit theorem for the Hamming statistic under the null hypothesis. We have that H is asymptotically normal with mean

$$\frac{1}{N_r}\sum_{i=1}^{N_r}\sum_{j=1}^{N_r} a_{ij} = 1$$

and variance

$$\frac{1}{N_r - 1}\sum_{i=1}^{N_r}\sum_{j=1}^{N_r} d_{ij}^2 = \frac{(r-1)}{N_r - 1}, \tag{6.5}$$

where

$$d_{ij} = a_{ij} - \bar{a}_{i.} - \bar{a}_{ij} + \bar{a}_{..} = a_{ij} - \frac{1}{N_r} - \frac{1}{N_r} + \frac{1}{N_r} = a_{ij} - \frac{1}{N_r}.$$

The result follows from the fact that

$$\frac{\max_{1\le i.j\le N_r} d_{ij}^2}{\frac{1}{N_r}\sum_{i=1}^{N_r}\sum_{j=1}^{N_r} d_{ij}^2} \le \frac{\frac{1}{N_r^2}\max_{1\le k\le r} t_k^{-2}}{\frac{r-1}{N_r}} \to 0.$$

Example 6.2. The statistic in the two-sample ordered location problem based on Hamming's distance becomes

$$H = \sum_{i=n_1+1}^{n_1+n_2} a_{i\mu(i)}$$

where

$$a_{ij} = \begin{cases} \frac{1}{n_1} & i,j \in \{1,\ldots,n_1\} \\ \frac{1}{n_2} & i,j \in \{n_1+1,\ldots,n_1+n_2\} \\ 0 & \text{otherwise.} \end{cases}$$

It follows from Hoeffding's combinatorial central limit theorem (see Theorem 3.6), H is asymptotically normal with mean equal to 1 and variance

$$\frac{1}{n_1+n_2-1}$$

as $\min(n_1,n_2) \to \infty$.

6.1.3. The Multi-Sample Problem in the Unordered Case

The solution to the multi-sample problem in the unordered case may be derived by considering the set of all possible orderings of the alternative. In particular, consider the following specific ordered testing problem

$$H_0 : F_1(x) = \ldots = F_r(x), \text{ for all } x$$

against

$$H_1^h : F_{e_{hr}}(x) \le \ldots \le F_{e_{h1}}(x),$$

where we have a fixed $h, 1 \le h \le r!$ and permutations $[e_{h1},\ldots,e_{hr}]$. We may now phrase the unordered testing hypothesis problem as that of testing

$$H_0 : F_1(x) = \ldots = F_r(x), \text{ for all } x$$

against

$$H_1 : \bigcup_{i=1}^{r!} H_1^h.$$

Let T_h be a test statistic for testing H_0 against H_1^h with rejection region $\{T_h < c\}$. We define the test statistic T_M for testing H_0 against H_1 using the following three steps:

1. *Linearization*: Let $\tilde{\boldsymbol{\alpha}} = (\alpha_1, \ldots, \alpha_{r!})$ and put

$$T_L(\tilde{\boldsymbol{\alpha}}) = \sum_{h=1}^{r!} \alpha_h T_h.$$

2. *Normalization*: Put

$$T_N(\tilde{\boldsymbol{\alpha}}) = \frac{T_L(\tilde{\boldsymbol{\alpha}}) - E_0[T_L(\tilde{\boldsymbol{\alpha}})]}{\sqrt{Var_0 T_L(\tilde{\boldsymbol{\alpha}})}}.$$

3. *Minimization*: Put

$$T_M = \min_{\tilde{\boldsymbol{\alpha}}} T_N(\tilde{\boldsymbol{\alpha}}).$$

The use of this approach in the case of the Spearman distance leads to the well-known Kruskal-Wallis test statistic for the one-way analysis of variance. Details of the proof of the next two theorems may be found in Alvo and Pan (1997).

Theorem 6.1 (Spearman Case). *The test statistic in the unordered case based on Spearman distance rejects the null hypothesis for large values of* $T_M = \bar{\boldsymbol{\mu}} \boldsymbol{\Sigma}_S^{-1} \bar{\boldsymbol{\mu}}$ *where*

$$\bar{\boldsymbol{\mu}} = (\bar{\mu}_1 - E_0[\bar{\mu}_1], \ldots, \bar{\mu}_{r-1} - E_0[\bar{\mu}_{r-1}])$$

with

$$\bar{\mu}_k = \sum_{i=N_{k-1}+1}^{N_k} \mu(i)$$

and its expectation under null

$$E_0[\bar{\mu}_k] = n_k \left(\frac{N_r + 1}{2} \right).$$

Also, the $(r-1) \times (r-1)$ *covariance matrix* $\boldsymbol{\Sigma}_S$ *has components*

$$Cov(\bar{\mu}_k, \bar{\mu}_{k'}) = \begin{cases} n_k (N_k - n_k)(N_r + 1)/12 & k = k' \\ -n_k n_{k'} (N_r + 1)/12 & k \neq k' \end{cases}$$

and inverse

$$\Sigma_S^{-1} = \frac{12}{N_r(N_r+1)} \left(diag\left\{ \frac{1}{n_1}, \ldots \frac{1}{n_r} \right\} + J/N_r \right)$$

Moreover, under the null hypothesis as $\min\{n_1, \ldots, n_r\} \to \infty$,

$$T_M \xrightarrow{\mathcal{L}} \chi^2_{r-1}.$$

The statistic T_M coincides with the well-known Kruskal-Wallis test statistic

$$\frac{12}{N_r\left(N_r+1\right)}\sum_{k=1}^{r} n_k\left(\frac{\bar{\mu}_k}{n_k}-\frac{N_r+1}{2}\right)^2.$$

Proof. The proof is a direct consequence of the multivariate central limit theorem. □

Theorem 6.2 (Hamming Case). *The test statistic in the unordered case based on Hamming distance rejects the null hypothesis for large values of $T_M = \bar{\boldsymbol{\mu}}\boldsymbol{\Sigma}_H^{-1}\bar{\boldsymbol{\mu}}$ where*

$$\bar{\boldsymbol{\mu}} = \left(\bar{\mu}_{11} - E_0\bar{\mu}_{11}, \ldots, \bar{\mu}_{r-1,r-1} - E_0\bar{\mu}_{r-1,r-1}\right)$$

with

$$\bar{\mu}_{kp} = \sum_{j=N_{k-1}+1}^{N_k} a_{p\mu(j)}$$

and

$$a_{ij} = \begin{cases} 1 & j \in \{N_{i-1}, \ldots, N_i\} \\ 0 & otherwise \end{cases}$$

expectation

$$E_0\left[\bar{\mu}_{kp}\right] = n_k$$

and covariance

$$\boldsymbol{\Sigma}_H = Cov\left(\bar{\mu}_{kp}, \bar{\mu}_{k'p'}\right) = \begin{cases} \frac{n_k n_p (N_r-n_k)(N_r-n_p)}{N_r^2(N_r-1)} & k=k', p=p' \\ -\frac{n_k n_p (N_r-n_k) n_{p'}}{N_r^2(N_r-1)} & k=k', p\neq p' \\ -\frac{n_k n_p (N_r-n_p) n_{k'}}{N_r^2(N_r-1)} & k\neq k', p=p' \\ \frac{n_k n_p n_{k'} n_{p'}}{N_r^2(N_r-1)} & k\neq k', p\neq p' \end{cases}$$

Moreover, under the null hypothesis as $\min\{n_1, \ldots, n_r\} \to \infty$,

$$T_M \xrightarrow{\mathcal{L}} \chi^2_{(r-1)^2}.$$

Equivalently, $\bar{\mu}_{kp}$ may be interpreted as the number of rankings $\{\mu\left(N_{k-1}+1\right), \ldots, \mu\left(N_k\right)\}$ in the set $\{N_{p-1}+1, \ldots, N_p.\}$

Proof. The proof is a direct consequence of the multivariate central limit theorem. □

6.1.4. Tests for Umbrella Alternatives

A second application of the unified theory is to the case of testing for trend when there is the possibility of an umbrella type alternative. As an example, consider the data on intelligence scores in Appendix A. Weschler Adult Intelligence Scale scores were recorded on 12 males listed by age groups. A look at the data reveals that the peak is located in the $35-54$ age group. In general, we would like to test the null hypothesis that there is no difference due to age against the alternative that the scores rise monotonically prior to the peak and decrease thereafter. Two situations arise: when the location of the peak is known and when it is unknown.

Let $F_1(x), \ldots, F_r(x)$ be r continuous distributions and suppose we wish to test

$$H_0 : F_1(x) = \ldots = F_r(x), \text{ for all } x$$

against the umbrella alternative

$$H_1 : F_1(x) \geq \ldots \geq F_p(x) \leq \ldots \leq F_r(x)$$

with strict inequality for some x. Suppose that there are m_i observations from F_i, $i = 1, \ldots, r$ and that $n = m_1 + \ldots + m_r$.

In the case when the location of the peak is known, Alvo (2008) obtained the test statistic corresponding to both Spearman and Kendall distance functions and showed that they were asymptotically equivalent under the condition that

$$\min_i (m_i) \longrightarrow \infty$$

in such a way that $\frac{m_i}{n} \longrightarrow \lambda_i > 0, n \longrightarrow \infty$. Moreover, he derived the asymptotic distribution of the test statistics under both the null and alternative hypotheses. For the Spearman distance in the special case $m_i = m$, it was shown that when the location of the peak is known, the test statistic is

$$S_p = mn \left\{ \sum_{i \leq p} \frac{i}{p} \left[\bar{\mu}_i - \frac{n+1}{2} \right] + \sum_{i > p} \frac{(k+1-i)}{(k+1-p)} \left[\bar{\mu}_i - \frac{n+1}{2} \right] \right\}$$

where $\bar{\mu}_i$ represents the average of the ranks in the ith population. Under the null hypothesis, S_p is asymptotically normal with mean 0 and variance

$$\sigma_p^2 = \frac{m\left[n(n+1)\right]^2}{12} \left[\sum_{i=1}^{p} \left(\frac{i}{p} - \frac{r+1}{2r} \right)^2 + \sum_{i=p+1}^{r} \left(\frac{r+1-i}{r+1-p} - \frac{r+1}{2r} \right)^2 \right].$$

Alvo (2008) also developed a test when the location of the peak is unknown. This is based on the statistic

$$S_{max} = \max_p \left(\frac{S_p}{\sigma_p} \right)$$

whose asymptotic distribution is that of $\boldsymbol{Bz}$ where $\boldsymbol{z}$ has a standard multivariate normal distribution and the matrix $\boldsymbol{B}$ derives from the relation

$$Cov\left(\frac{S_1}{\sigma_1},\ldots,\frac{S_r}{\sigma_r}\right)=\boldsymbol{BB}'.$$

The distribution of $\boldsymbol{Bz}$ is generally obtained by simulation.

6.2. Test for Dispersion

In this section, we shall be concerned with detecting differences in variability between two populations. We shall assume that the two populations of interest have the same median and are described by cumulative distribution functions $F_X(x)$ and

$$F_Y(x)=F_X(\mu+(x-\mu)/\gamma),$$

for some dispersion parameter $\gamma>0$. We would like to test

$$H_0:\gamma\geq 1$$

$$H_1:\gamma<1$$

The alternative states that the second population is less spread out than the first. We shall assume for simplicity of the presentation that both sample sizes are even numbers with

$$n_i=2m_i, i=1,2.$$

Using the unified theory to the dispersion problem, rank all the observations together and let items $1,\ldots,n_1$ be from the first population and items $n_1+1,\ldots,n_1+n_2$ be from the second. The equivalence class $\{\boldsymbol{\mu}\}$ consists of all permutations obtained by permuting the labels assigned to the items in each population respectively. Moreover, in view of the assumption that the medians are the same, we also transpose the items ranked in positions i and $n+1-i$. The extremal class E consists of all permutations which rank the items from the second population in the middle and the items from the first population at the two ends. This is because the first population is more diverse. Using the Spearman distance, it can be shown that the test statistic takes the form

$$S=\sum_{i=1}^{n_1}\left|R_i-\frac{n+1}{2}\right|$$

which is precisely the Freund-Ansari-Bradley test (Hájek and Sidak (1967), p. 95).

6.3. Tests of the Equality of Several Independent Samples

Alvo (2016) developed an alternative approach to test for the equality of r distributions which exploits the smooth model directly. Let $\{X_{ij}\}$ $i = 1, \ldots, r; j = 1, \ldots, n_i$ be independent random variables such that for fixed i, X_{ij} have a common cdf F_i. We wish to test the hypotheses

$$H_0 : F_i = F, \; i = 1, \ldots, r$$

against

$$H_1 : F_i \neq F_j, \text{ for some } i \neq j.$$

We are concerned with differences in location. Consider first a single observation from each population. Select the kernel to be the sign function, denoted sgn, which compares a pair of observations, one from each population and define for the ith population the smooth alternative density as proportional to

$$\pi(x_1, \ldots, x_r; \theta_i) \sim \left\{ \theta_i \sum_{l \neq k} \text{sgn}(x_k - x_l) - K(\theta_i) \right\}.$$

Suppose now that we observe a random sample from each of the r populations: $\{X_{ij}, j = 1, \ldots, n_i\}$, $i = 1, \ldots, r$. The composite log likelihood function for the ith population is proportional to

$$\theta_i \sum_{l \neq i} \sum_{j=1}^{n_i} \sum_{j'=1}^{n_l} \text{sgn}(x_{ij} - x_{lj'}) - \sum_{i=1}^{r} \sum_{l \neq i}^{r} n_i n_l K(\theta_i)$$

and hence the composite log likelihood function taking into account all the populations is proportional to

$$l(\boldsymbol{\theta}) \sim \sum_{i=1}^{r} \theta_i \sum_{l \neq i} \sum_{j=1}^{n_i} \sum_{j'=1}^{n_l} \text{sgn}(x_{ij} - x_{lj'}) - \sum_{i=1}^{r} \sum_{l \neq i}^{r} n_i n_l K(\theta_i).$$

The null hypothesis may now be defined as

$$H_0 : \theta_i = \theta, i = 1, \ldots, r \tag{6.6}$$

$$H_1 : \theta_i \neq \theta_j, \text{ for some } i \neq j. \tag{6.7}$$

Under the null hypothesis,

$$\sum_{i=1}^{r} \theta_k \sum_{l \neq i}^{r} \operatorname{sgn}(x_i - x_l) = \theta \sum_{i=1}^{r} \sum_{l \neq i}^{r} \operatorname{sgn}(x_i - x_l) = 0$$

which shows that the model can actually be specified by using only $(r-1)$ parameters.

Consequently, we may redefine the parameters so that $\sum \theta_i = 0$, and we wish to test $H_0 : \theta_i = 0, i = 1, \ldots, r$. It follows that $K(0) = 0$.

Suppose now that we observe a random sample from each of the r populations: $X_{ij}, j = 1, \ldots, n_i, i = 1, \ldots, r$. The composite log likelihood function is proportional to

$$l(\boldsymbol{\theta}) \sim \sum_{i=1}^{r} \theta_i \sum_{l \neq i} \sum_{j=1}^{n_i} \sum_{j'=1}^{n_l} \operatorname{sgn}(x_{ij} - x_{lj'}) - \sum_{i=1}^{r} \sum_{l \neq i}^{r} n_i n_l K(\theta)$$

Since,

$$\left(R_{ij} - \frac{n+1}{2}\right) = \frac{1}{2} \sum_{l \neq i}^{r} \sum_{j'=1}^{n_l} \operatorname{sgn}(x_{ij} - x_{lj'})$$

we have

$$\begin{aligned} l(\boldsymbol{\theta}) &\sim 2 \sum_{i=1}^{r} \theta_i \sum_{j=1}^{n_i} \left(R_{ij} - \frac{n+1}{2}\right) - \sum_{i=1}^{r} \sum_{l \neq i}^{r} n_i n_l K(\theta) \\ &= 2 \sum_{i=1}^{r} n_i \theta_i \left(\bar{R}_i - \frac{n+1}{2}\right) - \sum_{i=1}^{r} \sum_{l \neq i}^{r} n_i n_l K(\theta) \end{aligned}$$

where $\bar{R}_i$ is the average of the overall ranks assigned to the kth population. The score function is given by the $r \times 1$ vector

$$\boldsymbol{U} = \left[n_1 \left(\bar{R}_1 - \frac{n+1}{2}\right), \ldots, n_r \left(\bar{R}_r - \frac{n+1}{2}\right)\right]'$$

The g-inverse of the corresponding covariance matrix is given by the diagonal

$$\boldsymbol{I}^{-1} = diag\left(\frac{12}{n(n+1)n_k}\right)$$

Kruskal (1952) and it follows that the Rao score test statistic, denoted KW, is

$$\begin{aligned} KW &= \boldsymbol{U}'\boldsymbol{I}^{-1}\boldsymbol{U} \\ &= \frac{12}{n(n+1)} \sum_{i=1}^{r} n_i \left(\bar{R}_i - \frac{n+1}{2}\right)^2 \end{aligned} \tag{6.8}$$

which is the usual Kruskal-Wallis statistic. Under the null hypothesis,

$$KW \xrightarrow{\mathcal{L}} \chi^2_{r-1}.$$

We reject the null hypothesis for large values of KW. The Kruskal-Wallis test is thus locally most powerful for testing (6.6).

When there are ties in the data, we may adjust the ranks by using the mid-ranks for the tied data as we did for the Wilcoxon test. The permutation method can also be applied to the KW statistic. However, in order to maintain the chi-square approximation, the KW statistic should be modified. Recall that under the one-way ANOVA model, $SSB/\sigma^2 \sim \chi^2(r-1)$ under H_0, and the mean of the ranks is still $\frac{n+1}{2}$ no matter whether there are ties or not. It is thus natural to assume that for some constant C

$$KW_{ties} \equiv C\sum_{i=1}^{r} n_i(\bar{R}_i - \frac{n+1}{2})^2$$

where to fulfill the chi-square approximation, we must have

$$E(KW_{ties}) = r - 1.$$

According to the properties of sampling from finite distributions, we have

$$E(\bar{R}_i - \frac{n+1}{2})^2 = Var(\bar{R}_i) = \frac{(n-n_i)\sigma^2}{(n-1)n_i}$$

where

$$\sigma^2 = \frac{n^2-1}{12} - \frac{\sum_{i=1}^{g}(t_i^3 - t_i)}{12n}$$

is the population variance of the combined ranks or adjusted ranks and g is the number of tied groups. Hence. we have,

$$E(KW_{ties}) = E[C\sum_{i=1}^{r} n_i(\bar{R}_i - \frac{n+1}{2})^2] = C(r-1)\frac{n\sigma^2}{n-1}$$

with

$$C = \frac{n-1}{n\sigma^2}.$$

Thus, a more general form of the KW statistic in the case of ties is given by

$$KW_{ties} = \frac{KW}{1 - \frac{\sum_{i=1}^{g}(t_i^3 - t_i)}{n^3 - n}}.$$

Remark. Note that this is only an intuitive derivation of the KW statistic in the case of ties. For a more formal proof, see Kruskal and Wallis (1952). The Kruskal-Wallis test can be implemented using either of the R functions below:

- If no ties, use `kruskal.test` function in the base R installation.
- If ties exist, use `kruskal_test` function in the R add-on package `coin`.

The parametric paradigm allows for a further investigation into possible differences among the populations. We may conduct a bootstrap study by sampling with replacement n_i observations from the ith population and computing the overall average ranking $\bar{R}_i$. We note that the maximum composite likelihood equation for each θ_i is

$$\begin{aligned}\left(\bar{R}_i - \frac{n+1}{2}\right) &= \frac{\partial K(\theta_i)}{\partial \theta_i} \\ &= \frac{1}{n_i}\sum_{l \neq i}\left[P_\theta\left(X_i > X_l\right) - \frac{1}{2}\right] \qquad (6.9)\end{aligned}$$

Hence after each bootstrap iteration, we may obtain an estimate of the left-hand side of (6.9). The histogram of bootstrapped values under the null hypothesis should be centered around 0. An example illustrates this computation.

Example 6.3. Eighteen lobsters of the same size in a species are divided randomly into three groups and each group is prepared by a different cook using the same recipe. Each prepared lobster is then rated on each of the criteria of aroma, sweetness, and brininess by professional taste testers. The following shows the combined scores for the lobsters prepared by the three cooks. A higher score represents a better taste of the lobster.

Cook A	Cook B	Cook C
7.03	8.63	7.75
9.97	6.51	4.37
8.25	5.93	5.07
7.99	9.01	6.31
8.35	8.59	4.59
6.79	7.65	6.63

Based on the data, apply the Kruskal-Wallis test to test the null hypothesis that median scores for all three cooks are the same at the 5% level of significance.

Solution. Using the function `kruskal.test` in R, $KW = 6.9825$, $df = 2$, p-value $= 0.03046 < 0.05$. There is enough evidence to conclude that the median scores for all three cooks are not the same.

6.4. Tests for Interaction

In this section, we consider the general two-factor design with equal numbers of replications in each cell. Such designs are utilized in statistics to test for main effects and for interactions in a variety of experiments. In more recent times, they have been applied in a genetics environment in order to understand the underlying biological mechanisms (Gao and Alvo, 2005b). In the gene expression data of Drosophila melanogaster (Jin et al., 2001) for example, there are 24 cDNA micro arrays, 6 for each combination of two genotypes (Oregon R and Samarkand) and two sexes. As each array used two different dyes, there were in total 48 separate labeling reactions. Focusing on the individual expression level of a gene and its relationship with genotypes and sexes, the objective of the study was to identify genes whose expression levels are affected by the interaction between the two factors. For such data, the assumption of normality for the error terms is not warranted and consequently, nonparametric procedures are needed. We shall consider a nonparametric test for interaction based on the row ranks and column ranks of the data.

6.4.1. Tests for Interaction: More Than One Observation per Cell

Suppose now we have data from a double array with I rows and J columns with an equal number $n > 1$ of observations in each cell, $X_{ijk}, i = 1,,,.I, j = 1,\ldots,J, k = 1,\ldots,n$. We first note that

$$\frac{1}{nJ+1}\sum_{j'=1}^{J}\sum_{k'=1}^{n}\operatorname{sgn}\left(x_{ijk}-x_{ij'k'}\right)=\left(\frac{R_{ijk}}{nJ+1}-\frac{1}{2}\right),$$

where R_{ijk} is the rank of X_{ijk} among the ith row. Similarly,

$$\frac{1}{nI+1}\sum_{i'=1}^{I}\sum_{k'=1}^{n}\operatorname{sgn}\left(x_{ijk}-x_{i'jk'}\right)=\left(\frac{C_{ijk}}{nI+1}-\frac{1}{2}\right),$$

where C_{ijk} is the rank of X_{ijk} among the jth column. Placing this problem within the context of a smooth model, the composite log likelihood function viewed from the perspective of rows only is proportional to

$$\begin{aligned}
l(\boldsymbol{\theta}) &\sim \sum_{ij}\theta_{ij}\sum_{k=1}^{n}\sum_{j'=1}^{J}\sum_{k'=1}^{n}\operatorname{sgn}\left(x_{ijk}-x_{ij'k'}\right)-\left(n^2J\right)K(\boldsymbol{\theta})\\
&\sim \sum_{ij}\theta_{ij}\sum_{k=1}^{n}\left(R_{ijk}-\frac{(nJ+1)}{2}\right)-\left(n^2J\right)K(\boldsymbol{\theta})\\
&\sim \sum_{ij}\theta_{ij}\sum_{k=1}^{n}\frac{R_{ijk}}{nJ+1}-nK(\theta).
\end{aligned}$$

Similarly, the composite log likelihood function viewed from the perspective of columns only is proportional to

$$l(\boldsymbol{\theta}) \sim \sum_{ij} \theta_{ij} \sum_{k=1}^{n} \frac{C_{ijk}}{nI+1} - nK(\theta).$$

Let

$$x_{ijk} = \frac{R_{ijk}}{nJ+1} + \frac{C_{ijk}}{nI+1}.$$

Consequently, the composite log likelihood from the perspective of both rows and columns is proportional to the sum

$$\begin{aligned}\sum_{ij} \theta_{ij} \sum_{k=1}^{n} \left(\frac{R_{ijk}}{nJ+1} + \frac{C_{ijk}}{nI+1} \right) - K(\boldsymbol{\theta}) &\sim \sum_{ij} \theta_{ij} \sum_{k=1}^{n} x_{ijk} - nK(\boldsymbol{\theta}) \\ &\sim n\left[\sum_{ij} \theta_{ij} \bar{x}_{ij.} - K(\boldsymbol{\theta}) \right].\end{aligned}$$

Using dots to denote an average over that index, set the parameter space

$$\theta_{ij} = \theta_{..} + \alpha_i + \beta_j + \gamma_{ij}$$

where

$$\alpha_i = (\theta_{i.} - \theta_{..}), \beta_j = (\theta_{.j} - \theta_{..}), \theta_{..} = \frac{1}{IJ} \sum \theta_{ij}$$

and

$$\gamma_{ij} = \theta_{ij} - \theta_{i.} - \theta_{.j} + \theta_{..}$$

Here, $\alpha_i, \beta_j, \gamma_{ij}$ represent respectively the row, column, and interaction effects. It can be seen that

$\sum_{ij} \theta_{ij} \bar{x}_{ij.} =$

$$\sum \theta_{..} \bar{x}_{...} + J \sum_i \alpha_i (\bar{x}_{i..} - \bar{x}_{...}) + I \sum_j \beta_j (\bar{x}_{.j.} - \bar{x}_{...}) + \sum_{ij} \gamma_{ij} (\bar{x}_{ij.} - \bar{x}_{i..} - \bar{x}_{.j.} + \bar{x}_{...})$$

We wish to test the null hypothesis of no interaction, that is

$$H_0 : \gamma_{ij} = 0, \text{ for all } i.j.$$

The score vector $\boldsymbol{U}$ has ij component equal to

$$\frac{\partial l(\boldsymbol{\theta})}{\partial \gamma_{ij}} = [\bar{x}_{ij.} - \bar{x}_{i..} - \bar{x}_{.j.} + \bar{x}_{...}] - \frac{\partial K(\boldsymbol{\theta})}{\partial \gamma_{ij}}$$

Consequently, the test statistic to test the hypothesis of no interaction is given by

$$n^{-1}\boldsymbol{U}'\boldsymbol{\Sigma}^{-}\boldsymbol{U} \tag{6.10}$$

where $\boldsymbol{\Sigma}^{-}$ represents the generalized inverse of the variance-covariance matrix of $\boldsymbol{U}$. The asymptotic distribution of this statistic under the null hypothesis was shown to be a $\chi^2_{(I-1)(J-1)}$ (Gao and Alvo (2005a)). As well, the distribution under contiguous alternatives was shown to be $\chi^2_{(I-1)(J-1)}(\delta)$ where δ is the noncentrality parameter. The demonstration of these results makes use of results in Hajek (1968).

Example 6.4. Consider the gene expression data of Drosophila melanogaster of Jin et al. (2001). The gene fs(1) $k10$ is known to be expressed in reproductive systems and its expression level was reportedly affected by the gender and genotype interaction. The row-column statistic was applied to this data to account for the genotype, the gender, and the genotype-gender interaction. It was found that the interaction effect was statistically significant with a p-value equal to 0.004. The parametric F statistic and the aligned rank transform using the residuals yielded similar results. In order to illustrate the robustness of the nonparametric procedures, the analyses were redone with the first observation changed to an arbitrarily large number. The performance of the F statistic was severely affected and yielded a nonsignificant result. On the other hand, the nonparametric procedures were unaffected.

Next, we recall that in an example of a 3×4 factorial design considered by Box and Cox (1964) it was claimed that only after application of a nonlinear transformation can the error term be stabilized and the data made suitable for standard statistical analysis. We applied the row-column procedure to the un-transformed data and obtained a p-value of 0.44. Thus the hypothesis of no interaction was not rejected, a finding that concurs with Box and Cox. The aligned test on the other hand yielded a p-value of 0.02 which indicates the presence of interaction. However, for the transformed data, the aligned test with a p-value of 0.45 did not reject the null hypothesis.

Chapter Notes

1. Gao and Alvo (2005b) provide a brief historical look at the analysis of unbalanced two-way layout with interaction effects. Using the notion of a weighted rank, they present tests for both main effects and interaction effects. In addition, there is a discussion of the asymptotic relative efficiency of the proposed tests relative to the parametric F test. Various simulations further exemplify the power of the proposed tests. In a specific application, it is shown that the test statistic is the most robust in the presence of extreme outliers compared to other procedures.

2. Gao et al. (2008) also consider nonparametric multiple comparison procedures for unbalanced one-way factorial designs whereas Gao and Alvo (2008) treat nonparametric multiple comparison procedures for unbalanced two-way layouts.

3. Alvo and Pan (1997) considered the two-way layout with an ordered alternative using the tools of the unified theory. It was seen that the Spearman's distance induced the Page statistic (Page, 1963). The statistic induced by Hamming's distance was new (see also Schach (1979)). Alvo and Cabilio (1995) considered the two-way layout with ordered alternatives when the data within blocks may be incomplete and they obtained generalizations of the Page and Jonckheere statistics. Cabilio and Peng (2008) considered a multiple comparison procedure for ordered alternatives when the data are incomplete which maintains the experimentwise error rate at a preassigned level.

6.5. Exercises

Exercise 6.1. Consider the randomized block experiment given by the model

$$X_{ij} = b_i + \tau_j + e_{ij},\ i = 1, \ldots, n; j = 1, \ldots, t$$

where b_i is a block effect, τ_j is a treatment effect, and $\{e_{ij}\}$ are independent identically distributed error terns having a continuous distribution. Use the unified theory of hypothesis testing to obtain the statistic that corresponds to the Spearman measure of distance in order to test

$$H_0 : \tau_1 = \tau_2 = \ldots = \tau_t$$

against the ordered alternative

$$H_1 : \tau_1 \leq \tau_2 \leq \ldots \leq \tau_t$$

with at least one strict inequality. (See Alvo and Cabilio (1995).)

Exercise 6.2. Suppose that one observes at times $t_1 < t_2 < \ldots < t_k$ a random sample of n_i binary variables $\{y_{ij}\}$ taking values 1 or 0 with unknown probabilities $\theta_i, 1 - \theta_i$ respectively. Use the unified theory of hypothesis testing to obtain the statistic that corresponds to the Spearman measure of distance in order to test the null hypothesis of homogeneity

$$H_0 : \theta_1 = \theta_2 = \ldots = \theta_k$$

against the ordered alternative

$$H_1 : \theta_1 \leq \theta_2 \leq \ldots \leq \theta_k$$

with at least one strict inequality. (See Alvo and Berthelot (2012).)

Exercise 6.3. Using the following coded data on drug toxicity during 5 hours, test for an umbrella alternative when the peak toxicity is assumed to be during the 12:00–13:00 time period.

10:00–11:00	11:00–12:00	12:00–13:00	13:00–14:00	14:00–15:00
3.8	4.6	6.3	5.1	1.5
5.5	7.0	8.2	6.6	3.0
4.9	5.9	7.4	7.7	4.1

Exercise 6.4. We are interested in comparing 3 energy drinks in terms of endurance. The response is time to exhaustion on a treadmill. Test for differences in the mean between the three drinks.

Drink 1	Drink 2	Drink 3
42	48	62
36	34	48
54	56	75
44	46	52
28	32	44
45	50	65

7. Tests for Trend and Association

In this chapter, we consider additional applications of the smooth model paradigm described earlier in Chapter 4. We begin by considering tests for trend. We then proceed with the study of the one-sample test for a randomized block design. We obtain a different proof of the asymptotic distribution of Friedman's statistic based on Alvo (2016) who developed a likelihood function approach for the analysis of ranking data. Further, we derive a test statistic for the two-sample problem as well as for problems involving various two-way experimental designs. We exploit the parametric paradigm further by introducing the use of penalized likelihood in order to gain further insight into the data. Specifically, if judges provide rankings of t objects, penalized likelihood enables us to focus on those objects which exhibit the greatest differences.

7.1. Tests for Trend

Suppose that a series of observations of a random variable (concentration, unit well yield, biologic diversity, etc.) have been collected over some period of time. We would like to determine if their values exhibit either an increasing or a decreasing trend. In statistical terms this translates into a determination of whether the probability distribution from which they arise has changed over time. We would also like to describe the amount or rate of that change, in terms of changes in some central value of the distribution such as a mean or a median. The null hypothesis is that there is no trend. However, any given test brings with it a precise mathematical definition of what is meant by "no trend," including a set of background assumptions usually related to a type of distribution and to the serial correlation.

It is often of interest to test for a monotone trend in a series of data. For example, one may be interested in testing whether or not the pH in a river is decreasing or increasing over time. In such a case, the underlying distribution of the data is seldom known especially for small samples. We may develop, without loss in generality, a nonparametric test for a monotone increasing trend.

Let $X_1, \ldots, X_n$ be a random sample from some continuous distribution and consider the model in (5.7):

$$\pi\left(h\left(x,y\right);\theta\right) = \exp\left[\theta\, I(x < y) - K(\theta)\right] g(h(x,y))$$

M. Alvo, P. L. H. Yu, *A Parametric Approach to Nonparametric Statistics*, Springer Series in the Data Sciences, https://doi.org/10.1007/978-3-319-94153-0_7

where $g(h(x,y)) = \frac{1}{2}, x \neq y$ and $I(A) = 1$ if the event A occurs and $= 0$ otherwise with

$$h(x,y) = I(x < y)$$

$$e^{K(\theta)} = \frac{e^{\theta}+1}{2}.$$

The kernel $h(x,y)$ may be used to compare any two values along the sequence of the observations and is a measure of the slope. The case when $\theta = 0$ corresponds to the situation when there is trend. We may construct the composite likelihood function

$$L(\theta_i) = \prod_{j \neq i} \pi((h(x_i, x_j)); \theta_i) = \exp^{\left[\theta_i \sum_{j \neq i} I(x_i < x_j) - (n-1)K(\theta_i)\right]} \prod_{j \neq i} (g(h(x_i, x_j))).$$

The choice of kernel function is motivated by the fact that in testing for an increasing trend, we should focus on observations to the right of the present observation. It is seen that

$$\sum_{j \neq i} I(X_i < X_j) = R_i - 1,$$

where R_i is the rank of X_i among $X_1, \ldots, X_n$. Hence, the log of the composite likelihood is proportional to

$$\begin{aligned} \ell(\boldsymbol{\theta}) &= \log \prod_{i=1}^{n} L_i(\theta_i) \\ &\sim \sum_{i=1}^{n} \theta_i R_i - \sum_{i=1}^{n} \theta_i - (n-1) \sum K(\theta_i). \end{aligned}$$

Suppose now that we define

$$\theta_i = \beta \left(i - \frac{n+1}{2} \right).$$

It follows that

$$\ell(\boldsymbol{\theta}) \sim \beta \sum_{i=1}^{n} \left(i - \frac{n+1}{2} \right) R_i - (n-1) \sum K(\theta_i)$$

so that the test of no trend

$$H_0 : \theta_i = 0 \text{ for all } i$$

is now a test that $\beta = 0$. Differentiating with respect to β and setting it equal to 0 lead equivalently to the Rao score test statistic

$$S_n = \sum_{i=1}^{n} \left(i - \frac{n+1}{2} \right) R_i \tag{7.1}$$

which rejects for large absolute values. It should be noted that with this same approach we can also test for quadratic trends by choosing

$$\theta_i = \beta \left(i - \frac{n+1}{2} \right)^2 .$$

The statistic in (7.1) is a well-known test of trend. Since it is a linear rank statistic, its asymptotic distribution can be easily obtained from Theorem 3.2. In fact, we have that for large n

$$\frac{\sum_{i=1}^{n} \left(i - \frac{n+1}{2} \right) \frac{R_i}{n+1}}{\sigma} \xrightarrow{\mathcal{L}} \mathcal{N}(0,1)$$

where

$$\sigma^2 = \frac{1}{12} \sum \left(i - \frac{n+1}{2} \right)^2 = \frac{n(n^2-1)}{12}$$

Yu et al. (2002) obtained a generalization of the trend statistic in the presence of ties.

Example 7.1. In Appendix A.7, precipitation data for Saint John, New Brunswick, Canada was analyzed for the period 1894–1991 using the Spearman statistic (Alvo and Cabilio (1994)). The Z-score for St John was calculated to be 2.08 indicating there is an increasing trend.

7.2. Problems of Concordance

We recall Section 4.3 and consider the one-sample ranking problem whereby a group of judges are each asked to rank a set of t objects in accordance with some criterion. Let $\mathcal{P} = \{\boldsymbol{\nu}_j, j = 1, \ldots, t!\}$ be the space of all $t!$ permutations of the integers $1, 2, \ldots, t$ and let the probability mass distribution defined on $\mathcal{P}$ be given by

$$\boldsymbol{p} = (p_1, \ldots, p_{t!}),$$

where $p_j = \Pr(\boldsymbol{\nu}_j)$. Conceptually, each judge selects a ranking $\boldsymbol{\nu}$ in accordance with the probability mass distribution p. We are interested in testing the null hypothesis that each of the rankings is selected with equal probability, that is

$$H_0 : \boldsymbol{p} = \boldsymbol{p}_0 \text{ vs } H_1 : \boldsymbol{p} \neq \boldsymbol{p}_0. \tag{7.2}$$

where $\boldsymbol{p}_0 = \frac{1}{t!}\mathbf{1}$.

Define a k-dimensional vector score function $\boldsymbol{X}(\boldsymbol{\nu})$ on the space $\mathcal{P}$ and let its smooth probability mass function be given as

$$\pi(\boldsymbol{x}_j; \boldsymbol{\theta}) = \exp\left(\theta' \boldsymbol{x}_j - K(\boldsymbol{\theta})\right) \frac{1}{t!}, \quad j = 1, \ldots, t! \tag{7.3}$$

where $K(\boldsymbol{\theta})$ is a normalizing constant and $\boldsymbol{\theta}$ is a k-dimensional vector. Since

$$\sum_{j=1}^{t1} \pi(\boldsymbol{x}_j; \boldsymbol{\theta}) = 1$$

it can be seen that $K(\mathbf{0}) = 0$ and hence the hypotheses in (7.2) are equivalent to

$$H_0 : \boldsymbol{\theta} = \mathbf{0} \text{ vs } H_1 : \boldsymbol{\theta} \neq \mathbf{0}. \tag{7.4}$$

It follows that the log likelihood function is proportional to

$$l(\boldsymbol{\theta}) \sim n\left[\boldsymbol{\theta}'\hat{\boldsymbol{\eta}} - K(\boldsymbol{\theta})\right],$$

where

$$\hat{\boldsymbol{\eta}} = \left[\sum_{j=1}^{t!} \boldsymbol{x}_j \hat{p}_{nj}\right], \hat{p}_{nj} = \frac{n_j}{n}$$

and n_j represents the number of observed occurrences of the ranking $\boldsymbol{\nu}_j$. The Rao score statistic evaluated at $\boldsymbol{\theta} = \mathbf{0}$ is

$$\begin{aligned} U(\boldsymbol{\theta}; \boldsymbol{X}) &= n\frac{\partial}{\partial\theta}\left[\boldsymbol{\theta}'\hat{\boldsymbol{\eta}} - K(\mathbf{0})\right] \\ &= n\left[\hat{\boldsymbol{\eta}} - \frac{\partial}{\partial\theta}K(\mathbf{0})\right] \end{aligned}$$

whereas the information matrix is

$$\boldsymbol{I}(\boldsymbol{\theta}) = -n\left[\frac{\partial^2}{\partial\theta^2}K(\mathbf{0})\right].$$

The test then rejects the null hypothesis whenever

$$n^2\left[\hat{\boldsymbol{\eta}} - \frac{\partial}{\partial\boldsymbol{\theta}}K(\mathbf{0})\right]' \boldsymbol{I}^{-1}(\mathbf{0})\left[\hat{\boldsymbol{\eta}} - \frac{\partial}{\partial\boldsymbol{\theta}}K(\mathbf{0})\right] > \chi_f^2(\alpha),$$

where $\chi_f^2(\alpha)$ is the upper $100(1-\alpha)\%$ critical value of a chi-square distribution with $f = \text{rank}(\boldsymbol{I}(\boldsymbol{\theta}))$ degrees of freedom. We note that the test just obtained is the locally most powerful test of H_0. In the next section, we specialize this test statistic and consider the score functions of Spearman and Kendall.

7.2.1. Application Using Spearman Scores

In this section, we consider the Spearman score function which is defined to be the t-dimensional random vector of adjusted ranks

$$\boldsymbol{x}_j = \left(\nu_j(1) - \frac{t+1}{2}, \ldots, \nu_j(t) - \frac{t+1}{2}\right)', \; j = 1, \ldots, t!$$

Let $\boldsymbol{T}_S = (\boldsymbol{x}_j)$ be the $t \times t!$ matrix of possible values of $\boldsymbol{X}$. In the next theorem, we consider properties of the Spearman scores.

Theorem 7.1. *Under the null hypothesis for the Spearman scores,*

(a) the covariance function of $\mathbf{X}$ is given by

$$Cov(\mathbf{X}) = \frac{1}{t!}\boldsymbol{T}_S\boldsymbol{T}'_S = \frac{t+1}{12}\left[t\boldsymbol{I} - \boldsymbol{J}_t\right]. \tag{7.5}$$

(b) a generalized inverse of the covariance function is given by

$$\left(\boldsymbol{T}_S\boldsymbol{T}'_S\right)^- = \frac{12}{t(t+1)}\left[\boldsymbol{I} + \boldsymbol{J}_t\right]. \tag{7.6}$$

Proof. We refer the reader to (Alvo and Yu (2014), Chapter 4). □

Next, we demonstrate that the Rao score statistic is the well-known Friedman test (Friedman, 1937).

Theorem 7.2. *Under the null hypothesis, the Rao score statistic is asymptotically χ^2_{t-1} and is given by*

$$W = \frac{12n}{t(t+1)}\sum_{i=1}^{t}\left[\bar{R}_i - \frac{t+1}{2}\right]^2, \tag{7.7}$$

where $\bar{R}_i$ is the average of the ranks assigned to the ith object.

Proof. It can be seen that

$$\begin{aligned}\hat{\boldsymbol{\eta}} &= \left[\sum_{j=1}^{t!}\boldsymbol{x}_j\hat{p}_{nj}\right] \\ &= \boldsymbol{T}_S\hat{\boldsymbol{p}}_n.\end{aligned}$$

where

$$\hat{\boldsymbol{p}}_n = (\hat{p}_{nj})$$

is the $t! \times 1$ vector of relative frequencies. Since under the null hypothesis, $\boldsymbol{\theta} = \mathbf{0}$,

$$E_\theta [\boldsymbol{X}] = 0,$$

it follows that $U(\boldsymbol{\theta}; \boldsymbol{X}) = n\boldsymbol{T}_S \hat{\boldsymbol{p}}_n$ and

$$\begin{aligned} I(\boldsymbol{\theta}) &= \frac{n}{t!} \boldsymbol{T}_S \boldsymbol{T}'_S \\ &= n \left(\frac{t+1}{12}\right) [t\boldsymbol{I} - \boldsymbol{J}_t] . \end{aligned}$$

Consequently, the test statistic becomes

$$\begin{aligned} n^2 [\boldsymbol{T}_S \hat{\boldsymbol{p}}_n]' \boldsymbol{I}^{-1}(\boldsymbol{\theta}) [\boldsymbol{T}_S \hat{\boldsymbol{p}}_n] &= n^2 (\boldsymbol{T}_S \hat{\boldsymbol{p}}_n)' \left(\frac{12}{nt(t+1)} [\boldsymbol{I} + \boldsymbol{J}_t]\right) (\boldsymbol{T}_S \hat{\boldsymbol{p}}_n) \\ &= \left(\frac{12n}{t(t+1)}\right) (\boldsymbol{T}_S \hat{\boldsymbol{p}}_n)' (\boldsymbol{T}_S \hat{\boldsymbol{p}}_n) \\ &= \left(\frac{12n}{t(t+1)}\right) \sum_{i=1}^{t} \left(\bar{R}_i - \frac{t+1}{2}\right)^2 . \end{aligned} \tag{7.8}$$

For the last equality, we used the fact that

$$\boldsymbol{T}_S \hat{\boldsymbol{p}}_n = \left(\bar{\boldsymbol{R}} - \frac{t+1}{2} \mathbf{1}\right),$$

where $\bar{\boldsymbol{R}} = (\bar{R}_1, \ldots, \bar{R}_t)$ is the vector of the average ranks assigned to the objects. □

In the next section, we consider the Kendall scores as a second application.

7.2.2. Application Using Kendall Scores

Suppose now that the random vector $\boldsymbol{X}$ takes values $(t_K(\boldsymbol{\nu}))_q$ where the qth element is given by

$$(t_K(\boldsymbol{\nu}))_q = \operatorname{sgn}[\nu(j) - \nu(i)]$$

for $q = (i-1)\left(t - \frac{i}{2}\right) + (j-i), 1 \le i < j \le t$. This is the Kendall score function whose $\binom{t}{2} \times t!$ matrix of possible values is given by

$$\boldsymbol{T}_K = (t_K(\boldsymbol{\nu}_1), \ldots, t_K(\boldsymbol{\nu}_{t!}))'$$

Theorem 7.3. *Under the null hypothesis for the Kendall scores,*

(a) the covariance function of $\mathbf{X}$ is given by

$$Cov\,(\boldsymbol{X}) = \frac{1}{t!}\boldsymbol{T}_K \boldsymbol{T}'_K$$

whose entries $A\,(s,s',t,t') = \frac{1}{t!}\Sigma_{\boldsymbol{\nu}}\,\text{sgn}\,(\nu\,(s) - \nu\,(t))\,\text{sgn}\,(\nu\,(s') - \nu\,(t'))$ are given by

$$A\,(s,s',t,t') = \begin{cases} 0 & s \neq s', t \neq t' \\ 1 & s = s', t = t' \\ \frac{1}{3} & s = s', t \neq t' \\ -\frac{1}{3} & s = t', s' \neq t \end{cases}.$$

Moreover, the eigenvalues of $Cov\,(\boldsymbol{X})$ are $\frac{1}{3}, \frac{t+1}{3}$ with multiplicities $\binom{t-1}{2}, (t-1)$ respectively.

(b) the inverse matrix has entries of the form

$$\boldsymbol{B}\,(s,s',t,t') = (t-1)\begin{cases} 0 & s \neq s', t \neq t' \\ \frac{3}{t+1} & s = s', t = t' \\ -1 & s = s', t \neq t' \\ 1 & s = t', s' \neq t \end{cases}.$$

Proof. Part 1 follows from Lemma 4.1 in (Alvo and Yu (2014), p. 58). Part 2 follows by direct calculation. □

As an example, consider the case $t = 3$. Then,

$$\boldsymbol{X}(\nu) = \begin{pmatrix} \text{sgn}(\nu(2) - \nu(1)) \\ \text{sgn}(\nu(3) - \nu(1)) \\ \text{sgn}(\nu(3) - \nu(2)) \end{pmatrix}$$

It can be seen that,

$$\boldsymbol{T}_K \boldsymbol{T}'_K = \begin{pmatrix} 6 & 2 & -2 \\ 2 & 6 & 2 \\ -2 & 2 & 6 \end{pmatrix}$$

and

$$(\boldsymbol{T}_K \boldsymbol{T}'_K)^{-1} = \begin{pmatrix} \frac{1}{4} & -\frac{1}{8} & \frac{1}{8} \\ -\frac{1}{8} & \frac{1}{4} & -\frac{1}{8} \\ \frac{1}{8} & -\frac{1}{8} & \frac{1}{4} \end{pmatrix}.$$

The inverse matrix can be readily computed even for values of $t = 10$.

Theorem 7.4. *Under the null hypothesis, the Rao score statistic for the Kendall scores is asymptotically* $\chi^2_{\binom{t}{2}}$ *as* $n \to \infty$ *and is given by*

$$n\left(\boldsymbol{T}_K\hat{\boldsymbol{p}}_n\right)'\left(\boldsymbol{T}_K\boldsymbol{T}_K'\right)^{-1}\left(\boldsymbol{T}_K\hat{\boldsymbol{p}}_n\right). \tag{7.9}$$

Proof. The proof follows as for the Spearman statistic. □

An alternate form of the asymptotic distribution is given in Alvo et al. (1982):

$$n\left(\boldsymbol{T}_K\hat{\boldsymbol{p}}_n\right)'\left(\boldsymbol{T}_K\hat{\boldsymbol{p}}_n\right) \xrightarrow{L} \frac{1}{3\binom{t}{2}}\left\{(t+1)\,\chi^2_{t-1} + \chi^2_{\binom{t-1}{2}}\right\} - 1$$

where the left-hand side can be calculated as

$$\frac{\sum\left(2\#_i - n\right)^2}{n\binom{t}{2}}.$$

The summation is over all $\binom{t}{2}$ pairs of objects and $\#_i$ is the number of judges whose ranking of the pair i of objects agrees with the ordering of the same pair in a criterion ranking such as the natural ranking. The distribution of the Kendall statistic (7.9) is simpler though its form is somewhat more complicated. In the alternate form, the reverse is true.

In this section, we have seen that we can derive some well-known statistics through the parametric paradigm and that these are locally most powerful. We proceed next to show that a similar result can be obtained using Hamming score functions.

7.2.3. Application Using Hamming Scores

Suppose now that the random vector $\boldsymbol{X}$ takes values $\left(t_H(\nu)\right)_q$ where the qth element is given by

$$\left(t_H(\boldsymbol{\nu})\right)_q = I\left(\left[\nu(i) \le j\right] - \frac{j}{t}\right)$$

for $q = (i-1)\left(t - \frac{i}{2}\right) + (j-i), 1 \le i < j \le t$. This is the Hamming score function whose $t^2 \times t!$ matrix of possible values is

$$\boldsymbol{T}_H = \left(t_H(\boldsymbol{\nu}_1), \ldots, t_H(\boldsymbol{\nu}_{t!})\right)'.$$

Theorem 7.5. *Under the null hypothesis for the Hamming scores,*

(a) the covariance function of $\mathbf{X}$ is given by

$$\boldsymbol{\Gamma} = Cov(\boldsymbol{X}) = \frac{1}{t-1}\left(\boldsymbol{I} - \frac{\boldsymbol{J}}{t}\right) \otimes \left(\boldsymbol{I} - \frac{\boldsymbol{J}}{t}\right)$$

and $\boldsymbol{\Gamma}$ has a distinct eigenvalue $\frac{1}{t-1}$ with multiplicity $(t-1)^2$.

(b) The asymptotic distribution of the test statistic is

$$(t-1)\,n\,(\boldsymbol{T}_H\hat{\boldsymbol{p}}_n)'\,(\boldsymbol{T}_H\hat{\boldsymbol{p}}_n) \xrightarrow{L} \chi^2_{(t-1)^2}.$$

Moreover, the statistic can be calculated as

$$(\boldsymbol{T}_H\hat{\boldsymbol{p}}_n)'\,(\boldsymbol{T}_H\hat{\boldsymbol{p}}_n) = \sum_l \sum_i D_i^2\,(l) - n^2$$

where $D_i\,(l)$ is the number of rankers who assign rank l to object i.

This test was first introduced by Anderson (1959) and rediscovered by Kannemann (1976). Schach (1979) obtained the asymptotic distribution of the statistic based on Hamming distance under the null hypothesis and under contiguous alternatives making use of Le Cam's third lemma. Alvo and Cabilio (1998) extended the statistic to include various block designs.

7.3. The Two-Sample Ranking Problem

We may now consider the two-sample ranking problem. For simplicity, we shall make use of the Spearman scores throughout. Let $\boldsymbol{X}_1, \boldsymbol{X}_2$ be two independent random vectors whose distributions, as in the one-sample case, are expressed for simplicity as

$$\pi\,(\boldsymbol{x}_j;\boldsymbol{\theta}_l) = \exp\left\{\boldsymbol{\theta}_l'\boldsymbol{x}_j - K\,(\boldsymbol{\theta}_l)\right\} p_l\,(j)\,,\ j=1,\ldots,t!, l=1,2,$$

where $\boldsymbol{\theta}_l = (\theta_{l1},\ldots,\theta_{lt})'$ represents the vector of parameters for population l. We are interested in testing

$$H_0:\boldsymbol{\theta}_1=\boldsymbol{\theta}_2 \text{ vs } H_1:\boldsymbol{\theta}_1\neq\boldsymbol{\theta}_2.$$

The probability distribution $\{p_l\,(j)\}$ represents an unspecified null situation. Define

$$\hat{\boldsymbol{p}}_l = \left(\frac{n_{l1}}{n_l},\ldots,\frac{n_{lt!}}{n_l}\right)',$$

where n_{ij} represents the number of occurrences of the ranking $\boldsymbol{\nu}_j$ in sample l.

Also, for $l=1,2$, set $\sum_j n_{ij} \equiv n_l$, $\boldsymbol{\gamma}=\boldsymbol{\theta}_1-\boldsymbol{\theta}_2$ and

$$\boldsymbol{\theta}_l \;=\; \boldsymbol{m}+b_l\boldsymbol{\gamma},$$

where

$$\boldsymbol{m}=\frac{n_1\boldsymbol{\theta}_1+n_2\boldsymbol{\theta}_2}{n_1+n_2}, b_1=\frac{n_{2,}}{n_1+n_2}, b_2=-\frac{n_1}{n_1+n_2}.$$

Let $\boldsymbol{\Sigma}_l$ be the covariance matrix of $\boldsymbol{X}_l$ under the null hypothesis defined as

$$\boldsymbol{\Sigma}_l = \boldsymbol{\Pi}_l - \boldsymbol{p}_l \boldsymbol{p}_l',$$

where $\boldsymbol{\Pi}_l = diag\left(p_l\left(1\right), \ldots, p_l\left(t!\right)\right)$ and $\boldsymbol{p}_l = \left(p_l\left(1\right), \ldots, p_l\left(t!\right)\right)'$. The logarithm of the likelihood L as a function of $(\boldsymbol{m}, \boldsymbol{\gamma})$ is proportional to

$$\log\, L\left(\boldsymbol{m}, \boldsymbol{\gamma}\right) \quad \sim \quad \sum_{l=1}^{2} \sum_{j=1}^{t!} n_{lj} \left\{ \left(\boldsymbol{m} + b_l \boldsymbol{\gamma}\right)' \boldsymbol{x}_j - K\left(\boldsymbol{\theta}_l\right) \right\}.$$

Theorem 7.6. *Consider the two-sample ranking problem whereby we wish to test*

$$H_0 : \boldsymbol{\theta}_1 = \boldsymbol{\theta}_2 \ vs \ H_1 : \boldsymbol{\theta}_1 \neq \boldsymbol{\theta}_2.$$

The Rao score test statistic is given by

$$n \left(\boldsymbol{T}_S \hat{\boldsymbol{p}}_1 - \boldsymbol{T}_S \hat{\boldsymbol{p}}_2\right)' \hat{\boldsymbol{D}} \left(\boldsymbol{T}_S \hat{\boldsymbol{p}}_1 - \boldsymbol{T}_S \hat{\boldsymbol{p}}_2\right). \tag{7.10}$$

It has asymptotically a χ^2_f whenever $n_l/n \rightarrow \lambda_l > 0$ as $n \rightarrow \infty$, where $n = n_1 + n_2$. Here $\hat{\boldsymbol{D}}$ is the Moore-Penrose inverse of $\boldsymbol{T}_S \hat{\boldsymbol{\Sigma}} \boldsymbol{T}_S'$ and $\hat{\boldsymbol{\Sigma}}$ is a consistent estimator of $\boldsymbol{\Sigma} = \frac{\boldsymbol{\Sigma}_1}{\lambda_1} + \frac{\boldsymbol{\Sigma}_2}{\lambda_2}$ and f is the rank of $\hat{\boldsymbol{D}}$.

Proof. The Rao score vector evaluated under the null hypothesis is given by

$$\left(\frac{\partial \log\, L\left(\boldsymbol{m}, \boldsymbol{\gamma}\right)}{\partial \gamma_r}\right) = \frac{n_1 n_2}{n_1 + n_2} \left(\boldsymbol{T}_S \hat{\boldsymbol{p}}_1 - \boldsymbol{T}_S \hat{\boldsymbol{p}}_2\right)$$

and consequently, the Rao score test statistic becomes

$$n \left(\boldsymbol{T}_S \hat{\boldsymbol{p}}_1 - \boldsymbol{T}_S \hat{\boldsymbol{p}}_2\right)' \hat{\boldsymbol{D}} \left(\boldsymbol{T}_S \hat{\boldsymbol{p}}_1 - \boldsymbol{T}_S \hat{\boldsymbol{p}}_2\right).$$

The result follows from the general theory in Section 7.1. □

The result of this section was first obtained in Feigin and Alvo (1986) using notions of diversity. It is derived presently through the parametric paradigm. See Chapter 4.2 of Feigin and Alvo (1986) for a discussion on the efficient calculation of the test statistic.

The parametric paradigm can also be used to deal with the two-sample mixture problem using a distribution expressed as

$$\pi\left(\boldsymbol{X}_1, \boldsymbol{X}_2; \boldsymbol{\theta}_1, \boldsymbol{\theta}_2\right) = \lambda \pi\left(\boldsymbol{X}_1; \boldsymbol{\theta}_1\right) + \left(1 - \lambda\right) \pi\left(\boldsymbol{X}_2; \boldsymbol{\theta}_2\right), 0 < \lambda < 1.$$

In that case, the use of the EM algorithm can provide estimates of the parameters (see Casella and George (1992)).

7.4. The Use of Penalized Likelihood in Tests of Concordance: One and Two Group Cases

In the previous sections, it was possible to derive well-known test statistics for the one- and two-sample ranking problems through the parametric paradigm. In this section, we make use of the parametric paradigm to obtain new results for the ranking problems. Specifically, we consider a negative penalized likelihood function defined to be the negative likelihood function subject to a constraint on the parameters which is then minimized with respect to the parameter. This approach yields further insight into ranking problems.

For the one-sample ranking problem, let

$$\Lambda(\boldsymbol{\theta}, c) = -\boldsymbol{\theta}' \left[\sum_{j=1}^{t!} n_j \boldsymbol{x}_j\right] + nK(\boldsymbol{\theta}) + \lambda(\sum_{i=1}^{t} \theta_i^2 - c) \tag{7.11}$$

represent the penalizing function for some prescribed values of the constant c. We shall assume for simplicity that $\|\boldsymbol{x}_j\| = 1$. When t is large (say $t \geq 10$), the computation of the exact value of the normalizing constant $K(\boldsymbol{\theta})$ involves a summation of $t!$ objects. McCullagh (1993) noted the resemblance of (7.3) to the continuous von Mises-Fisher density

$$f(\boldsymbol{x}; \boldsymbol{\theta}) = \frac{\|\boldsymbol{\theta}\|^{\frac{t-3}{2}}}{2^{\frac{t-3}{2}} t! I_{\frac{t-3}{2}}(\|\boldsymbol{\theta}\|)\Gamma(\frac{t-1}{2})} \exp(\boldsymbol{\theta}' \boldsymbol{x}),$$

where $\|\boldsymbol{\theta}\|$ is the norm of $\boldsymbol{\theta}$, $\boldsymbol{x}$ is on the unit sphere, and $I_v(z)$ is the modified Bessel function of the first kind given by

$$I_v(z) = \sum_{k=0}^{\infty} \frac{1}{\Gamma(k+1)\Gamma(v+k+1)} \left(\frac{z}{2}\right)^{2k+\nu}.$$

This seems to suggest the approximation of the constant $K(\boldsymbol{\theta})$ by

$$\exp(-K(\boldsymbol{\theta})) \approx \frac{1}{t!} \cdot \frac{\|\boldsymbol{\theta}\|^{\frac{t-3}{2}}}{2^{\frac{t-3}{2}} I_{\frac{t-3}{2}}(\|\boldsymbol{\theta}\|)\Gamma(\frac{t-1}{2})}.$$

The accuracy of this approximation is very good as discussed further in Chapter 11 in connection with Bayesian models for ranking data.

We proceed to find the maximum penalized likelihood estimation for $\boldsymbol{\theta}$ using algorithms implemented in MATLAB that converge very fast. Following the estimation of $\boldsymbol{\theta}$, we apply the basic bootstrap method in order to determine the distribution of $\boldsymbol{\theta}$. We sample n rankings with replacement from the observed data. Then we find the maximum

Table 7.1.: Combined data on leisure preferences

Rankings	(123)	(132)	(213)	(231)	(312)	(321)
Frequencies	1	1	1	5	7	12

likelihood estimate of $\boldsymbol{\theta}$ from each bootstrap sample. Repeating this procedure 10,000 times leads to the bootstrap distribution for $\boldsymbol{\theta}$. In this way, we can draw useful inference from the distribution $\boldsymbol{\theta}$ and in particular construct two-sided confidence intervals for its components. We applied this to a data set with $t = 3$. Define the probabilities of the rankings

$$p_1 = P\{123\}, p_2 = P\{132\}, p_3 = P\{213\}, p_4 = P\{231\}, p_5 = P\{312\}, p_6 = P\{321\}$$

and consider the interpretation of the $\boldsymbol{\theta}$ vector. Since

$$\boldsymbol{\theta}'\boldsymbol{T}_S\boldsymbol{p} = \begin{bmatrix}\theta_1 & \theta_2 & \theta_3\end{bmatrix}\begin{bmatrix}p_5 + p_6 - p_1 - p_2 \\ p_2 + p_4 - p_3 - p_5 \\ p_1 + p_3 - p_4 - p_6\end{bmatrix},$$

we note that θ_1 weights the difference between

$$p_5 + p_6 = \Pr(\text{giving rank 3 to item 1})$$

and

$$p_1 + p_2 = \Pr(\text{giving rank 1 to item 1})$$

which compares the probabilities of assigning the lowest and highest rank to object 1. The other components make similar comparisons for the other objects.

Example 7.2. Sutton Data (One-Sample Case)
C. Sutton considered in her 1976 thesis, the leisure preferences and attitudes on retirement of the elderly for 14 white and 13 black females in the age group 70–79 years. Each individual was asked: with which sex do you wish to spend your leisure time? Each female was asked to rank the three responses: male(s), female(s), or both, assigning rank 1 for the most desired and 3 for the least desired. The first object in the ranking corresponds to "male," the second to "female," and the third to "both." To illustrate the approach in the one-sample case, we combined the data from the two groups as in Table 7.1.

We applied the method of penalized likelihood in this situation and the results are shown in Table 7.2. To better illustrate our result, we rearrange our result (unconstrained $\boldsymbol{\theta}$, $c = 1$) and data as Table 7.3. It can be seen that θ_1 is the largest coefficient and object 1 (Male) shows the greatest difference between the number of judges choosing rank 1 or rank 3 which implies that the judges dislike spending leisure time with males

Table 7.2.: Penalized likelihood for the combined data

c	θ_1	θ_2	θ_3	$\Lambda(\boldsymbol{\theta}, c)$
0.5	0.53	-0.06	-0.47	50.00
1	0.75	-0.09	-0.66	50.36
2	1.06	-0.12	-0.93	54.62

Table 7.3.: The combined data reexpressed

Object	Number of judges	Action	Difference		$c = 1$
Male	2	assign rank 1	-17	θ_1	0.75
	19	assign rank 3			
Female	8	assign rank 1	2	θ_2	-0.09
	6	assign rank 3			
Both	17	assign rank 1	15	θ_3	-0.66
	2	assign rank 3			

the most. For object 3 (Both), the greater value of negative θ_3 implies the judges prefer to spend leisure time with both sexes the most. θ_2 is close to zero and we deduce the judges show no strong preference on Female. This is consistent with the hypothesis that $\boldsymbol{\theta}$ close to zero means randomness. To conclude, the results also show that θ_i weights the difference in probability between assigning the lowest and the top rank to object i. A negative value of θ_i means the judges prefer object i more whereas a positive θ_i means the judges are more likely to assign a lower rank to object i.

We plot the bootstrap distribution of $\boldsymbol{\theta}$ in Figure 7.1. For H_0: $\theta_i = 0$, we see that θ_1 and θ_3 are significantly different from 0 whereas θ_2 is not. We also see that the bootstrap distributions are not entirely bell shaped leading us to conclude that a traditional t-test method may not be appropriate in this case.

For the Kendall score, we consider once again the Sutton data ($t = 3$) and apply a penalized likelihood approach. The results are exhibited in Table 7.4.

We rearrange the Sutton data focusing on paired comparisons and the results ($c = 1$) are displayed in Table 7.5. First, we note that all the $\theta_i's$ are negative. This is consistent

Table 7.4.: Penalized likelihood using the Kendall score function for the Sutton data

Paired comparison			Choice of c				no constraint
object i	object j		$c = 0.5$	$c = 1$	$c = 2$	$c = 10$	
1	2	θ_1	-0.35	-0.49	-0.70	-1.56	-0.60
1	3	θ_2	-0.56	-0.80	-1.13	-2.53	-0.97
2	3	θ_3	-0.24	-0.34	-0.48	-1.08	-0.41
$\Lambda(\boldsymbol{\theta}, c)$			42.79	40.17	40.20	127.76	39.59

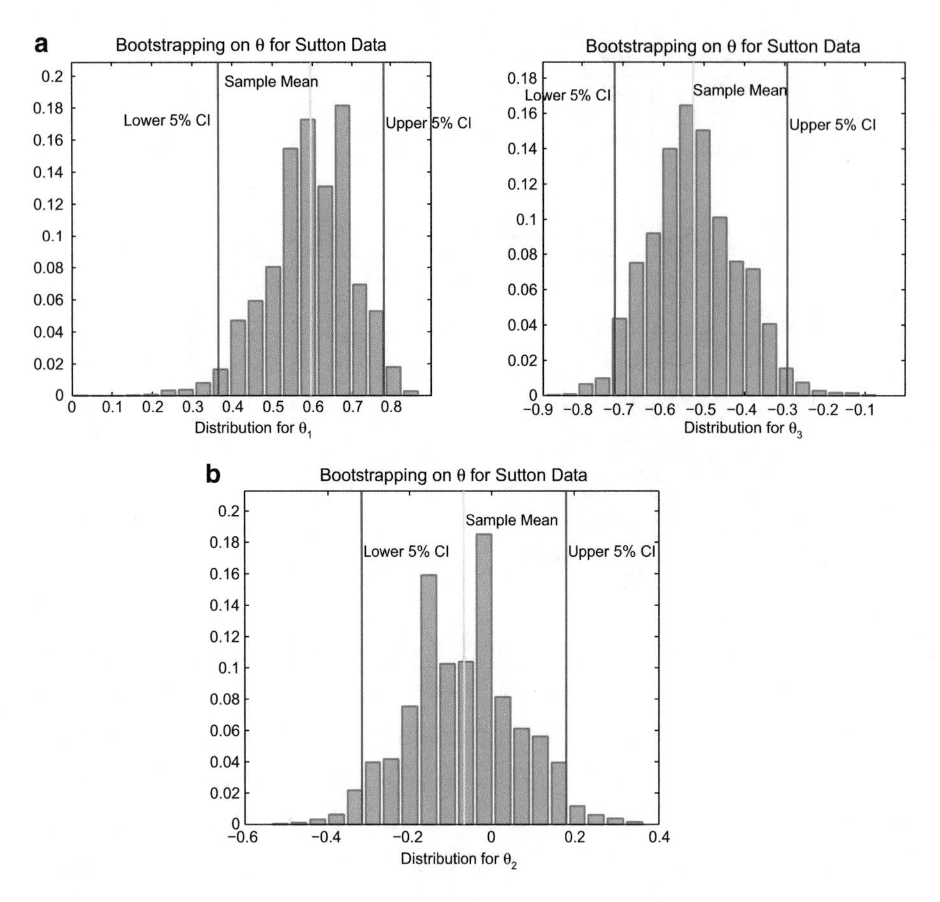

Figure 7.1.: The distribution of θ for Sutton data by bootstrap method

Table 7.5.: Paired comparison for the Sutton data and the estimation of $\boldsymbol{\theta}$

object i	object j	Number of judges	Paired comparison	θ
1	2	7	more prefer 1	-0.49
		20	more prefer 2	
1	3	3	more prefer 1	-0.80
		24	more prefer 3	
2	3	9	more prefer 2	-0.34
		18	more prefer 3	

Table 7.6.: Penalized likelihood results for the Sutton data

c	γ_1	γ_2	γ_3	m_1	m_2	m_3	$\Lambda(\boldsymbol{m},\gamma)$
0.5	0.34	-0.57	0.24	0.59	-0.07	-0.52	46.88
1	0.48	-0.81	0.34	0.58	-0.06	-0.52	46.38
2	0.67	-1.15	0.48	0.57	-0.06	-0.51	46.46
10	1.50	-2.57	1.07	0.47	-0.04	-0.43	58.73

Table 7.7.: The Sutton data on leisure preferences

Rankings	(123)	(132)	(213)	(231)	(312)	(321)
Frequencies for white females	0	0	1	0	7	6
Frequencies for black females	1	1	0	5	0	6

with our interpretations. The judges show a strong preference for Males to Both and Males to Females. They least prefer Females to Both. We may conclude that the $\theta_i's$ represent well the paired preferences among the judges. We applied penalized likelihood in this situation and the results are shown in Table 7.6.

Example 7.3. Sutton Data (Two-Sample Case)
For the two-sample case, we consider a penalized likelihood approach to determine those components of γ which most separate the populations. Hence, we consider minimizing with respect to parameters $\boldsymbol{m}$ and $\boldsymbol{\gamma}$ the function:

$$\Lambda(\boldsymbol{m},\boldsymbol{\gamma}) = -\sum_{l=1}^{2}(\boldsymbol{m}+b_l\boldsymbol{\gamma})\sum_{j=1}^{t!} n_{lj}x_{lj} + \sum_{l=1}^{2} n_l K(\boldsymbol{m}+b_l\boldsymbol{\gamma}) + \lambda(\sum_{i=1}^{t}\gamma_i^2 - c)$$

for some prescribed values of the constant c and λ. We continue to use the approximation to the normalizing constant from the von Mises-Fisher distribution to approximate $K(\boldsymbol{\theta})$.

Here γ_i shows the difference between the two population's preference on object i. A negative γ_i means that population 1 shows more preference on object i compared to population 2. A positive γ_i means that population 2 shows more preference on object i compared to population 1. For γ_i close to zero, there is no difference between the two populations on that object. As we shall see, this interpretation is consistent with the results in the real data applications. From the definition of $\boldsymbol{m}$, we know that $\boldsymbol{m}$ is the common part of $\boldsymbol{\theta}_1$ and $\boldsymbol{\theta}_2$. More specifically, $\boldsymbol{m}$ is the weight average of $\boldsymbol{\theta}_1$ and $\boldsymbol{\theta}_2$ taking into account the sample sizes of the populations.

As an application consider the Sutton data ($t = 3$) found in Table 7.7. Rearranging the results for $c = 1$ we have the original data in Table 7.8. First, it is seen that $\boldsymbol{m}$ is just like the $\boldsymbol{\theta}$'s in the one-sample problem. For example, m_3 is the smallest value and the whole population prefers object "Both" best. m_3 is the largest and the whole

Table 7.8.: The Sutton data and the estimation of $\boldsymbol{m}, \gamma$

Object:	#white females	#black females	Sum	Action	γ	$\boldsymbol{m}$
Male	0	2	2	give rank 1	0.48	0.58
	13	6	19	give rank 3		
Female	8	0	8	give rank 1	-0.81	-0.06
	0	6	6	give rank 3		
Both	6	11	17	give rank 1	0.34	-0.52
	1	1	2	give rank 3		

population mostly dislikes object "Male." This is not surprising since we know that $\boldsymbol{m}$ is the common part of $\boldsymbol{\theta}_1$ and $\boldsymbol{\theta}_2$. For the parameter $\boldsymbol{\gamma}$, we note that white females prefer to spend leisure time with Females (8 individuals assign rank 1) whereas black females do not (6 individuals assign rank 3). We see that γ_2 is negative and largest in absolute value. There is a significant difference of opinion with respect to object 2, Female. For objects "Male" and "Both," black females prefer them to white females. To conclude, the results are consistent with the interpretation of $\boldsymbol{m}$ and $\boldsymbol{\gamma}$.

We conclude that the use of the parametric paradigm provided more insight on the objects being ranked. Further details for the penalized likelihood approach are found in Alvo and Xu (2017).

7.5. Design Problems

We begin by defining the notion of compatibility introduced by Alvo and Cabilio (1999) which will be useful for the study of incomplete block designs. We then recall some asymptotic results found in Alvo and Yu (2014) and conclude by relating those to the score statistic obtained by embedding the problem in a parametric framework. Illustrations are given only for the Spearman distance.

7.5.1. Compatibility

Compatibility is a concept that was introduced in connection with incomplete rankings. Specifically, suppose that $\boldsymbol{\mu} = (\mu(1), \ldots, \mu(t))$ represents a complete ranking of t objects and that $\boldsymbol{\mu}^* = (\mu^*(o_1), \ldots, \mu^*(o_k))$ represents an incomplete ranking of a subset k of these objects where $o_1, \ldots, o_k$ represent the labels of the objects ranked. Alternatively, for an incomplete ranking, we may retain the $t-$dimensional vector notation and indicate by a " $-$ " an unranked object. Hence, the incomplete ranking $\boldsymbol{\mu}^* = (2, -, 3, 4, 1)$ indicates that among the $t = 5$ objects, only object "2" is not ranked. In this notation, complete and incomplete rankings are of the same length t.

Definition 7.1. A complete ranking $\boldsymbol{\mu}$ of t objects is said to be compatible with an incomplete ranking $\boldsymbol{\mu}^*$ of a subset k of these objects, $2 \leq k \leq t$, if the relative ranking of every pair of objects ranked in $\boldsymbol{\mu}^*$ coincides with their relative ranking in μ.

As an example, the incomplete ranking $\boldsymbol{\mu}^* = (2, -, 3, 4, 1)$ gives rise to a class of order preserving complete compatible rankings. Denoting by $\boldsymbol{C}(\boldsymbol{\mu}^*)$ the set of complete rankings compatible with $\boldsymbol{\mu}^*$, we have that

$$\boldsymbol{C}(\boldsymbol{\mu}^*) = \{(2,5,3,4,1), (2,4,3,5,1), (2,3,4,5,1), (3,2,4,5,1), (3,1,4,5,2)\}.$$

The total number of complete rankings of t objects compatible with an incomplete ranking of a subset of k objects is given by $\frac{t!}{k!}$. This follows from the fact that there are $\binom{t}{k}$ ways of choosing k integers for the ranked objects, one way in placing them to preserve the order and then $(t-k)!$ ways of rearranging the remaining integers. The product is thus

$$a \equiv \binom{t}{k}(t-k)! = \frac{t!}{k!}.$$

In general, we may define a matrix of compatibility $\boldsymbol{C}$ so that each column contains exactly a ones and each row contains exactly one 1.

For a given pattern of $t - k_h$ missing observations, each permutation of the k_h objects has its own distinct set of $t!/k_h!$ compatible t-rankings, so that each column of $\boldsymbol{C}_h$ contains exactly $t!/k_h!$ 1's, and each row exactly one 1. For any incomplete k_h−ranking $\boldsymbol{\mu}^*$, this definition can be shown to lead to an analogue of $\boldsymbol{T}$ given by

$$\boldsymbol{T}_h^* = \frac{k_h!}{t!}\boldsymbol{T}\boldsymbol{C}_h \tag{7.12}$$

whose columns are the score vectors of

$$\boldsymbol{\mu}^* = (\mu^*(1), \mu^*(2), \ldots, \mu^*(k_h))$$

as $\boldsymbol{\mu}^*$ ranges over each of the $k_h!$ permutations (Alvo and Cabilio, 1991).

Example 7.4. Let $t = 3$, $k_h = 2$. The complete rankings associated with the rows are in the order (123), (132), (213), (231), (312), (321). For the incomplete rankings (12_), (21_) indexing the columns, the associated compatibility matrix $\boldsymbol{C}_h$ is

$$\boldsymbol{C}_h = \begin{bmatrix} 1 & 0 \\ 1 & 0 \\ 0 & 1 \\ 1 & 0 \\ 0 & 1 \\ 0 & 1 \end{bmatrix}.$$

In the Spearman case, for the incomplete rankings (12_), (21_)

$$\boldsymbol{T}_h^* = \frac{1}{3}\begin{bmatrix} -1 & -1 & 0 & 0 & 1 & 1 \\ 0 & 1 & -1 & 1 & -1 & 0 \\ 1 & 0 & 1 & -1 & 0 & -1 \end{bmatrix}\begin{bmatrix} 1 & 0 \\ 1 & 0 \\ 0 & 1 \\ 1 & 0 \\ 0 & 1 \\ 0 & 1 \end{bmatrix} = \begin{bmatrix} -\frac{2}{3} & \frac{2}{3} \\ \frac{2}{3} & -\frac{2}{3} \\ 0 & 0 \end{bmatrix}.$$

For completeness, we may also extend the notion of compatibility to tied rankings defined as follows.

Definition 7.2. A tied ordering of t objects is a partition into e sets, $1 \leq e \leq t$, each containing d_i objects, $d_1 + d_2 + \ldots + d_e = t$, so that the d_i objects in each set share the same rank i, $1 \leq i \leq e$. Such a pattern is denoted $\delta = (d_1, d_2, \ldots, d_e)$. The ranking denoted by $\boldsymbol{\mu}_\delta = (\mu_\delta(1), \ldots, \mu_\delta(t))$ resulting from such an ordering is called a tied ranking and is one of $\frac{t!}{d_1!, d_2!, \ldots, d_e!}$ possible permutations.

For example, if $t = 3$ objects are ranked, it may happen that objects 1 and 2 are equally preferred to object 3. Consequently, the rankings $(1, 2, 3)$ and $(2, 1, 3)$ would both be plausible and should be placed in a "compatibility" class. The average of the rankings in the compatibility class which results from the use of the Spearman distance would then yield the ranking

$$\frac{1}{2}[(1, 2, 3) + (2, 1, 3)] = (1.5, 1.5, 3).$$

It is seen that this notion of compatibility for ties justifies the use of the mid-rank when ties are present. Associated with every tied ranking, we may define a $t! \times \frac{t!}{d_1!, d_2!, \ldots, d_e!}$ matrix of compatibility. Yu et al. (2002) considered the problem of testing for independence between two random variables when the pattern for ties and for missing observations are fixed.

7.5.2. General Block Designs

The parametric paradigm may be extended to the study of more general block designs. We shall restrict attention to the Spearman distance. Consider the situation in which t objects are ranked k_h at a time, $2 \leq k_h \leq t$ by b judges (blocks) independently and in such a way that each object is presented to r_i judges and each pair of objects (i, j) is compared by λ_{ij} judges, $h = 1, \ldots, b$, $i, j = 1, \ldots, t$. We would like to test the hypothesis of no treatment effect, that is:

H_0 : *each judge, when presented with the specified k_h objects, picks the ranking at random from the space of $k_h!$ permutations of $(1, 2, \ldots, k_h)$.*

In the study of the asymptotic behavior of various statistics for such problems, we consider n replications of such basic designs. For any incomplete k_h-ranking $\boldsymbol{\mu}^*$, define the score vector of

$$\boldsymbol{\mu}^* = (\mu^*(1), \mu^*(2), \ldots, \mu^*(k_h))$$

as $\boldsymbol{\mu}^*$ ranges over each of the $k_h!$ permutations (Alvo and Cabilio, 1991). From Alvo and Cabilio (1991), for a given permutation of $(1, 2, \ldots, k_h)$, indexed by $s = 1, 2, \ldots, k_h!$, define the vector $\boldsymbol{x}_{h(s)}$ whose ith entry is given by

$$\frac{(t+1)}{(k_h+1)}\left(\mu^*_{h(s)}(i) - \frac{k_h+1}{2}\right)\delta_h(i) = \left(\frac{(t+1)}{(k_h+1)}\mu^*_{h(s)}(i) - \frac{t+1}{2}\right)\delta_h(i), \tag{7.13}$$

$\delta_h(i)$ is either 1 or 0 depending on whether the object i is, or is not, ranked in block h, and $\mu^*_{h(s)}(i)$, as defined above, is the rank of object i for the permutation indexed by s for block pattern h. This is also the corresponding ith row element of column s of

$$\boldsymbol{T}^*_h = \frac{k_h!}{t!}\boldsymbol{T}\boldsymbol{C}_h.$$

An (i, j) element of $\left(\frac{k_h!}{t!}\boldsymbol{T}_S\boldsymbol{C}_h\right)\left(\frac{k_h!}{t!}\boldsymbol{T}_S\boldsymbol{C}_h\right)'$ is thus of the form

$$\sum_{s=1}^{k_h!}\left(\frac{(t+1)}{(k_h+1)}\mu^*_{h(s)}(i) - \frac{t+1}{2}\right)\left(\frac{(t+1)}{(k_h+1)}\mu^*_{h(s)}(j) - \frac{t+1}{2}\right)\delta_h(i)\delta_h(j),$$

For a specific pattern of missing observations for each of the b blocks, the matrix of scores is given by

$$\boldsymbol{T}^* = (\boldsymbol{T}^*_1 \mid \boldsymbol{T}^*_2 \mid \ldots \mid \boldsymbol{T}^*_b) = \boldsymbol{T}\left(\frac{k_1!}{t!}\boldsymbol{C}_1 \mid \frac{k_2!}{t!}\boldsymbol{C}_2 \mid \ldots \mid \frac{k_b!}{t!}\boldsymbol{C}_b\right),$$

where the index of each block identifies the pattern of missing observations, if any, in that block. It was shown in Alvo and Cabilio (1991) that the proposed test rejects H_0 for large values of

$$G \equiv (\boldsymbol{T}^*\boldsymbol{f})'(\boldsymbol{T}^*\boldsymbol{f}) = \boldsymbol{f}'(\boldsymbol{T}^*)'\boldsymbol{T}^*\boldsymbol{f}, \tag{7.14}$$

where the $\sum_{h=1}^{b} k_h!$ dimensional vector $\boldsymbol{f}$ is the vector of frequencies, for each of the b patterns of the observed incomplete rankings. That is, $\boldsymbol{f} = (\boldsymbol{f}_1 \mid \boldsymbol{f}_2 \mid \cdots \mid \boldsymbol{f}_b)'$, where $\boldsymbol{f}_h$ is the $k_h!$ dimensional vector of the observed frequencies of each of the $k_h!$ ranking permutations for the incomplete pattern $h = 1, 2, \ldots, b$. Using the fact that for these distance measures the matrix $\boldsymbol{T}^*$ is orthogonal to the vector of 1's, and proceeding in a manner analogous to Alvo and Cabilio (1991) gives the following. Moreover, we have

Theorem 7.7. *Under H_0 as $n \to \infty$,*

$$\frac{1}{\sqrt{n}} \boldsymbol{T}^* \boldsymbol{f} \xrightarrow{\mathcal{L}} \mathcal{N}(\boldsymbol{0}, \boldsymbol{\Gamma}), \tag{7.15}$$

where $\mathcal{N}(\boldsymbol{0}, \boldsymbol{\Gamma})$ is a multivariate normal with mean 0 and covariance matrix

$$\boldsymbol{\Gamma} = \sum_{h=1}^{b} \frac{1}{k_h!} \left(\frac{k_h!}{t!} \boldsymbol{T} \boldsymbol{C}_h \right) \left(\frac{k_h!}{t!} \boldsymbol{T} \boldsymbol{C}_h \right)'. \tag{7.16}$$

Thus

$$n^{-1} G = n^{-1} (\boldsymbol{T}^* \boldsymbol{f})' (\boldsymbol{T}^* \boldsymbol{f}) \xrightarrow{\mathcal{L}} \sum \alpha_i z_i^2, \tag{7.17}$$

where $\{z_i\}$ are independent identically distributed normal variates, and $\{\alpha_i\}$ are the eigenvalues of $\boldsymbol{\Gamma}$.

In the **Spearman case**, for a given permutation of $(1, 2, \ldots, k_h)$, indexed by $s = 1, 2, \ldots, k_h!$, the corresponding ith row element of column s of $\frac{k_h!}{t!} \boldsymbol{T}_S \boldsymbol{C}_h$ is found to be

$$\frac{(t+1)}{(k_h+1)} \left(\mu^*_{h(s)}(i) - \frac{k_h+1}{2} \right) \delta_h(i) = \left(\frac{(t+1)}{(k_h+1)} \mu^*_{h(s)}(i) - \frac{t+1}{2} \right) \delta_h(i), \tag{7.18}$$

$\delta_h(i)$ is either 1 or 0 depending on whether the object i is, or is not, ranked in block h, and $\mu^*_{h(s)}(i)$, as defined above, is the rank of object i for the permutation indexed by s for block pattern h.

Lemma 7.1. *In the Spearman case we have that the $t \times t$ matrix*

$$\boldsymbol{\Gamma} = \sum_{h=1}^{b} \frac{1}{k_h!} \left(\frac{k_h!}{t!} \boldsymbol{T}_S \boldsymbol{C}_h \right) \left(\frac{k_h!}{t!} \boldsymbol{T}_S \boldsymbol{C}_h \right)' = \sum_{h=1}^{b} \gamma_h^2 \boldsymbol{A}_h$$

where

$$\boldsymbol{A}_h = \begin{cases} \frac{k_h-1}{k_h} \delta_h(j) & \text{on the diagonal,} \\ -\frac{1}{k_h} \delta_h(j) \delta_h(j') & \text{off the diagonal,} \end{cases}$$

and $\gamma_h^2 = \frac{1}{(k_h-1)} \sum_{j=1}^{k_h} \left(\frac{(t+1)}{(k_h+1)} j - \frac{(t+1)}{2} \right)^2$. The elements of $\boldsymbol{\Gamma}$ are thus

$$\begin{cases} (t+1)^2 \left[\frac{1}{12} \sum_{h=1}^{b} \frac{k_h-1}{k_h+1} \delta_h(j) \right] & \text{on the diagonal,} \\ (t+1)^2 \left[-\frac{1}{12} \sum_{h=1}^{b} \frac{1}{k_h+1} \delta_h(j) \delta_h(j') \right] & \text{off the diagonal.} \end{cases} \tag{7.19}$$

Note that the elements of each row of $\boldsymbol{\Gamma}$ sum to 0, so that rank $(\boldsymbol{\Gamma}) \le t - 1$.

The matrix $\mathbf{\Gamma}$ with elements given in Lemma 7.1 is closely related to the *information matrix* of a block design. John and Williams (1995) details how this matrix occurs in the least squares estimation of treatment effects and the role its eigenvalues play in determining optimality criteria for choosing between different designs. This information matrix $\boldsymbol{A}$ has components as follows:

$$\begin{cases} \sum_{h=1}^{b} \frac{k_h - 1}{k_h} \delta_h(i) & \text{on the diagonal,} \\ \\ -\sum_{h=1}^{b} \frac{1}{k_h} \delta_h(i) \delta_h(j) & \text{off the diagonal.} \end{cases} \tag{7.20}$$

Note that $\boldsymbol{A}$ and $\mathbf{\Gamma}$ share the same rank.

In view of the independence of the observations in each block, we may define a smooth alternative for each block:

$$\pi\left(\boldsymbol{x}_{h(s)}; \boldsymbol{\theta}_h\right) = \exp\left[\boldsymbol{\theta}_h' \boldsymbol{x}_{h(s)} - K\left(\boldsymbol{\theta}_h\right)\right] / \left(k_h!\right), \quad s = 1, \ldots, k_h!, h = 1, \ldots, b,$$

where $\boldsymbol{\theta}_h$ represents the $k_h \times 1$ vector of parameters associated with the k_h objects ranked in block h. We interpret the vectors $\boldsymbol{x}_{h(s)}$ as column vector with values taken by a random vector $\boldsymbol{X}^{(h)}$ with probability mass function

$$P\left(\boldsymbol{X}^{(h)} = \boldsymbol{x}_{h(s)}\right) = \pi\left(\boldsymbol{\theta}_h; \boldsymbol{x}_{h(s)}\right), s = 1, \ldots, k_h!.$$

The smooth alternative for the entire block design will be the product of the models for each block. We are interested in testing

$$\begin{aligned} H_0: \ \boldsymbol{\theta}_h &= \ 0, \text{ for all } h, \\ H_1: \ \boldsymbol{\theta}_h &\neq \ 0, \text{ for some } h. \end{aligned}$$

The log likelihood, derived from the joint multinomial probability function, is given by

$$\begin{aligned} l(\boldsymbol{\theta}) &\sim \sum_{h=1}^{b} \sum_{s=1}^{k_h!} n_{h(s)} \log \pi\left(\boldsymbol{\theta}_h; \boldsymbol{x}_{h(s)}\right) \\ &= \sum_{h=1}^{b} \boldsymbol{\theta}_h' \left[\sum_{s=1}^{k_h!} n_{h(s)} \boldsymbol{x}_{h(s)}\right] - \sum_{h=1}^{b} \sum_{s=1}^{k_h!} n_{h(s)} K\left(\boldsymbol{\theta}_h\right) \end{aligned}$$

where $n_{h(s)}$ represents the frequency of occurrence of the value $\boldsymbol{x}_{h(s)}$. Under this formulation, using a similar argument as in Section 7.2, we can show that the score test leads to the result in Theorem 7.7. Two examples of block designs are considered below.

Example 7.5. Balanced Incomplete Block Designs (BIBD)
In the complete ranking case $k_h = t, b = n$, so that the design becomes a randomized

block. An example of a test in such a situation is the Friedman Test with test statistic

$$G_S = \sum_{i=1}^{t} \left(R_i - \frac{n(t+1)}{2} \right)^2,$$

where R_i is the sum of the ranks assigned by the judges to object i. Under H_0, as $n \to \infty$,

$$\frac{1}{n} G_S \xrightarrow{\mathcal{L}} \left\{ \frac{t(t+1)}{12} \right\} \chi^2_{t-1}.$$

An interpretation of the Friedman statistic is that it is essentially the average of the Spearman correlations between all pairs of rankings. In the balanced incomplete block design, we have $k_h = k, r_i = r, \lambda_{ij} = \lambda, bk = rt$, and $\lambda(t-1) = r(k-1)$. An example of a test in such a case is due to Durbin whose test statistic is

$$G_S = \sum_{i=1}^{t} \left(\frac{(t+1)}{(k+1)} R_i - \frac{nr(t+1)}{2} \right)^2.$$

Under H_0, as $n \to \infty$,

$$\frac{1}{n} G_S \xrightarrow{\mathcal{L}} \left\{ \frac{\lambda t(t+1)^2}{12(k+1)} \right\} \chi^2_{t-1}.$$

Example 7.6. Group Divisible and Cyclic Design
It may not always be possible to construct balanced incomplete block designs. For example, the smallest number of replications for a BIBD is $r = \frac{\lambda(t-1)}{(k-1)}$. When $t = 6, k = 4$, the balanced design would require $r = 10$ replications and hence $rt = 60$ experimental units. Such a large number of units may either not be available or too costly for the researcher to acquire. The partially balanced group divisible design helps to reduce the number of experimental units required at the cost of forcing some comparisons between treatments to be more precise than others. In this design, the t objects occur in g groups of d objects, $t = gd$. Within each group pairs of objects are compared by λ_1 judges, whereas each pair of objects between groups is compared by λ_2 judges. Such designs must satisfy the additional conditions

$$bk = rt, r(k-1) = (d-1)\lambda_1 + d(g-1)\lambda_2.$$

An example of a group divisible design is given in Table 7.9. Here, $t = 6, b = 3, k = 4, r = 2, g = 3, d = 2, \lambda_1 = 2, \lambda_2 = 1$. Treatments $(1,4), (2,5), (3,6)$ are compared $\lambda_1 = 2$ times whereas all the other pairs are compared only once. The number of experimental units required here is $rt = 12$ compared to the 60 for a BIBD.

Incomplete block designs often require the use of tables which may not always be available in the field. Moreover, care must be taken to record data correctly for such designs. Cyclic designs on the other hand are easily constructed from an initial block and the treatments can be easily assigned. Such designs are obtained by cyclic development of

Table 7.9.: Example of a group divisible design

Treatments	1	2	3	4	5	6
Block 1	X	X		X	X	
Block 2		X	X		X	X
Block 3	X		X	X		X

Table 7.10.: Example of a cyclic design

Treatments	1	2	3	4	5	6
Block 1	X		X	X		
Block 2		X		X	X	
Block 3			X		X	X
Block 4	X			X		X
Block 5	X	X			X	
Block 6		X	X			X

an initial block or combinations of such sets. Let $\lambda_0 = r$, and $\lambda_{j-1} = \lambda_{1j}, j = 2, \ldots, t$, be the number of judges that compare object 1 with object j. The matrix $\boldsymbol{A}$ is a circulant related to the matrix derived by the cyclic development of $(\lambda_0, \lambda_1, \ldots, \lambda_{t-1})$, and its eigenvalues are given in John and Williams (1995). The eigenvalues of $\boldsymbol{\Gamma}$ are

$$\alpha_i = \frac{(t+1)^2}{12}\left\{\frac{r(k-1)}{(k+1)} - \frac{1}{(k+1)}\sum_{h=1}^{t-1}\lambda_h \cos\left(\frac{2\pi i h}{t}\right)\right\}, \; i = 1, 2, \ldots, t-1.$$

An example of a cyclic design with $t = 6, k = r = 3, b = 6$ is given in Table 7.10. Note that this is not a BIBD since for example, $\lambda_{14} = 2 \neq \lambda_{13} = 1$.

7.6. Exercises

Exercise 7.1. 1. Calculate $E[W]$ under the null hypothesis.

Hint: First show that $W = \frac{12}{nt(t+1)}\sum_{i=1}^{t}\left(\sum_{j=1}^{n} R_{ij}^2\right) - 3n(t+1)$ and then use the properties of the ranks from Lemma 3.1 under the null hypothesis.

Exercise 7.2. Obtain for $t = 4$, an interpretation for the components of $\boldsymbol{T_{SP}}$.

Exercise 7.3. Some evidence suggests that anxiety and fear can be differentiated from anger according to feelings along a dominance-submissiveness continuum. In order to determine the reliability of the ratings on a sample of animals, the ranks of the ratings given by two observers, Ryan and Jacob, were tabulated below. Perform a suitable test at the 5% significance level whether Ryan and Jacob agree on their ratings.

Animal	1	2	3	4	5	6	7	8	9	10	11	12	13	14	15	16
Ryan's Ranks	4	7	13	3	15	12	5	14	8	6	16	2	10	1	9	11
Jacob's Ranks	2	6	14	7	13	15	3	11	8	5	16	1	12	4	9	10

Exercise 7.4. It is widely believed that the Earth is getting warmer year by year, caused by the increasing amount of greenhouse gases emitted by human, which are trapping more heat in the Earth's atmosphere. However, some people cast doubts on that, claiming that the average global temperature actually remains stable over time whereas extreme weathers occur more and more frequently. In order to test these claims, the data on the annual mean temperature ($MEAN$, in °C), the annual maximum (MAX) and minimum (MIN) temperatures (in °C) in Hong Kong for the years ($YEAR$) 1961 to 2016 are extracted from the Hong Kong Observatory website, see Appendix A.8.

(a) Test whether the annual mean temperature $MEAN$ shows an increasing trend at the 5% significance level using (i) Spearman test and (ii) Kendall test.

(b) Test whether the annual temperature range (defined as $RANGE \equiv MAX - MIN$) shows a monotone trend at the 5% significance level using (i) Spearman test and (ii) Kendall test.

(c) Based on the results found in (a) to (b), draw your conclusions and discuss the limitation of your analysis.

Exercise 7.5. Eight subjects were given one, two, and three alcoholic beverages at different widely space times. After each drink, the closest distance in feet they would approach an alligator was measured. Test the hypothesis that there is no difference between the different amounts of drinks consumed.

	1	2	3	4	5	6	7	8
One drink	19.0	14.4	18.2	15.6	14.6	11.2	13.9	11.6
Two drinks	6.3	11.6	9.7	5.9	13.0	9.8	4.8	10.7
Three drinks	3.3	1.2	3.7	7.1	2.6	1.9	5.2	6.4

Exercise 7.6. There are 6 psychiatrists who examine 10 patients for depression. Each patient is seen by 3 psychiatrists who provide a score as shown below. Analyze the data using a nonparametric balanced incomplete block design. Are there differences among the psychiatrists?

Patient	1	2	3	4	5	6
1	10	14	10			
2	3	2		1		
3	7		12			9
4	3			8	5	
5	20				26	20
6		20	14		20	
7		5		8		14
8		14			18	15
9			12	17	12	
10			18	19		13

8. Optimal Rank Tests

Lehmann and Stein (1949) and Hoeffding (1951b) pioneered the development of an optimal theory for nonparametric tests, parallel to that of Neyman and Pearson (1933) and Wald (1949) for parametric testing. They considered nonparametric hypotheses that are invariant under permutations of the variables in multi-sample problems[1] so that rank statistics are the maximal invariants, and extended the Neyman-Pearson and Wald theories for independent observations to the joint density function of the maximal invariants. Terry (1952) and others subsequently implemented and refined Hoeffding's approach to show that a number of rank tests are locally most powerful at certain alternatives near the null hypothesis. We shall first consider Hoeffding's change of measure formula and derive some consequences with respect to the two-sample problem. This formula assumes knowledge of the underlying distribution of the random variables and leads to an optimal choice of score functions and subsequently to locally most powerful tests. Hence, for any given underlying distributions, we may obtain the optimal test statistic.

In previous chapters, we did not assume knowledge of the underlying distributions of the random variables. Instead, we derived test statistics through a parametric embedding based on either a kernel or score function. The connection between the two approaches is now evident. Using the results of the present chapter, it will then be possible to calculate the efficiency of previously derived test statistics with respect to the optimal for specific underlying distribution of the variables. We postpone the discussion of efficiency to Chapter 9 of this book.

In Section 8.1 we see that the locally most powerful tests can be included in the unified theory discussed in Section 6.1. In Section 8.2 we discuss the regression problem, whereas in Section 8.3 we present the optimal choice of score function in the two-way layout.

[1]Lehmann & Stein considered the case for two samples and Hoeffding for multi-samples.

M. Alvo, P. L. H. Yu, *A Parametric Approach to Nonparametric Statistics*,
Springer Series in the Data Sciences, https://doi.org/10.1007/978-3-319-94153-0_8

8.1. Locally Most Powerful Rank-Based Tests

Rank-based tests are invariant to monotone transformations and consequently do not depend on the scale of measurement. This is not always the case. For example, the t-test is altered if x is replaced by $\log x$ In this section, we shall consider a change of measure formula (Hoeffding, 1951b), known in the literature as Hoeffding's Formula, in connection with the distribution of the ranks of the data. In this approach, we are able to incorporate a priori knowledge of the underlying distributions of the random variables to determine the locally most powerful test. Let $Z_1, \ldots, Z_n$ be a sequence of independent random variables with densities $f_1, \ldots, f_n$ respectively. Suppose that the null hypothesis postulates that the variables are identically distributed. Let $R = (R_1, \ldots, R_n)$ denote the vector of ranks where R_i is the rank of Z_i among $Z_1, \ldots, Z_n$. The following lemma is due to Hoeffding.

Lemma 8.1 (Hoeffding's Formula). *Assume that $V_1 < \ldots < V_n$ are the order statistics of a sample of size n from a density h. Then*

$$P(R = r) = \int \cdots \int_{(v_1 < \ldots v_n)} f_1(v_1) \ldots f_n(v_n)\, dv_1 \ldots dv_n \tag{8.1}$$

$$= \int \cdots \int_{(v_1 < \ldots v_n)} \frac{f_1(v_1) \ldots f_n(v_n)}{n! h(v_1) \ldots h(v_n)} n!\, h(v_1) \ldots h(v_n)\, dv_1 \ldots dv_n \tag{8.2}$$

$$= \frac{1}{n!} E_h \left[\frac{f_1\left(V_{(r_1)}\right) \ldots f_n\left(V_{(r_n)}\right)}{h\left(V_{(r_1)}\right) \ldots h\left(V_{(r_n)}\right)} \right]. \tag{8.3}$$

To illustrate the use of Hoeffding's lemma, we consider the two-sample problem where $X_1, \ldots, X_m$ constitute a random sample from a cdf F having density f and $Y_1, \ldots, Y_{n-m}$ are a random sample from a cdf G having density g. As well, assume that $f = 0$ implies $g = 0$. Substitute in Hoeffding's formula (8.3)

$$f_i = \begin{cases} f & i = 1, \ldots, m \\ g & i = m+1, \ldots, n \end{cases}$$

and $h = g$, we have

$$P(R = r) = \frac{1}{n!} E_g \left[\frac{f\left(V_{(r_1)}\right) \ldots f\left(V_{(r_m)}\right)}{g\left(V_{(r_1)}\right) \ldots g\left(V_{(r_m)}\right)} \right].$$

It follows that if $R_1, \ldots, R_m$ denote the ranks of the first population,

$$
\begin{aligned}
P\{R_1 = r_1, \ldots, R_m = r_m\} &= \sum P\left(R_1 = r_1, \ldots, R_m = r_m, R_{m+1}, \ldots, R_{m+n}\right) \\
&= \frac{1}{n!} E_g \left[\frac{f\left(V_{(r_1)}\right) \ldots f\left(V_{(r_m)}\right)}{g\left(V_{(r_1)}\right) \ldots g\left(V_{(r_m)}\right)} \right] \left[\sum 1\right] \\
&= \frac{m!\,(n-m)!}{n!} E_g \left[\frac{f\left(V_{(r_1)}\right) \ldots f\left(V_{(r_m)}\right)}{g\left(V_{(r_1)}\right) \ldots g\left(V_{(r_m)}\right)} \right] \\
&= \frac{1}{\binom{n}{m}} E_g \left[\frac{f\left(V_{(r_1)}\right) \ldots f\left(V_{(r_m)}\right)}{g\left(V_{(r_1)}\right) \ldots g\left(V_{(r_m)}\right)} \right], \qquad (8.4)
\end{aligned}
$$

where $V_{(1)} < \ldots < V_{(N)}$ are the order statistics from a random sample of n uniform variables on $(0, 1)$ and the second summation is over all possible permutations of the first and second samples among themselves. The expectation E_g is with respect to the probability measure under which the n observations are i.i.d. with common density function g, assuming that g is positive whenever f is.

Corresponding to a density function f with cdf F, Hájek and Sidak (1967) considered a general score function for location problems defined as

$$
\begin{aligned}
a_n(i, f) &= E\varphi\left(V_n^{(i)}, f\right), 1 \le i \le n \qquad (8.5) \\
&= n\binom{n-1}{i-1} \int_0^1 \varphi(v, f)\, v^i (1-v)^{n-i}\, dv \qquad (8.6)
\end{aligned}
$$

where $V_n^{(1)} < \ldots < V_n^{(n)}$ are the order statistics from a uniform distribution on $(0, 1)$ and

$$\varphi(v, f) = \frac{f'\left(F^{-1}(v)\right)}{f\left(F^{-1}(v)\right)}, 0 < v < 1.$$

Theorem 8.1 (Location Alternatives). *Assume $\int |f'(x)|\, dx < \infty$. Suppose that we have a sample of size m from the first population and a sample of size n from the second. The test with critical region*

$$\sum_{i=1}^{m} a_n(R_i f) \ge k$$

is the locally most powerful test for testing

$$H_0 : f(x_1, \ldots, x_n) = \prod_{i=1}^{m+n} f_X(x_i) \qquad (8.7)$$

against

$$H_1 : f(x_1, \ldots, x_n) = \prod_{i=1}^{m} f_X(x_i - \Delta) \prod_{i=m+1}^{m+n} f_X(x_i), \Delta > 0.$$

Proof. See page 67 of Hájek and Sidak (1967). □

We shall consider two examples to illustrate this theorem which results from an application of Hoeffding's formula. These include the Wilcoxon test when $g(x) = e^x/(1+e^x)^2$ is the logistic density and the Fisher-Yates test when g is the standard normal density.

Example 8.1 (Wilcoxon Test). Suppose that we have a random sample of size m from the logistic distribution whose density is given by

$$f(x) = \frac{e^{-x}}{(1+e^{-x})^2}, x > 0.$$

Then, $F(x) = \frac{1}{1+e^{-x}}$ and

$$\log f(x) = -x - 2\log\left(1+e^{-x}\right)$$

from which

$$\frac{\partial}{\partial x}\log f(x) = \frac{f'((x))}{f((x))} = -1 + 2\left[1 - F(x)\right].$$

Hence,

$$\frac{f'(F^{-1}(V))}{f(F^{-1}(V))} = -1 + 2\left[1 - V\right].$$

Recall from Lemma 2.1 that the ith order statistic from a uniform has density

$$f_{V_{(i)}}(x) = \frac{m!}{(i-1)!\,(m-i)!}v^{i-1}(1-v)^{m-i}$$

Consequently,

$$\begin{aligned} E\left[1 - V_{(i)}\right] &= \int \frac{m!}{(i-1)!\,(m-i)!}v^{i-1}(1-v)^{m+1-i}dv \\ &= \frac{m!}{(i-1)!\,(m-i)!}\frac{(i-1)!\,(m+1-i)!}{(m+2)!} \\ &= \frac{i}{m+1}. \end{aligned}$$

It follows that the locally most powerful test when the underlying density is logistic is given by

$$\begin{aligned} -\sum_{i=1}^{m} E\left[\frac{f'(F^{-1}(V))}{f(F^{-1}(V))}\right] &= -\sum_{i=1}^{m} E\left[-1 + 2\left\{1 - V_{(r_i)}\right\}\right] \\ &= 2\sum_{i=1}^{m}\frac{R_i}{m+1} - m \end{aligned}$$

which is recognized as the Wilcoxon statistic.

Example 8.2 (Fisher-Yates Test). Suppose that we have a random sample of size m from the normal distribution with mean μ and variance $\sigma^2 > 0$ given by

$$f(x) = \frac{1}{\sqrt{2\pi}\sigma} e^{-\frac{1}{2}\left(\frac{x-\mu}{\sigma}\right)^2} - \infty < x, \mu < \infty.$$

Then,

$$\frac{\partial}{\partial x} \log f(x) = -\frac{1}{\sigma}\left(\frac{x-\mu}{\sigma}\right).$$

Once again, using the density of the ith order statistic, we see that

$$\begin{aligned} E\left[\frac{X_{(i)}-\mu}{\sigma}\right] &= \int \frac{m!}{(i-1)!\,(m-i)!} z\left[\Phi(z)\right]^{i-1} f(x)\left[1-\Phi(z)\right]^{m-i} dz \\ &= \int_0^1 \frac{m!}{(i-1)!\,(m-i)!} \Phi^{-1}(u)\left[u\right]^{i-1}\left[1-u\right]^{m-i} du \\ &= E\left[\Phi^{-1}\left(V_{(i)}\right)\right], \end{aligned}$$

where $V_{(i)}$ is the ith order statistic from a random sample of m uniform random variables on the interval $(0, 1)$. There is no closed form for this last expectation, though there are various approximations.[2] In fact, using the delta method first order approximation,

$$E\left[\Phi^{-1}\left(V_{(i)}\right)\right] \approx \Phi^{-1}\left(E\left[V_{(i)}\right]\right) = \Phi^{-1}\left(\frac{i}{m+1}\right).$$

This then yields the normal scores or Fisher-Yates test statistic

$$\sum_{i=1}^{m} E\left[\Phi^{-1}\left(V_{(R_i)}\right)\right]$$

which may be approximated by

$$\sum_{i=1}^{m} \Phi^{-1}\left(\frac{R_i}{m+1}\right).$$

The latter is called the van der Warden test.

[2] See also Royston (1982) who provides the approximation $E\left[X_{(i)}\right] = \mu + \sigma\Phi^{-1}\left(\frac{i}{m+1}\right)\left[1 + \frac{\left(\frac{i}{m+1}\right)\left(1-\frac{i}{m+1}\right)}{2(m+2)\left[\phi\left[\Phi^{-1}\frac{i}{m+1}\right]\right]^2}\right]$.

A parametric embedding argument can be used to give an alternative derivation of the local optimality of the Fisher-Yates and Wilcoxon tests. In the two-sample problem, define

$$\pi\left(\boldsymbol{x}_{1j},\boldsymbol{x}_{2j};\boldsymbol{\theta}_1,\boldsymbol{\theta}_2\right)=\exp\left\{\sum_{\ell=1}^{2}\left[\boldsymbol{\theta}_{\ell}^{\prime}\boldsymbol{x}_{\ell j}-K\left(\boldsymbol{\theta}_{\ell}\right)\right]\right\}p_{0j},\quad j=1,\ldots,n!, \tag{8.8}$$

where $\boldsymbol{\theta}_{\ell}=\left(\theta_{\ell 1},\ldots,\theta_{\ell k}\right)^{\prime}$ represents the parameter vector for sample $\ell(=1,2)$ and x_{1j}, x_{2j} are the data from sample 1 and sample 2 with respective sizes m and $n-m$ that are associated with the ranking (permutation) ν_j, $j=1,\ldots,n!$.

Under the null hypothesis $H_0:\boldsymbol{\theta_1}=\boldsymbol{\theta}_2$, we can assume without loss of generality that the underlying $V_1,\ldots,V_n$ from the combined sample are i.i.d. uniform (by considering $G(V_i)$, where G is the common distribution function, assumed to be continuous, of the V_i) and that all rankings of the V_i are equally likely. Hence (12.5) represents an exponential family constructed by exponential tilting of the baseline measure (i.e., corresponding to H_0) on the rank-order data. This has the same spirit as Neyman's smooth test of the null hypothesis that the data are i.i.d. uniform against alternatives in the exponential family (4.1). The Neyman-Pearson lemma can be applied to show that the score tests have maximum local power at the alternatives that are near $\boldsymbol{\theta}=\mathbf{0}$. The Neyman parametric embedding in (4.1) makes these results directly applicable to the rank-order statistics. In particular, this shows that the two-sample Wilcoxon test of H_0 is locally most powerful for testing the uniform distribution against the truncated exponential distribution for which the data are constrained to lie in the range $(0,1)$ of the uniform distribution. Note that these exponential tilting alternatives differ from the location alternatives in the preceding paragraph not only in their distributional form (truncated exponential instead of logistic) but also in avoiding the strong assumption of the preceding paragraph that the data have to be generated from the logistic distribution even under the null hypothesis (8.7).

Similar results are also valid for tests against scale alternatives. Define the scores

$$a_{1n}\left(i,f\right)\cong\varphi_1\left(EV_n^{(i)},f\right)=\varphi_1\left(\frac{i}{n+1},f\right),$$

where

$$\varphi_1\left(v,f\right)=-1-F^{-1}\left(v\right)\frac{f^{\prime}\left(F^{-1}\left(v\right)\right)}{f\left(F^{-1}\left(v\right)\right)},\quad 0<v<1.$$

Theorem 8.2 (Scale Alternatives). *Assume* $\int_{-\infty}^{\infty}\left|xf^{\prime}\left(x\right)dx\right|<\infty$. *The test with critical region*

$$\sum_{i=1}^{n}c_i a_{1n}\left(R_i f\right)\geq k$$

is the locally most powerful test for testing

$$H_0 : f(x_1, \ldots, x_n) = \prod_{i=1}^{n} f_X(x_i)$$

against

$$H_1 : f(x_1, \ldots, x_n) = \exp\left(-\Delta \sum_{i=1}^{n} c_i\right) \prod_{i=1}^{n} f_X(x_i - \mu) e^{-\Delta c_i}, \Delta > 0,$$

where the $\{c_i\}$ *are regression constants.*

Proof. See page 69 of Hájek and Sidak (1967). □

Example 8.3 (Locally Most Powerful Test for Scale)**.** Suppose that F has an absolutely continuous density f for which $\int |xf'(x)| \, dx < \infty$. We would like to test

$$H_0 : F(x) = G(x) \text{ versus } H_1 : G(x) = F(x/\theta), \theta > 1.$$

Let

$$S_m = \sum_{j=1}^{m} E_F\left(-1 - V_{(q_j)} \frac{f'(V_{(r_j)})}{f(V_{(r_j)})}\right),$$

where $V_{(1)} < \ldots < V_{(n)}$ are the order statistics of a random sample of N from F. Then the locally most powerful rank test is given by

$$\phi(q) = \begin{cases} 1 & S_m > k_\alpha \\ \gamma & S_m = k_\alpha \\ 0 & S_m < k_\alpha \end{cases}$$

In the special case where F is a normal distribution with mean 0 and variance σ^2, then the locally most powerful rank test is based on the sum

$$S_m = \sum_{j=1}^{m} E\left[Z_{(i)}^2\right],$$

where $Z_{(1)} < \ldots < Z_{(N)}$ are the order statistics of a random sample of N from a standard normal distribution. On the other hand, if $f(x) = e^{-x}, x > 0$, then the locally most powerful rank test is based on the sum of the Savage scores given by

$$S_m = \sum_{j=1}^{m} \left[\log\left(1 - \frac{j}{N+1}\right) - 1\right].$$

One may well ask whether the unified theory described in Chapter 6 produces locally most powerful tests. Indeed we recall the locally most powerful tests are a subset of the class of general linear rank statistics. Consider the two-sample problem for a location alternative. The locally most powerful test rejects for small values of

$$\sum_{i=1}^{n_1} E_g\left[-\frac{g'(V_{(\mu(i))})}{g(V_{(\mu(i))})}\right],$$

where $V_{(1)} < \ldots < V_{(n)}$ are the order statistics of a random sample of size n from g. Let

$$h\left(j\right) = E_g\left[-\frac{g'(V_{(j)})}{g(V_{(j)})}\right]$$

and consider the Euclidean distance function

$$d\left(\mu,\nu\right) = \sum_{i=1}^{n}\left[h\left(\mu\left(i\right)\right) - h\left(\nu\left(i\right)\right)\right]^2.$$

Expanding this sum, we see that the test statistic induced by this distance function can be shown to be

$$\sum_{i=1}^{n_1} h\left(\mu\left(i\right)\right) = \sum_{i=1}^{n_1} E_g\left[-\frac{g'(V_{(\mu(i))})}{g(V_{(\mu(i))})}\right].$$

The demonstration is identical to the one using the permutations directly.

8.2. Regression Problems

We may also obtain locally most powerful tests for regression problems. Consider the general regression model

$$\boldsymbol{Y} = \boldsymbol{X\beta} + \boldsymbol{\epsilon}, \tag{8.9}$$

where $\boldsymbol{Y}$ is an n-dimensional vector response variable, $\boldsymbol{X}$ is an $n \times p$ design matrix, $\boldsymbol{\beta}$ is a p-dimensional vector of regression parameters, and $\boldsymbol{\epsilon}$ is a vector random error term with independent identically distributed components having continuous distribution $F\left(\cdot\right)$ and density f_ϵ. Let $Y_1, \ldots, Y_n$ be a sample of n observed responses and let

$$\boldsymbol{x}_i = \left(x_{i1}, \ldots, x_{ip}\right)'$$

be the corresponding covariates. Denote the order statistics $Y_{(1)} < \ldots < Y_{(n)}$ and let their ranks be denoted by $\boldsymbol{R} = \left(R_1, \ldots, R_n\right)$. The probability distribution of $\boldsymbol{R}$ is given by

$$f\left(\boldsymbol{r};\Delta\right) = \int \prod_{i=1}^{n} f_\epsilon\left(u_i - x_i'\boldsymbol{\beta}\right) du_i \tag{8.10}$$

where integration is over the set $\{(u_1, \ldots, u_n) \,|u_1 < \ldots < u_n\}$. Kalbfleisch (1978) arrived at (8.10) through group considerations. Specifically, it was argued that the regression model (8.9) conditional on $\boldsymbol{x}$ is invariant under the group of increasing differentiable transformations acting on the response. In the parameter space, this leaves $\boldsymbol{\beta}$ invariant. We may obtain the locally most powerful test of the hypothesis $H_0 : \boldsymbol{\beta} = \mathbf{0}$ by computing the score function U at $\boldsymbol{\beta} = \mathbf{0}$

$$U = \left. \frac{\partial \log\, f(\boldsymbol{r}; \boldsymbol{\beta})}{\partial \boldsymbol{\beta}} \right|_{\boldsymbol{\beta}=\mathbf{0}}. \tag{8.11}$$

The lth component for $l = 1, \ldots, p$ is

$$\begin{aligned} U_l &= \frac{1}{f(\boldsymbol{r}; \mathbf{0})} \frac{\partial f(\boldsymbol{r}; \boldsymbol{\beta})}{\partial \boldsymbol{\beta}} \\ &= n! \int \prod_{i=1}^{n} f_\epsilon(u_i) \left(- \sum_{j=1}^{n} x_{(j)l} a(j) \right) du_i \\ &= - \sum_{j=1}^{n} x_{(j)l} a(j) \\ &= - \sum_{j=1}^{n} x_{jl} a(R_j), \end{aligned} \tag{8.12}$$

where

$$a(i) = E \left[\frac{\partial \log\, f_\epsilon(Z_{(i)})}{\partial Z_{(i)}} \right] \tag{8.13}$$

and $Z_{(i)}$ is the ith order statistic of a sample of size n from f_ϵ. Hence, when $p > 1$ we obtain a vector of linear rank statistics.

If $p = 1$, then

$$\sum_{j=1}^{n} x_j a(R_j)$$

is the usual simple linear rank statistic. There is a close connection in that case with the usual Pearson correlation coefficient as we saw in Section 3.1 which we recall in the next example.

Example 8.4. Consider the simple linear regression model

$$Y_i = \alpha + \Delta x_i + \varepsilon_i, i = 1, \ldots, n$$

where $\{\varepsilon_i\}$ are independent identically distributed random variables from a cumulative distribution function have median 0. We wish to test

$$H_0 : \Delta = \Delta_0$$

against

$$H_1 : \Delta \neq \Delta_0.$$

Assume $x_1 < \ldots < x_n$ and let R_i denote the rank of the residual $Y_i - \Delta_0 x_i$ among $\{Y_j - \Delta_0 x_j\}$. The usual Pearson correlation coefficient is given by

$$\rho_n = \frac{\sum_{i=1}^n (x_i - \bar{x}_n)(a(R_i) - \bar{a})}{\left[\sum_{i=1}^n (x_i - \bar{x}_n)^2 \sum_{i=1}^n (a(R_i) - \bar{a}_n)^2\right]^{1/2}},$$

where the score function is an increasing function with $a(1) \neq a(n)$ and

$$\bar{a} = \frac{1}{n} \sum_{i=1}^n a(R_i).$$

It is seen that ρ_n is linearly equivalent to the linear rank statistic

$$\sum_{i=1}^n x_i a(R_i).$$

We may determine the optimal regression score statistic corresponding to specific error distributions using (8.13) as in the following examples.

(i) Suppose f_ϵ is the standard normal distribution. Then, $a(i) = -E\left[Z_{(i)}\right]$ which leads to the Fisher-Yates normal scores test.

(ii) If f_ϵ is the logistic density

$$f_\epsilon(z) = \frac{e^{-z}}{(1 + e^{-z})^2},$$

then

$$a(i) = E\left[\frac{-1 + e^{-Z_{(i)}}}{1 + e^{-Z_{(i)}}}\right] = 1 - 2E\left[\frac{1}{1 + e^{-Z_{(i)}}}\right].$$

Since $Y = \frac{1}{1+e^{-Z}}$ has a uniform distribution on the interval $(0, 1)$, it follows that

$$a(i) = 1 - 2E\left[Y_{(i)}\right] = 1 - 2\left[\frac{i}{n+1}\right]$$

which leads to the Wilcoxon statistic.

(iii) If f_ϵ is the extreme value density

$$f_\epsilon(z) = e^{(z-e^z)}$$

then

$$a(i) = E\left[1 - e^{Z_{(i)}}\right].$$

We may obtain a general form for the score test in the regression model. The score function U evaluated at $\Delta = 0$ has variance-covariance matrix

$$I_0 = E\{UU'\} = E\sum_{ii'} x_{il}x_{i'l'}a(R_i)\,a(R_{i'}).$$

Suppose that $\sum_{i=1}^n x_{il} = 0$ and $\sum_{ii'} x_{il}x_{i'l'} = 0$. Then,

$$\begin{aligned}
E\sum_{ii'} x_{il}x_{i'l'}a(R_i)\,a(R_{i'}) &= \sum_{ii'} x_{il}x_{i'l'}E\left(a(R_i)\,a(R_{i'})\right)\\
&= \sum_{i} x_{il}x_{il'}Ea^2(R_i) + \sum_{i\neq i'} x_{il}x_{i'l'}E\left(a(R_i)\,a(R_{i'})\right)\\
&= \sum_{i} x_{il}x_{il'}Ea^2(R_i) - \sum_{i} x_{il}x_{il'}E\left(a(R_i)\,a(R_{i'})\right)\\
&= \left(\sum_{i} x_{il}x_{il'}\right)\left(Ea^2(R_i) - E\left(a(R_i)\,a(R_{i'})\right)\right)
\end{aligned}$$

Now note

$$\begin{aligned}
Ea^2(R_i) - E\left(a(R_i)\,a(R_{i'})\right) &= \frac{1}{n}\sum_{i=1}^n a^2(i) - \frac{1}{n(n-1)}\sum_{i\neq i'} a(i)\,a(i')\\
&= \frac{1}{n}\sum_{i=1}^n a^2(i) - \frac{1}{n(n-1)}\left[\left(\sum a(i)\right)^2 - \sum a^2(i)\right]\\
&= \frac{1}{(n-1)}\sum_{i=1}^n (a(i) - \bar{a})^2
\end{aligned}$$

where $\bar{a} = \frac{\sum_{i=1}^n a(i)}{n}$.

Now set the design matrix $X = (x_{ij})$. It follows that the ll' term of $X'X$ is given by

$$\left(\sum_{i=1}^n x_{il}x_{il'}\right),$$

and hence the variance-covariance matrix I_0 becomes

$$I_0 = (X"X)\left(\frac{1}{n-1}\left(\sum_{i=1}^{n}(a(i)-\bar{a})^2\right)\right).$$

Assuming that $\bar{a}=0$, we have

$$I_0 = (X"X)\left(\frac{1}{n-1}\left(\sum_{i=1}^{n}a^2(i)\right)\right)$$

and hence using (8.12) and (8.13), the score test takes the form

$$U'I_0^{-1}U.$$

8.3. Optimal Tests for the Method of n-Rankings

Sen (1968a) considered the problem of finding efficient tests by the method of n-rankings. Specifically, he considered the two-way layout in situations where block effects are not additive or when the method of ranking after alignment is not practicable. Alvo and Cabilio (2005) extended his results to situations involving more general experimental designs. In what follows, we begin with a review of the complete ranking case and then consider the incomplete ranking situation.

Suppose that we have n judges (blocks) who rank t objects labeled $(1,2,\ldots,t)$. Each of the $1\leq i\leq n$ judges is presented with $2\leq k_i\leq t$ objects which are ranked independently by the judges. This ranking is a permutation of $(1,2,\ldots,k_i)$ and in order to indicate which objects are ranked by judge i, this is represented by

$$\boldsymbol{R}_i = (R_i(o_1), R_i(o_2), \ldots, R_i(o_{k_i})),$$

where the o_j with $1\leq o_1<o_2<\ldots<o_{k_i}\leq t$ are the labels of the actual objects being ranked. Occasionally we may wish to represent this as a t-vector in which missing ranks are represented by the symbol "_." If $k_i=t$, the ranking is said to be complete, otherwise it is incomplete. We wish to test the hypothesis H_0 : each judge, when presented with the specified k_i objects, picks the ranking at random from the space of $k_i!$ permutations of $(1,2,\ldots,k_i)$.

Instead of basing our statistic directly on the ranks themselves, we may wish to replace the ranks assigned by each judge by real valued score functions

$$a(j,k_i), 1\leq j\leq k_i\leq t.$$

In order to motivate the discussion, we begin in the next section with the complete case where $k_i=t$.

8.3.1. Complete Block Designs

Consider the situation where each of the n independent rankings is complete and suppose we wish to test

H_0^* : Each judge picks a complete ranking at random from the space of $t!$ permutations of $(1, \ldots, t)$

For a given function $a(j,t)$, $1 \leq j \leq t$, and a ranking $\boldsymbol{R}$ define the vector of adjusted scores

$$\boldsymbol{a}(\boldsymbol{R}) = (a(R(1),t) - \overline{a}_t, a(R(2),t) - \overline{a}_t, \ldots, a(R(t),t) - \overline{a}_t)', \text{where } \overline{a}_t = t^{-1} \sum_{r=1}^{t} a(r,t). \tag{8.14}$$

Under H_0^*, $E(a(R(j),t)) = \overline{a}_t$, and the covariance matrix of $\boldsymbol{a}(\boldsymbol{R})$ is given by

$$\boldsymbol{\Sigma}_0 = \frac{t}{t-1} \sigma_0^2 \left(\boldsymbol{I} - \frac{1}{t} \boldsymbol{J} \right) \tag{8.15}$$

where

$$\sigma_0^2 \equiv Var(a(R(j),t)) = t^{-1} \sum_{r=1}^{t} (a(r,t) - \overline{a}_t)^2, \tag{8.16}$$

$\boldsymbol{I}$ is the identity $(t \times t)$ matrix, and $\boldsymbol{J}$ is the $(t \times t)$ matrix of 1's. A measure of similarity between the complete rankings $\boldsymbol{R}_1$ and $\boldsymbol{R}_2$ is given by

$$\mathcal{A}(\boldsymbol{R}_1, \boldsymbol{R}_2) = \boldsymbol{a}(\boldsymbol{R}_1)' \boldsymbol{a}(\boldsymbol{R}_2) = \sum_{j=1}^{t} (a(R_1(j),t) - \overline{a}_t)(a(R_2(j),t) - \overline{a}_t). \tag{8.17}$$

Let the $(t \times t!)$ matrix $\boldsymbol{T}$ represent the collection of adjusted score vectors $\boldsymbol{a}(\boldsymbol{R})$ as $\boldsymbol{R}$ ranges over all its $t!$ possible values, and let $\boldsymbol{f}$ be the $t!$ vector of frequencies of the observed rankings. The $(t! \times t!)$ matrix $\boldsymbol{T}'\boldsymbol{T}$ has components $\mathcal{A}(\boldsymbol{R}_1, \boldsymbol{R}_2)$ with $\boldsymbol{R}_1$ and $\boldsymbol{R}_2$ ranging over all $t!$ permutations of $(1, 2, \ldots, t)$. With $\boldsymbol{R}_i$, $i = 1, \ldots, n$, representing the observed rankings, and

$$S_n(j) = \sum_{i=1}^{n} a(R_i(j),t),$$

let

$$\boldsymbol{S}_n = \left(S_n\left(1\right), S_n\left(2\right), \ldots, S_n\left(t\right)\right)',$$

so that $\boldsymbol{Tf} = \left(\boldsymbol{S}_n - n\bar{a}_t\mathbf{1}\right)$, where $\mathbf{1}$ is the $t-$vector of 1's. Proceeding as in (Alvo and Cabilio (1991)), the proposed statistic is the quadratic form

$$n^{-1}\boldsymbol{f}'\left(\boldsymbol{T}'\boldsymbol{T}\right)\boldsymbol{f} = \left(\frac{1}{\sqrt{n}}\boldsymbol{Tf}\right)'\left(\frac{1}{\sqrt{n}}\boldsymbol{Tf}\right) = n^{-1}\sum_{j=1}^{t}\left(S_n\left(j\right) - n\bar{a}_t\right)^2. \tag{8.18}$$

Large values of this statistic are inconsistent with H_0. Under H_0^*, as $n \to \infty$, an application of the central limit theorem for the multinomial shows that $n^{-1/2}\boldsymbol{Tf}$ is asymptotically normal with mean vector $\mathbf{0}$ and covariance matrix

$$\frac{1}{t!}\boldsymbol{T}\left(\boldsymbol{I} - \frac{1}{t!}\boldsymbol{J}\right)\boldsymbol{T}' = \frac{1}{t!}\boldsymbol{TT}'.$$

Some manipulation shows this covariance to be equal to $\boldsymbol{\Sigma}_0$. Consequently, since $\mathbf{I}-\frac{1}{t}\mathbf{J}$ is an idempotent of rank $\left(t-1\right)$, as $n \to \infty$ the statistic in (8.18) has asymptotically the same distribution as $t\left(t-1\right)^{-1}\sigma_0^2\chi^2_{(t-1)}$, where χ^2 is a random variable distributed with a chi-square distribution with $(t-1)$ degrees of freedom. This leads to the same statistic as defined in Hájek and Sidak (1967) and Sen (1968a), namely

$$Q_n = \frac{t-1}{n\sum_{r=1}^{t}\left(a\left(r,t\right) - \bar{a}_t\right)^2}\sum_{j=1}^{t}\left(S_n\left(j\right) - n\bar{a}_t\right)^2 \tag{8.19}$$

which, under H_0^*, as $n \to \infty$ has asymptotically a chi-square distribution with $(t-1)$ degrees of freedom.

8.3.2. Incomplete Block Designs

Consider the situation described earlier, in which the ith judge ranks $1 \leq k_i \leq t$ objects. In the development that follows, we will at first consider the situation in which a basic design of b blocks is defined, and this basic design is replicated n times, so that a total of nb judges rank the specified subsets of the t objects. The indicator function $\delta_i(j)$ is either 1 or 0 depending on whether the object $1 \leq j \leq t$ is, or is not, ranked in block i. The index set of objects ranked in block $1 \leq h \leq b$ of this basic design is denoted by

$$\Omega_h = \left\{1 \leq j \leq t \,|\, \delta_h(j) = 1\right\},$$

and the basic design is defined by

$$\Omega^* = \left\{\Omega_h \,|\, 1 \leq h \leq b\right\}.$$

Let $r_j = \sum_{h=1}^{b} \delta_h(j)$ denote the total number of blocks in the basic design that include object j, and $\lambda_{jj'} = \sum_{h=1}^{b} \delta_h(j)\delta_h(j')$ the number of such blocks which include both objects $1 \leq j \neq j' \leq t$, so that nr_j judges rank object j, and $n\lambda_{jj'}$ judges rank both objects j and j'. For an incomplete ranking pattern Ω_h, let

$$\{\boldsymbol{R}_{h(s)}, s = 1, 2, \ldots, k_h!\}$$

represent the set of possible k_h-rankings, that is the permutations of $(1, 2, \ldots, k_h)$ within the specified incomplete pattern. For any such incomplete k_h-ranking with a given pattern of missing observations, associate a matrix of compatibility $\boldsymbol{C}_h$ whose $s = 1, 2, \ldots, k_h!$ columns are indicators identifying which of the $t!$ complete rankings indexing its rows, are compatible with the particular k_h-permutation $\boldsymbol{R}_{h(s)}$. The analogue of the adjusted matrix $\boldsymbol{T}$, the collection of adjusted score vectors $\boldsymbol{a}(\boldsymbol{R})$ is given by

$$\boldsymbol{T}^* = (\boldsymbol{T}_1^* \mid \boldsymbol{T}_2^* \mid \ldots \mid \boldsymbol{T}_b^*)$$

where

$$\boldsymbol{T}_h^* = \frac{k_h!}{t!}\boldsymbol{T}\boldsymbol{C}_h.$$

Denote by $\mathcal{C}\left(\boldsymbol{R}_{h(s)}\right)$ the class of complete rankings compatible with the specified k_h-permutation $\boldsymbol{R}_{h(s)}$ indexed by column s of $\boldsymbol{C}_h$. Under H_0, the columns of $\boldsymbol{T}_h^*$ are the conditional expected adjusted scores

$$E\left(\boldsymbol{a}(\boldsymbol{R}) \left|\mathcal{C}\left(\boldsymbol{R}_{h(s)}\right)\right.\right).$$

This conditional expectation provides the appropriate weighting for scores in an incomplete design. As Theorem 8.3 below indicates, this weighted score is given by the following definition.

Definition 8.1. For a given score function $a(j,t)$, $1 \leq j \leq t$, if object j is ranked in a given block $1 \leq h \leq b$, and if $R_h(j) = r$, the weighted score is given by

$$a^*(r, k_h) = \sum_{q=r}^{t-k_h+r} a(q,t) \binom{q-1}{r-1}\binom{t-q}{k_h-r}\binom{t}{k_h}^{-1}, \quad r = 1, 2, \ldots, k_h. \tag{8.20}$$

Theorem 8.3. *For an incomplete ranking pattern Ω_h, $1 \leq h \leq b$, with $\boldsymbol{R}_{h(s)}$ $s = 1, 2, \ldots, k_h!$ denoting the permutations of $(1, 2, \ldots, k_h)$ within the specified incomplete pattern, the row j, column s element of $\boldsymbol{T}_h^* = \frac{k_h!}{t!}\boldsymbol{T}\boldsymbol{C}_h$ is*

$$\left(a^*\left(R_{h(s)}(j), k_h\right) - \bar{a}_t\right)\delta_h(j).$$

Proof. For an incomplete ranking pattern Ω_h, $1 \leq h \leq b$, the row j, column s element of $\boldsymbol{T}_h^* = \frac{k_h!}{t!}\boldsymbol{T}\boldsymbol{C}_h$ is the average of $(a(R(j), t) - \bar{a}_t)$ over all $t!/k_h!$ complete rankings $\boldsymbol{R}$

compatible with $\boldsymbol{R}_{h(s)}$. If object j is ranked in pattern Ω_h, and if $R_h(j) = r$, then there are exactly

$$\binom{q-1}{r-1}\binom{t-q}{k_h-r}(t-k_h)!$$

complete compatible rankings $\boldsymbol{R}$, for which

$$R(j) = q, r \leq q \leq t - k_h + r,$$

so that the average of such scores $a(R(j), t)$ is given by (8.20). If on the other hand object j is not ranked in pattern Ω_h, there are $(t-1)!/k_h!$ complete rankings $\boldsymbol{R}$ compatible with $\boldsymbol{R}_{h(s)}$ for which $R(j) = q$, $1 \leq q \leq t$, so that the sum of such scores is

$$(t-1)!/k_h! \sum_{q=1}^{t} a(q,t),$$

and thus the average score is $\overline{a}_t$. □

Note that if k_h, the number of objects ranked is the same for each block, the weighted scores in (8.20) are all the same function of r. Note further that

$$\sum_{q=r}^{t-k+r} \binom{q-1}{r-1}\binom{t-q}{k-r}\binom{t}{k}^{-1} = 1.$$

The matrix $(\boldsymbol{T}^*)'\boldsymbol{T}^*$ contains measures of similarity between any two incomplete rankings with patterns in Ω^*, given as averages of the measures of similarity in (8.17) between all the complete rankings compatible with each of the specified incomplete rankings. Under H_0, for an incomplete ranking $\boldsymbol{R}_h$,

$$k_h^{-1}\sum_{r=1}^{k_h} a^*(r,k_h) = E(a^*(R_h(j),k_h)) = E(a(R(j),t)) = \overline{a}_t = t^{-1}\sum_{r=1}^{t} a(r,t),$$

and the vector of adjusted weighted scores,

$$\begin{aligned} \boldsymbol{a}^*(\boldsymbol{R}_h) &= [(a^*(R_h(1),k_h)-\overline{a}_t)\delta_h(1), (a^*(R_h(2),k_h)-\overline{a}_t)\delta_h(2), \ldots, \\ &\quad (a^*(R_h(t),k_h)-\overline{a}_t)\delta_h(t)]' \end{aligned} \tag{8.21}$$

has the covariance matrix $\boldsymbol{\Sigma}_h = \gamma_h^2 \boldsymbol{A}_h$, where the $(t \times t)$ matrix is given by

$$\boldsymbol{A}_h = \begin{cases} \frac{k_h-1}{k_h}\delta_h(j) & \text{on the diagonal} \\ -\frac{1}{k_h}\delta_h(j)\delta_h(j') & \text{off the diagonal} \end{cases} \tag{8.22}$$

and

$$\gamma_h^2 = (k_h - 1)^{-1} \sum_{r=1}^{k_h} \left(a^*\left(r, k_h\right) - \bar{a}_t\right)^2.$$

Proceeding as in Alvo and Cabilio (1999), define the $\sum_{h=1}^{b} k_h!$ dimensional vector

$$\boldsymbol{f} = \left(\boldsymbol{f}_1 \mid \boldsymbol{f}_2 \mid \cdots \mid \boldsymbol{f}_b\right)',$$

where $\boldsymbol{f}_h$ is the $k_h!$ dimensional vector of the observed frequencies of each of the $k_h!$ ranking permutations for the incomplete pattern Ω_h, $h = 1, 2, \ldots, b$. In order to extend the notation for the weighted scores to all the nb rankings, define

$$a^*\left(r, k_i\right) = a^*\left(r, k_h\right) if\ i \equiv h \left(\operatorname{mod} b\right), b+1 \leq i \leq nb.$$

Analogous to the complete case statistic the test statistic in this more general setting is given by

$$\begin{aligned} n^{-1}\boldsymbol{f}'\left(\boldsymbol{T}^{*\prime}\boldsymbol{T}^*\right)\boldsymbol{f} &= \left(\frac{1}{\sqrt{n}}\boldsymbol{T}^*\boldsymbol{f}\right)'\left(\frac{1}{\sqrt{n}}\boldsymbol{T}^*\boldsymbol{f}\right) \\ &= n^{-1}\sum_{j=1}^{t}\left(\sum_{i=1}^{nb}\left(a^*\left(R_i\left(j\right), k_i\right) - \bar{a}_t\right)\delta_i\left(j\right)\right)^2. \end{aligned} \tag{8.23}$$

Set

$$\boldsymbol{S}_{nb} = \left(S_{nb}^*\left(1\right), S_{nb}^*\left(2\right), \ldots, S_{nb}^*\left(t\right)\right)', \boldsymbol{r} = \left(r_1, r_2, \ldots, r_t\right)',$$

where

$$S_{nb}^*\left(j\right) = \sum_{i=1}^{nb} a^*\left(R_i\left(j\right), k_i\right)\delta_i\left(j\right).$$

Since

$$\sum_{i=1}^{nb}\delta_i\left(j\right) = nr_j$$

it follows that

$$\left(\boldsymbol{S}_{nb} - n\bar{a}_t\boldsymbol{r}\right) = \sum_{i=1}^{nb}\boldsymbol{a}^*\left(\boldsymbol{R}_i\right).$$

The test statistic is then

$$G_n = n^{-1}\left(\boldsymbol{S}_{nb} - n\bar{a}_t\boldsymbol{r}\right)'\left(\boldsymbol{S}_{nb} - n\bar{a}_t\boldsymbol{r}\right) = n^{-1}\sum_{j=1}^{t}\left(S_{nb}^*\left(j\right) - nr_j\bar{a}_t\right)^2. \tag{8.24}$$

Under H_0, $n^{-1/2}\boldsymbol{T}^*\boldsymbol{f}$ has covariance matrix

$$\boldsymbol{\Sigma}_0 = \sum_{i=1}^{b} \gamma_i^2 \boldsymbol{A}_i, \tag{8.25}$$

so that an alternative form of the statistic in (8.24) is

$$n^{-1}(\boldsymbol{T}^*\boldsymbol{f})' \boldsymbol{\Sigma}_0^- \boldsymbol{T}^*\boldsymbol{f} = n^{-1}(\boldsymbol{S}_{nb} - n\bar{a}_t\boldsymbol{r})' \boldsymbol{\Sigma}_0^- (\boldsymbol{S}_{nb} - n\bar{a}_t\boldsymbol{r}) \tag{8.26}$$

where $\boldsymbol{\Sigma}_0^-$ is a generalized inverse of $\boldsymbol{\Sigma}_0$.

8.3.3. Special Cases: Wilcoxon Scores

When the score function is $a(j,t) = j$, the Wilcoxon score is simply the (unweighted) Spearman correlation. In this case, if object j is ranked in block i, and if $R_i(j) = r$, the weighted score in (8.20) becomes

$$a^*(r, k_i) = r \sum_{q=r}^{t-k_i+r} \binom{q}{r}\binom{t-q}{k_i - r}\binom{t}{k_i}^{-1} = r\binom{t+1}{t-k_i}\binom{t}{k_i}^{-1} = r\frac{t+1}{k_i+1}, \tag{8.27}$$

This form is derived in Alvo and Cabilio (1991). Further $\bar{a}_t = (t+1)/2$, so that

$$S_b^*(j) - r_j\bar{a}_t = (t+1)\sum_{i=1}^{b}(k_i+1)^{-1}(R_i(j) - (k_i+1)/2)\,\delta_i(j), \tag{8.28}$$

and the test statistic G_n becomes

$$\sum_{j=1}^{t}\left(S_b^*(j) - r_j\frac{t+1}{2}\right)^2 = (t+1)^2\sum_{j=1}^{t}\left(\sum_{i=1}^{b}(k_i+1)^{-1}\left(R_i(j) - \frac{k_i+1}{2}\right)\delta_i(j)\right)^2. \tag{8.29}$$

In this case, $\gamma_i^2 = \frac{1}{12}(t+1)^2 k_i (k_i+1)^{-1}$.

8.3.4. Other Score Functions

In the situation where the complete rankings $\boldsymbol{R}_i = (R_i(1), R_i(2), \ldots, R_i(t))$ arise from independent random variables $(X_{i1}, X_{i2}, \ldots, X_{it})$ with absolutely continuous distribution functions $F_{ij}(x) = F_i(x - \tau_j)$, where $\sum \tau_j = 0$, and $F_i(x)$ have continuous densities $f_i(x)$, for which $\int_{-\infty}^{\infty} f_i^2(x)\,dx < \infty$, Sen (1968b) shows that the asymptotic distribution under local translation alternatives for test statistics Q_n of the form in (8.19) is noncentral chi-squared with $(t-1)$ degrees of freedom and a specified noncentrality parameter $\Delta_n(a)$. An upper bound Δ_n^0 to this noncentrality parameter over all possible choices of score functions $a(r,t)$ is derived. For various distributions of $(X_{i1}, X_{i2}, \ldots, X_{it})$, identical over the blocks, Sen (1968b) derives the form of the optimal scores $a(r,t)$ in the sense that $\Delta_n(a) = \Delta_n^0$. In this sense, the Wilcoxon score statistic discussed above is optimal in the case that the rankings result from complete samples from the logistic distribution with density $f(x) = e^{-x}/(1+e^{-x})^2; -\infty < x < \infty$.

The score function for an incomplete k_i-block is of the form

$$a^*(r,k_i) = \sum_{q=r}^{t-k_i+r} a(q,t) \binom{q-1}{r-1} \binom{t-q}{k_i-r} \binom{t}{k_i}^{-1} = K(k_i,t)\, a(r,k_i), \tag{8.30}$$

where $K(k_i,t)$ depends only on the number of objects ranked in the block. Other score functions considered by Sen (1968) also have this property. In particular we have the following.

(a) With the score $a(1,t) = 1, a(t,t) = -1, a(r,t) = 0$ otherwise, Q_n is optimal when sampling from the uniform distribution with density $f(x) = 1; 0 \leq x \leq 1$. Direct substitution into (8.20) gives $a^*(r,k_i) = \frac{k_i}{t} a(r,k_i)$.

(b) With the score $a(1,t) = 1, a(r,t) = 0$ otherwise, Q_n is optimal when sampling from the exponential distribution with density $f(x) = e^{-x}; 0 \leq x < \infty$. Again, direct substitution into (8.20) gives $a^*(r,k_i) = \frac{k_i}{t} a(r,k_i)$.

(c) When sampling from the double exponential distribution with density $f(x) = e^{-|x|}$; $-\infty < x < \infty$, Q_n associated with the score function $a(r,t) = 1 - 2\sum_{i=0}^{r-1} \binom{t}{i} 2^{-t}$ is shown to be optimal. In order to derive the form of $a^*(r,k_i)$ in this case we make use of the following lemma.

Lemma 8.2. *Let $F(x;n) = \sum_{i=0}^{x} \binom{n}{i} p^i (1-p)^{n-i}, 0 \leq x \leq n$, be the cumulative binomial distribution function. Then, for all $1 \leq x \leq m \leq n$*

$$F(x-1;m) = \sum_{i=x}^{n-m+x} F(i-1;n) \binom{i-1}{x-1} \binom{n-i}{m-x} \binom{n}{m}^{-1}. \tag{8.31}$$

Proof. See Alvo and Cabilio (2005) for the proof of this lemma. □

Suppose that we have random variables $X_{i1}, \ldots, X_{it}$ which may or may not be observable but which underlie the ranks. We assume that the variables are independently distributed with continuous cdf $F_{i1}(x), \ldots, F_{it}(x)$ respectively for $i = 1, \ldots, n$. We would like to test the null hypothesis that

$$F_{i1}(x) = \cdots = F_{it}(x) \equiv F_i(x), i = 1, \ldots, n,$$

against the alternative that

$$F_{ij}(x) = F_i(x - \tau_j), j = 1, \ldots, t; \sum_{j=1}^{t} \tau_j = 0.$$

From (8.20)

$$a^*(r, k_i) = 1 - 2\left[\sum_{q=r}^{t-k_i+r} \sum_{i=0}^{q-1} \binom{t}{i} 2^{-t} \binom{q-1}{r-1} \binom{t-q}{k_i - r} \binom{t}{k_i}^{-1}\right],$$

we have that the quantity in square brackets is equal to $F(r-1; k_i) = \sum_{i=0}^{r-1} \binom{k_i}{i} 2^{-k_i}$, so that $a^*(r, k_i) = a(r, k_i)$.

The scores considered in this section have certain properties. Since $k_i^{-1} \sum_{r=1}^{k_i} a^*(r, k_i) = \overline{a}_t$, it follows that for all such scores, $\overline{a}_t = K(k, t)\, \overline{a}_k$. Further, if the design is such that the number of objects ranked in each block is constant, that is $k_i = k$ for all i, then the incomplete scores are equivalent to the complete case ones.

8.3.5. Asymptotic Distribution Under the Null Hypothesis

Consider the situation in which a basic design of b blocks is replicated n times. Under H_0, as $n \to \infty$, $n^{-1/2}\boldsymbol{T}^*\boldsymbol{f}$ is asymptotically normal with mean vector $\mathbf{0}$ and covariance matrix $\boldsymbol{\Sigma}_0$ given in (8.25), and thus the test statistic G_n is asymptotically distributed as $\sum \alpha_i z_i^2$, where $\{z_i\}$ are independent identically distributed normal variates, and $\{\alpha_i\}$ are the eigenvalues of $\boldsymbol{\Sigma}_0$. The matrix $\boldsymbol{A} = \sum_{i=1}^{b} \boldsymbol{A}_i$, with $\boldsymbol{A}_i$ given in (8.22) is the information matrix of the design, and $rank(\boldsymbol{\Sigma}_0) = rank(\boldsymbol{A}) \leq t - 1$. If the design is connected, that is $\lambda_{jj'} \geq 1$, for all $1 \leq j \neq j' \leq t$, so that every pair of objects is compared, then $rank(\boldsymbol{A}) = t - 1$. The critical values of the asymptotic distribution are easily calculated using the methods in Jensen and Solomon (1972) (see Alvo and Cabilio, 1995 for such an application.)

If the number of objects ranked in each block is the same, say $k_i = k$, the weighted scores are all $a^*(r, k)$. Similarly the variances σ_i^2 are all equal to $(k-1)\, k^{-1}\gamma^2$, where $\gamma^2 = (k-1)^{-1} \sum_{r=1}^{k} (a^*(r, k) - \overline{a}_t)^2$. The covariance matrix $\boldsymbol{\Sigma}_0 = \gamma^2 \boldsymbol{A}$,

$$\boldsymbol{A} = \begin{cases} \frac{k-1}{k} r_j & \text{on the diagonal,} \\ -\frac{1}{k}\lambda_{jj'} & \text{off the diagonal.} \end{cases} \tag{8.32}$$

For many designs, including balanced incomplete blocks, cyclic, and group divisible designs, the eigenvalues of $\boldsymbol{A}$ are found analytically. See Alvo and Cabilio (1999) for details.

The asymptotics and the null hypothesis may be recast in a different setting. Specifically consider the situation in which random variables $(X_{i1}, X_{i2}, \ldots, X_{it})$, which may or may not be observable, underlie the rankings

$$\boldsymbol{R}_i = (R_i(1), R_i(2), \ldots, R_i(t)), i = 1, \ldots, b.$$

These random variables are assumed independent with absolutely continuous distribution functions $F_{ij}(x) = F_i(x - \tau_j)$, where $\sum \tau_j = 0$, and $F_i(x)$ have continuous densities $f_i(x)$, for which $\int_{-\infty}^{\infty} f_i^2(x)\,dx < \infty$. The null hypothesis of random uniform selection of rankings becomes $H_0' : \tau = \mathbf{0}$, where $\tau' = (\tau_1, \tau_2, \ldots, \tau_t)$. If the asymptotics of interest are simply that the number of blocks b becomes large, the definitions and notation used earlier may be modified by setting $n = 1$ as appropriate. The test statistic may be rewritten as

$$G_b^* = (\boldsymbol{S}_b - \bar{a}_t \boldsymbol{r})' \boldsymbol{\Sigma}_0^- (\boldsymbol{S}_b - \bar{a}_t \boldsymbol{r}) \tag{8.33}$$

where $\boldsymbol{\Sigma}_0$ defined in (8.25) is the covariance matrix, under H_0, of

$$(\boldsymbol{S}_b - \bar{a}_t \boldsymbol{r}) = (S_b^*(1) - \bar{a}_t r_1, S_b^*(2) - \bar{a}_t r_2, \ldots, S_b^*(t) - \bar{a}_t r_t)'. \tag{8.34}$$

As $b \to \infty$, if the design is connected, G_b^* has an asymptotic χ^2 distribution with $t-1$ degrees of freedom.

8.3.6. Asymptotic Distribution Under the Alternative

In block design Ω_h with k_h objects being ranked, $h = 1, 2, \ldots, b$, let

$$p_{hr}^{(j)} = P(R_h(j) = r | \delta_h(j) = 1) \delta_h(j)$$

represent the probability that if object j is being ranked it will be assigned the rank $r, 1 \le r \le k_h$. Denote the mean score for this ranking pattern by

$$\mu_h(j) = \sum_{r=1}^{k_h} a^*(r, k_h)\, p_{hr}^{(j)}. \tag{8.35}$$

Note that according to our convention,

$$\mu_i(j) = \mu_h(j), k_i = k_h \; if \; i \equiv h, (\text{mod } b), rb + 1 \le i \le nb.$$

The variables

$$U_s = \sum_{j=1}^{t} c_j \sum_{i=(s-1)b+1}^{sb} \left[a^*(R_i(j), k_i)\, \delta_i(j) - \mu_i(j)\right], s = 1, 2, \ldots, n \tag{8.36}$$

are independent with zero means, and with $n^{-1}\sum_{s=1}^{n} E\left|U_s\right|^3$ bounded. Thus, by the Lindeberg-Feller Central Limit Theorem (Chapter 2.1) it follows that as $n \to \infty$, $n^{-1/2}\sum_{s=1}^{n} U_s$ is asymptotically normal for all $\boldsymbol{c} = (c_1, c_2, \ldots, c_t) \neq (1, 1, \ldots, 1)$. Thus we have that $n^{-1/2}\left(\boldsymbol{S}_{nb} - E\left(\boldsymbol{S}_{nb}\right)\right)$ has an asymptotic multivariate normal distribution. Note that the expected values of the elements of $\boldsymbol{S}_{nb}$ are

$$E\left(S^*_{nb}\left(j\right)\right) = n\mu_b\left(j\right) = n\sum_{h=1}^{b}\mu_h\left(j\right)\delta_h\left(j\right). \tag{8.37}$$

Specializing the situation described in the previous section, assume that the absolutely continuous distributions of the $\left(X_{i1}, X_{i2}, \ldots, X_{it}\right)$, are of the form $F_{ij}\left(x\right) = F\left(x - \tau_j\right)$, with continuous density $f\left(x\right)$ for which $\int_{-\infty}^{\infty} f^2\left(x\right)dx < \infty$. Consider the alternatives

$$H_{1n} : \tau = \tau_n = n^{-1/2}\theta,\ \theta = \left(\theta_1, \theta_2, \ldots, \theta_t\right).$$

In order to investigate the asymptotic distribution of the test statistic

$$n^{-1}\left(\boldsymbol{S}_{nb} - n\bar{a}_t\boldsymbol{r}\right)'\boldsymbol{\Sigma}_0^-\left(\boldsymbol{S}_{nb} - n\bar{a}_t\boldsymbol{r}\right)$$

under such local translation alternatives, we use arguments and notation similar to those in (Sen (1968b)). Thus let

$$\beta_s^{(h)} = \binom{k_h - 2}{s}\int_{-\infty}^{\infty} F\left(x\right)^{r-2}\left(1 - F\left(x\right)\right)^{k_h - r} f^2\left(x\right)dx,\ s = 0, 1, \ldots, k_h - 2$$

where $\beta_{-1}^{(h)} = \beta_{k_h - 1}^{(h)} = 0$. By definition, for all $j \in \Omega_h$,

$$p_{hr}^{(j)} = \sum_{S_j}\int_{-\infty}^{\infty} P\left(X_{s_1} \leq x, \ldots, X_{s_{r-1}} \leq x, X_{s_{r+1}} > x, \ldots, X_{s_{k_h}} > x | X_j = x\right) dF_j\left(x\right), \tag{8.38}$$

where the summation extends over all possible choices of $\left(s_1, \ldots, s_{r-1}\right)$ from $\Omega_h \backslash \left\{j\right\}$, with $\left(s_{r+1}, .., s_{k_h}\right)$ the complementary set. Use of the distributional assumptions, independence, and Taylor series expansion and some manipulation shows that the probability in (8.38) can be written as

$$p_{hr}^{(j)} = k_h^{-1} + n^{-1/2}k_h\left(\theta_j - \overline{\theta}_h\right)\left(\beta_{r-2}^{(h)} - \beta_{r-1}^{(h)}\right) + o\left(n^{-1/2}\right),$$

where $\overline{\theta}_h = k_h^{-1}\sum\theta_j$, and the summation is over all $j \in \Omega_h$. Hence we see that $\mu_b\left(j\right)$ defined in (8.35) is related to the null mean through

$$\mu_b\left(j\right) - r_j\bar{a}_t = \eta\left(j, a^*\right) + o\left(n^{-1/2}\right), \tag{8.39}$$

where

$$\eta(j, a^*) = \sum_{h=1}^{b} n^{-1/2} k_h \left(\theta_j - \overline{\theta}_h\right) \delta_h(j) \left(\sum_{r=1}^{k_h} a^*(r, k_h) \left(\beta_{r-2}^{(h)} - \beta_{r-1}^{(h)}\right)\right). \tag{8.40}$$

Turning to the covariance matrix, define

$$p_{h,rs}^{(j,j')} = P(R_h(j) = r, R_h(j') = s | \delta_h(j) \delta_h(j') = 1) \delta_h(j) \delta_h(j')$$

which can be shown to be expressible as

$$p_{h,rs}^{(j,j')} = \frac{1}{k_h(k_h - 1)} + o\left(n^{-1/2}\right). \tag{8.41}$$

The covariance matrix of the vector of scores

$$\begin{aligned} &[(a^*(R_h(1), k_h) - \mu_h(1)) \delta_h(1), (a^*(R_h(2), k_h) \\ &- \mu_h(2)) \delta_h(2), \ldots, (a^*(R_h(t), k_h) - \mu_h(t)) \delta_h(t)]' \end{aligned} \tag{8.42}$$

can be written as

$$\left[\gamma_h^2 \frac{k_h - 1}{k_h} + o\left(n^{-1/2}\right)\right] \delta_h(j)$$

on the diagonal and

$$\left[-\gamma_h^2 \frac{1}{k_h} + o\left(n^{-1/2}\right)\right] \delta_h(j) \delta_h(j')$$

off the diagonal. It follows that

$$Cov\left(n^{-1/2} \boldsymbol{S}_{nb}\right) - \boldsymbol{\Sigma}_0 \to 0 \text{ as } n \to \infty,$$

where $\boldsymbol{\Sigma}_0$ is defined in (8.25). Combining this result with the asymptotic normality of $n^{-1/2}\mathbf{S}_{nb}$ it follows that if the basic design is connected, the statistic

$$n^{-1} \left(\boldsymbol{S}_{nb} - n\bar{a}_t \boldsymbol{r}\right)' \boldsymbol{\Sigma}_0^- \left(\boldsymbol{S}_{nb} - n\bar{a}_t \boldsymbol{r}\right)$$

has for $n \to \infty$, a noncentral chi-squared distribution with $t - 1$ degrees of freedom and noncentrality parameter

$$[\eta(1, a^*), \ldots, \eta(t, a^*)] \boldsymbol{\Sigma}_0^- [\eta(1, a^*), \ldots, \eta(t, a^*)]'$$

where $\eta(j, a^*)$ is defined in (8.40).

The form of this noncentrality parameter is simplified if $k_h = k$ for all $h = 1, 2, \ldots, b$. In such a case, it becomes

$$k^2 (k-1) \frac{\left(\sum_{r=1}^{k} a^* (r,k) (\beta_{r-2} - \beta_{r-1}) \right)^2}{\sum_{r=1}^{k} (a^* (r,k) - \overline{a}_t)^2} \boldsymbol{\Theta}' \boldsymbol{A}^- \boldsymbol{\Theta}$$

where $\boldsymbol{\Theta}$ is the vector with components $\sum_{h=1}^{b} \left(\theta_j - \overline{\theta}_h \right) \delta_h (j), j = 1, 2, .., t,$ and $\boldsymbol{A}$ is defined in (8.32). This parameter may be written as a multiple of a correlation coefficient which is maximized when $a^* (r,k) - \overline{a}_t = c (\beta_{r-2} - \beta_{r-1})$. This relationship leads to the same conclusions as in Sen (1968b), which is that the score functions detailed in Section 4 are optimal for the stated underlying distributions.

8.4. Exercises

Exercise 8.1. Let X have an exponential distribution with mean 1. Find the optimal score statistic for the scale alternative.

Exercise 8.2. Let X have a Cauchy distribution. Find the optimal score statistic for the location alternative.

9. Efficiency

In this chapter, we consider the asymptotic efficiency of tests which requires knowledge of the distribution of the test statistics under both the null and the alternative hypotheses. In the usual cases such as for the sign and the Wilcoxon tests, the calculations are straightforward. However, the calculations become more complicated in the multi-sample situations and for these, we appeal to Le Cam's lemmas. This is illustrated in the case of test statistics involving both the Spearman and Hamming distances. The smooth embedding approach is useful in that for a given problem, it leads to test statistics whose power function can be determined. The latter is then assessed against the power function of the optimal test statistic derived from Hoeffding's formula for any given underlying distribution of the data.

9.1. Pitman Efficiency

Suppose that one is interested in testing a null hypothesis H_0 against an alternative H_1. Consider two competing tests $T_{n,1}, T_{n,2}$ both of which are at level α and are consistent, that is, the power function converges to 1 as the sample size gets large. We may choose between them only if we can define a suitable criterion that takes into account the size as well as the power function of the test. There are a number of different criteria as detailed in Chapter 10 of Serfling (2009). Here, we will only describe the criteria proposed by Pitman which is defined for local alternatives close to the null hypothesis. The criteria provides a simple measure of the comparison between two tests but it is not as informative as the power function would be since the latter looks at a full range of alternatives.

Let $X_1, \ldots, X_n$ be a random sample from some cdf F having density $f(x; \theta)$ for some parameter θ. Let Ω_0, Ω_1 denote the spaces where θ takes its values under the null and alternative hypotheses respectively. The power function of a test is defined to be the probability of rejecting the null hypothesis and is given by

$$\beta_n (T_n; \theta) = P_\theta (T_n \text{ rejects } H_0).$$

M. Alvo, P. L. H. Yu, *A Parametric Approach to Nonparametric Statistics*,
Springer Series in the Data Sciences, https://doi.org/10.1007/978-3-319-94153-0_9

The size of the test is defined to be

$$\alpha_n = \sup_{\theta \epsilon \Omega_0} \beta_n (T_n; \theta).$$

In the specific situation where

$$H_0 : \theta = 0$$

against

$$H_1 : \theta > 0.$$

we shall be concerned with smooth local alternatives of the form $\theta_n = \frac{h}{\sqrt{n}}$. These alternatives make it more difficult for a test to choose between the hypotheses as n gets large. We shall suppose that for both tests the following three conditions are satisfied:

(i) the test statistic T_n is asymptotically normal under P_{θ_n}

$$P_{\theta_n}\left(\frac{\sqrt{n}\,(T_n - \mu(\theta_n))}{\sigma(\theta_n)} \leq x\right) \longrightarrow \Phi(x)$$

where it is convenient to let $\mu(\theta_n)$ denote the mean of T_n and $\frac{\sigma^2(\theta_n)}{n}$ denote the variance of T_n.

(ii) $\mu(\theta_n)$ is differentiable at 0, and $\sigma(\theta_n) \longrightarrow \sigma(0)$ as $\theta_n \to 0$ and

(iii) the test rejects for large values of the test statistic, that is

$$\frac{\sqrt{n}\,(T_n - \mu(0))}{\sigma(0)} > z_\alpha$$

where z_α is the upper $100(1-\alpha)\,\%$ point of a standard normal distribution.

Under these conditions, the corresponding power function converges to

$$\begin{aligned} \beta_n(\theta_n) &= P_{\theta_n}\left(\frac{\sqrt{n}\,(T_n - \mu(\theta_n))}{\sigma(\theta_n)} \geq z_\alpha \frac{\sigma(0)}{\sigma(\theta_n)} - \frac{\sqrt{n}\,(\mu(\theta_n) - \mu(0))}{\sigma(\theta_n)}\right) \\ &= P_{\theta_n}\left(\frac{\sqrt{n}\,(T_n - \mu(\theta_n))}{\sigma(\theta_n)} \geq z_\alpha \frac{\sigma(0)}{\sigma(\theta_n)} - \frac{h}{\sigma(\theta_n)} \frac{(\mu(\theta_n) - \mu(0))}{\theta_n}\right) \\ &\longrightarrow \Phi(hc - z_\alpha), \end{aligned}$$

where

$$c = \lim_{n\to\infty}\left(\frac{1}{\sigma(\theta_n)} \frac{(\mu(\theta_n) - \mu(0))}{\theta_n}\right) = \frac{\mu'(0)}{\sigma(0)}.$$

The quantity c is called the slope of the test. The larger the slope, the more powerful the test. A comparison of two tests $T_{n,1}, T_{n,2}$ may then be made in terms of the ratio of the slopes denoted by c_1, c_2, respectively

$$\frac{c_1}{c_2} = \left[\frac{\mu_1'(0)}{\sigma_1(0)}\right] / \left[\frac{\mu_2'(0)}{\sigma_2(0)}\right]. \tag{9.1}$$

Hence, $T_{n,1}$ is preferred to $T_{n,2}$ if $\frac{c_1}{c_2} > 1$.

Example 9.1 (One-Sample Tests: Sign Test, t-Test, Wilcoxon Signed Rank Test). We first calculate the slope of the sign test. Suppose that we have a random sample of size n from some distribution having absolutely continuous cumulative distribution function $F(x - \theta)$ which is symmetric around its median θ. We would like to test the null hypothesis $H_0 : \theta = 0$ and the one-sided alternative $H_1 : \theta > 0$. Let $T_{n,1}$ be the proportion of $X_i > 0$. The sign test rejects the null hypothesis for large values of $T_{n,1}$. Set

$$p = P(X > 0) = 1 - F(-\theta)$$

Then, $T_{n,1}$ has a binomial distribution with parameters n and probability of success p. Hence,

$$\begin{aligned} \mu(\theta) = E_\theta[T_{n,1}] &= p, \\ Var_\theta[T_{n,1}] &= \frac{p(1-p)}{n}. \end{aligned}$$

It follows that $\sigma^2(\theta) = p(1-p)$ and the slope is

$$\frac{\mu'(0)}{\sigma(0)} = \frac{2f(0)}{\sigma}, \tag{9.2}$$

where $f(x) = F'(x)$.

Suppose now we consider the usual t-test for H_0 vs H_1 which rejects the null hypothesis for large values of the sample mean $T_{n,2} = \bar{X}_n$. Hence,

$$\begin{aligned} \mu(\theta) &= E[T_{n,2}] \\ &= \int x dF(x - \theta) \\ &= \theta \end{aligned}$$

Since the test statistic is given by

$$\frac{\sqrt{n}\bar{X}_n}{S_n},$$

where S_n^2 represents the sample variance, we see that as $n \to \infty$

$$S_n^2 \to \sigma^2 = \int x^2 dF(x).$$

It follows that

$$\frac{\mu'(0)}{\sigma(0)} = \frac{1}{\sigma}. \tag{9.3}$$

Hence, when $f(x)$ is the standard normal density, the asymptotic relative efficiency of the sign test to the t-test is

$$(2f(0))^2 = \left(\frac{2}{\sqrt{2\pi}}\right)^2 = 0.637.$$

The Wilcoxon signed rank statistic rejects for large values of

$$T_{n,3} = \sum_{i=1}^{n} R_i^+ I(X_i > 0),$$

where as in Section 5.2, R_i^+ is the rank of $|X_i|$ among the $\{|X_i|, i = 1, \ldots, n\}$. It follows that under the null hypothesis

$$\begin{aligned} E_0[T_{n,3}] &= \frac{n(n+1)}{4}, \\ Var_0[T_{n,3}] &= \frac{n(n+1)(2n+1)}{24}. \end{aligned}$$

The calculations of the mean and variance under the alternative for the Wilcoxon signed rank statistic are more involved and are given by the following theorem.

Theorem 9.1. *Let*

$$\begin{aligned} p_1 &= P(X > 0) \\ p_2 &= P(X_1 + X_2 > 0) \\ p_3 &= P(X_1 + X_2 > 0, X_1 > 0) \\ p_4 &= P(X_1 + X_2 > 0, X_1 + X_3 > 0). \end{aligned}$$

Then, we have,

$$E[T_{n,3}] = np_1 + \frac{n(n-1)}{2} p_2$$

and

$$Var\left[T_{n,3}\right] = np_1\left(1-p_1\right) + \frac{n\left(n-1\right)}{2}p_2\left(1-p_2\right)$$
$$+2n\left(n-1\right)\left(p_3 - p_1 p_2\right) + n\left(n-1\right)\left(n-2\right)\left(p_4 - p_2^2\right)$$

Proof. See page 47 of Hettmansperger (1994). Here $p_3 = \frac{\left(p_2+p_1^2\right)}{2}$. □

The probabilities in Theorem 9.1 can be approximated for values of θ close to 0 so that

$$\begin{aligned} p_1 &= 1 - F\left(-\theta\right) \\ &\approx 1 - \left[F\left(0\right) - \theta f\left(0\right)\right] \\ &= \frac{1}{2} - \theta f\left(0\right). \end{aligned}$$

Also,

$$\begin{aligned} p_4 &= P\left(X_1 + X_2 > 0, X_1 + X_3 > 0\right) \\ &= \left(\int 1 - F\left(-x\right)\right)^2 f\left(x\right) dx \\ &= \frac{1}{3} \end{aligned}$$

and

$$\begin{aligned} p_2 &= P\left(X_1 + X_2 > 0\right) \\ &\approx 1 - F^*\left(-2\theta\right) \\ &= \frac{1}{2} + 2\theta f^*\left(0\right) \end{aligned}$$

where F^* is the convolution distribution and f^* is its density. Using the symmetry of f whereby $f\left(x\right) = f\left(-x\right)$, we have that

$$f^*\left(0\right) = \int_{-\infty}^{\infty} f^2\left(x\right) dx.$$

It follows that for small values of θ

$$Var_0\left[T_{n,3}\right] \approx \frac{n\left(n+1\right)\left(2n+1\right)}{24}.$$

In order to compute the slope for the Wilcoxon signed rank test, we note that

$$\mu(\theta) = E[T_{n,3}] = np_1 + \frac{n(n-1)}{2}p_2$$

so that

$$\mu'(0) = nf(0) + n(n-1)\int_{-\infty}^{\infty} f^2(x)\,dx$$

and

$$\frac{\mu'(0)}{\sigma(0)} = \sqrt{12}\int_{-\infty}^{\infty} f^2(x)\,dx. \tag{9.4}$$

The calculations of the slopes in (9.2), (9.3), and (9.4) enable us to compute the asymptotic relative efficiencies.

Example 9.2 (Two-Sample Tests: Mann-Whitney-Wilcoxon Test and t-Test)**.** Suppose that we have two independent random samples of sizes n and m respectively from distributions $F(x)$,G$(x-\Delta)$ having densities f, g. Recalling the Mann-Whitney-Wilcoxon test in Section 5.3.2, it was seen that the test statistic was the counting form

$$T_{m,n} = \frac{1}{mn}\sum_{j=1}^{m}\sum_{i=1}^{n} I(X_i < Y_j)$$

which rejects the null hypothesis for large values. Then

$$\begin{aligned}\mu(\Delta) &= E[T_{m,n}] \\ &= P(X < Y) \\ &= \int_{-\infty}^{\infty} F(y)\,f(y-\Delta)\,dy \\ &= \int_{-\infty}^{\infty} F(y+\Delta)\,dF(y)\end{aligned}$$

and provided we can differentiate under the integral sign,

$$\mu'(0) = \int_{-\infty}^{\infty} f^2(y)\,dy.$$

The variance of $T_{m,n}$, given from Theorem 5.1, is asymptotically

$$\begin{aligned}Var[T_{m,n}] &= \frac{1}{(mn)^2}\left[mnq_1(1-q_1) + mn(n-1)(q_2-q_1^2) + mn(m-1)(q_3-q_1^2)\right] \\ &\approx \left[o(1) + m^{-1}(q_2-q_1^2) + n^{-1}(q_3-q_1^2)\right]\end{aligned}$$

so that for $\frac{m}{m+n} \to \lambda > 0$,

$$\begin{aligned}(m+n)\,Var\,[T_{m,n}] &\approx \lambda^{-1}(q_2 - q_1^2) + (1-\lambda)^{-1}(q_3 - q_1^2) \\ &= \frac{1}{12\lambda(1-\lambda)},\end{aligned}$$

using similar arguments following Theorem 9.1. It follows that the slope is given by

$$\frac{\mu'(0)}{\sigma(0)} = \sqrt{12\lambda(1-\lambda)} \int_{-\infty}^{\infty} f^2(x)\,dx. \tag{9.5}$$

By way of comparison, we may also compute the slope for the two-sample t-test using the statistic which rejects the null hypothesis for large values. It is given by the

Table 9.1.: Relative efficiency of the Wilcoxon two-sample test relative to the t-test

Distribution	Relative efficiency
Normal	$3/\pi$
Logistic	$\pi^2/9$
Uniform	1

ratio

$$\left(\frac{\bar{Y}_m - \bar{X}_n}{S_{n,m}\sqrt{m^{-1}+n^{-1}}}\right),$$

where $\bar{X}_n, \bar{Y}_m$ represent the sample means and $S^2_{n,m}$ represent the pooled sample variance estimate of the common variance σ^2. It is straightforward to show that the slope of the two-sample t-test is

$$\frac{\mu'(0)}{\sigma(0)} = \frac{\sqrt{\lambda(1-\lambda)}}{\sigma}. \tag{9.6}$$

We record in Table 9.1 the relative efficiency of the Wilcoxon test relative to the t-test for various underlying distributions.

9.2. Making Use of Le Cam's Lemmas

The calculation of the asymptotic distribution under the alternative is not always as straightforward as in the previous section. In what follows we shall consider examples where it is necessary to calculate the asymptotic relative efficiency of tests by making use of Le Cam's contiguity concepts and lemmas. We begin with a simple example obtained from Section 14.11 of van der Vaart (2007) involving the median test.

Example 9.3 (Median Test)**.** Consider the two-sample test in Example 9.2 with m observations in the first sample and n in the second. Set $N = m + n$. The median test rejects the null hypothesis that the medians in the two samples are the same for large values of

$$T_N = \frac{1}{N} \sum_{i=m+1}^{N} I\left(R_i \leq \frac{n+1}{2}\right)$$

where $R_1, \ldots, R_N$ are the ranks of the complete data and $I(A)$ is the indicator function of the set A. Under the null hypothesis,

$$\begin{aligned} \sqrt{N}\left(T_N - \frac{n}{2N}\right) &= \frac{1}{N\sqrt{N}}\left(-n\sum_{i=1}^{m} I\left(F(X_i) \leq \frac{1}{2}\right)\right. \\ &\quad + m\sum_{j=1}^{n} I\left(F(Y_i) \leq \frac{1}{2}\right) Bigg) + o_P(1) \qquad (9.7) \\ &\xrightarrow{L} \mathcal{N}\left(0, \frac{\lambda(1-\lambda)}{4}\right) \qquad (9.8) \end{aligned}$$

since the left-hand side of (9.7) can be expressed as a linear rank statistic. The right-hand side of (9.7) is the projection (see Exercise 3.8). On the other hand, for alternatives $\theta_N = \frac{h}{\sqrt{N}}$, the log likelihood ratio given by

$$\log \frac{\prod f(X_i) \prod g(Y_j - \theta_N)}{\prod f(X_i) \prod g(Y_j)} = -\frac{h\sqrt{1-\lambda}}{\sqrt{n}} \sum_{i=1}^{n} \frac{g'}{g}(Y_i) - \frac{h^2(1-\lambda) I_g}{2} + o_P(1) \qquad (9.9)$$

and hence is asymptotically normal. Moreover the joint distribution of the linear parts of the right-hand sides of (9.7) and (9.9) is multivariate normal. Slutsky's lemma implies a similar result for the right-hand sides. Hence by Le Cam's third lemma, under alternatives $\theta_N = \frac{h}{\sqrt{N}}$,

$$\sqrt{N}\left(T_n - \frac{n}{2N}\right) \xrightarrow{\mathcal{L}} \mathcal{N}\left(\tau(h), \frac{\lambda(1-\lambda)}{4}\right),$$

where

$$\tau(h) = -h\lambda(1-\lambda) \int_{F(y) \leq 1/2} \left(\frac{f'(y)}{f(y)}\right) dF(y)$$

is the asymptotic covariance. The slope of the median test is then given by

$$-2\sqrt{\lambda(1-\lambda)} \int_0^{1/2} \left(\frac{f'(v)}{f(v)}\right) \left(F^{-1}(v)\right) dv. \qquad (9.10)$$

9.2.1. Asymptotic Distributions Under the Alternative in the General Multi-Sample Problem: The Spearman Case

In this section, we derive the asymptotic efficiency for the Spearman statistic in the general multi-sample problem with ordered alternatives from Section 6.1.2. Let $f(x)$ be the corresponding density of $F(x)$ and let $\varphi(v)$ be a square integrable function on $(0,1)$. Assume

$$\begin{aligned} F^{-1}(v) &= \inf\{x : F(x \geq v)\} \\ I(f) &= \int_{-\infty}^{\infty} \left[\frac{f'(x)}{f(x)}\right]^2 f(x)\,dx < \infty \qquad (9.11) \\ \varphi(v,f) &= \frac{f'(F^{-1}(v))}{f(F^{-1}(v))}, 0 < v < 1 \end{aligned}$$

Let $F_1(x), \ldots, F_r(x)$ be r continuous distribution functions. Suppose we wish to test

$$H_0 : F_r(x) = \ldots = F_1(x) = F\left(x - \bar{d}\right)$$

against the alternative

$$H_1 : F_k(x) = F_k(x - \Delta_k), 1 \leq k \leq r,$$

where

$$\Delta_1 < \ldots < \Delta_r, \quad \bar{d} = \frac{1}{N_r}\sum_{i=1}^{N_r} d_i$$

and

$$d_i = \Delta_k, \; N_{k-1} < i \leq N_k. \qquad (9.12)$$

We shall also assume that as $min\{n_1, \ldots, n_r\} \to \infty$, the following regularity conditions hold under the alternative

$$N_r^{1/2}\Delta_k \to \delta_k \qquad (9.13)$$

$$\max_{1 \leq i \leq N_r} \left(d_i - \bar{d}\right)^2 \to 0 \qquad (9.14)$$

$$I(f)\sum_{i=1}^{N_r}\left(d_i - \bar{d}\right)^2 \to b^2, \text{ a finite constant.} \qquad (9.15)$$

Notation The notation $T_1 \sim T_2$ indicates that the statistics T_1, T_2 are asymptotically.

We may determine the locally most powerful test for this problem by using Hoeffding's lemma 8.1. In fact, provided

$$\int_{-\infty}^{\infty} |f'(x)|\, dx < \infty,$$

the locally most powerful test is given by the statistic

$$T_0 = -\sum_{i=1}^{N_r} \left(d_i - \bar{d}\right) \varphi\left(X_i - \bar{d}, f\right) \tag{9.16}$$

$$= -\sum_{i=1}^{N_r} \left(d_i - \bar{d}\right) \frac{f'\left(X_i - \bar{d}\right)}{f\left(X_i - \bar{d}\right)} \tag{9.17}$$

which under H_1 has an asymptotic normal distribution with mean 0 and variance

$$b^2 = I\left(f\right) \sum_{i=1}^{N_r} \left(d_i - \bar{d}\right)^2. \tag{9.18}$$

We shall make use of the following two theorems.

Theorem 9.2 (Hájek and Sidak (1967), Theorem a V.1.4). *Let* $\{V_i\}_1^n$ *be i.i.d. uniform random variables on the interval* $(0,1)$. *Suppose* $\varphi(v)$, $0 < v < 1$ *is a square integrable function such that*

$$\int_0^1 \left(\varphi\left(v\right)\right)^2 dv < \infty$$

and put

$$a_n^{\varphi}\left(i\right) = E\left\{\varphi\left(V_1\right) | \pi\left(1\right) = i\right\}, 1 \le i \le n.$$

Then

$$\lim_{n\to\infty} E\left\{a_n^{\varphi}\left(\pi\left(1\right)\right) - \varphi\left(V_1\right)\right\}^2 = 0.$$

Theorem 9.3 (Hájek and Sidak (1967), Theorem VI.2.4). *Let*

$$S_n = \sum_{i=1}^{n} \left(c_k - \bar{c}\right) a_n\left(\mu\left(i\right)\right),$$

where the scores $\{a_n(i)\}$ *satisfy*

$$\lim_{n\to\infty} \int_0^1 \left\{a_n\left(1 + [vn]\right) - \varphi\left(v\right)\right\}^2 dv = 0,$$

and the $\{c_k\}$ *satisfy Noether's condition. Then under the regression alternatives,*

$$\frac{S_n - m_n}{\sigma_n} \xrightarrow{\mathcal{L}} \mathcal{N}\left(0,1\right),$$

where

$$m_n = \sum_{i=1}^{n} (c_k - \bar{c}) \left(d_i - \bar{d}\right) \int_0^1 \varphi(v)\, \varphi(v, f)\, dv$$

$$\sigma_n^2 = \sum_{i=1}^{n} (c_k - \bar{c})^2 \int_0^1 (\varphi(v) - \bar{\varphi})^2\, dv.$$

Corollary 9.1. *Consider the Spearman statistic for the multi-sample ordered location problem*

$$S_r = \sum c(i) \frac{\mu(i)}{N_r + 1}$$

where for $1 \leq k \leq r$

$$c(i) = N_{k-1} + N_k, N_{k+1} < i \leq N_k.$$

Then as $\min\{n_1, \ldots, n_r\} \to \infty$*,* S_r *is, under the alternative, asymptotically normal with mean and variance given respectively by*

$$m_r = \frac{N_r^2}{2} + \sum_{i=1}^{N_r} (c_k - \bar{c}) \left(d_i - \bar{d}\right) \int_0^1 v\varphi(v, f)\, dv$$

$$\sigma_r^2 = \frac{N_r^3}{12} \sum w_k W_k W_{k-1}$$

with

$$\frac{n_k}{N_r} \to w_k, W_k = \sum_{i=1}^{k} w_i.$$

Proof. This is a direct application of Theorem 9.3. □

We are now in a position to compute the asymptotic efficiency of the Spearman statistic in the multi-sample ordered problem. We shall show that the asymptotic efficiency is given by the ratio

$$ARE = \frac{\left\{\sum_{k=1}^{r} w_k \delta_k (W_{k-1} + W_k - 1) \int_0^1 v\varphi(v, f)\, dv\right\}^2}{I(f) \left\{\frac{1}{12} \sum_{k=1}^{r} w_k W_{k-1} W_k\right\} \left\{\sum_{k=1}^{r} w_k \left(\delta_k - \sum_{k=1}^{r} w_k \delta_k\right)^2\right\}} \tag{9.19}$$

In fact, from Corollary 9.1,

$$\sum_{i=1}^{N_r} (c_k - \bar{c}) \left(d_i - \bar{d}\right) = N_r^{3/2} \sum_{i=1}^{r} w_k \delta_k (W_k + W_{k-1} - 1)$$

and

$$\sum_{i=1}^{N_r}\left(d_i-\bar{d}\right)^2=\sum_{i=1}^{N_r}w_k\left(\delta_k-\sum w_k\delta_k\right)^2$$

Hence, the asymptotic power is given by

$$\lim_{\min\{n_i\}\to\infty}P\left(\frac{S_r-E_0S_r}{\sqrt{Var_0S_r}}>k_\alpha\right)=1-\Phi\left(k_\alpha-\frac{\sum_{k=1}^r w_k\delta_k\left(W_{k-1}+W_k-1\right)\left\{\int_0^1 v\varphi\left(v,f\right)dv\right\}}{\sqrt{\frac{1}{12}\sum_{k=1}^r w_kW_{k-1}W_k}}\right) \quad (9.20)$$

On the other hand, the asymptotic power of the locally most powerful test defined by the optimal score function is given by

$$I\left(f\right)\left\{\sum_{k=1}^r w_k\left(\delta_k-\sum_{k=1}^r w_k\delta_k\right)^2\right\}.$$

In Table 9.2, we record various integrals which enable us to compute the efficiencies. To illustrate the calculation, note that for the standard normal, using a change of variable and integration by parts,

$$\begin{aligned}\int_0^{1/2}u\varphi\left(u\right)du &= \int_0^{1/2}u\Phi^{-1}\left(u\right)du\\ &= \frac{1}{\sqrt{2\pi}}\int_{-\infty}^0\Phi\left(v\right)ve^{-\frac{v^2}{2}}dv\\ &= \left[\Phi\left(u\right)\frac{1}{\sqrt{2\pi}}e^{-\frac{u^2}{2}}\right]_{-\infty}^0+\int_{-\infty}^0\left[\frac{1}{\sqrt{2\pi}}e^{-\frac{v^2}{2}}\right]^2dv\\ &= -\frac{1}{2\sqrt{2\pi}}+\frac{1}{4\sqrt{\pi}}.\end{aligned}$$

A similar calculation then shows

$$\begin{aligned}\int_0^1 u\varphi\left(u\right)du &= \int_0^{1/2}u\varphi\left(u\right)du+\int_{1/2}^1 u\varphi\left(u\right)du\\ &= =\left(-\frac{1}{2\sqrt{2\pi}}+\frac{1}{4\sqrt{\pi}}\right)+\left(\frac{1}{2\sqrt{2\pi}}+\frac{1}{4\sqrt{\pi}}\right)\\ &= \frac{1}{2\sqrt{\pi}}.\end{aligned}$$

9.2.2. Asymptotic Distributions Under the Alternative in the General Multi-Sample Problem: The Hamming Case

Theorem 9.3 provides the asymptotic distribution under the alternative for simple linear rank statistics in the multi-sample case with ordered alternatives. For other test statistics such as the one generated by Hamming distance, we need to consider generalized rank statistics. To this end, we considerthe following theorems. First, we exhibit

Table 9.2.: Computations of integrals

	Normal	Double exponential	Logistic
I(f)	σ^{-2}	1	$\frac{1}{3}$
$\int_{1/2}^{1} \varphi(u)\, du$	$\frac{1}{\sigma\sqrt{2\pi}}$	$\frac{1}{2}$	$\frac{1}{4}$
$\int_{1/2}^{1} u\varphi(u)\, du$	$\frac{1}{2\sqrt{2\pi}} + \frac{1}{4\sqrt{\pi}} \approx 0.3405$	$\frac{3}{8}$	$\frac{5}{24}$
$\int_{0}^{1/2} u\varphi(u)\, du$	$-\frac{1}{2\sqrt{2\pi}} + \frac{1}{4\sqrt{\pi}} \approx -0.0585$	$-\frac{1}{8}$	$-\frac{1}{24}$
$\int_{1/2}^{1} u^2\varphi(u)\, du$	$\frac{1}{4\sqrt{2\pi}} + \frac{1}{2(\pi)^{3/2}}\left(arctan^{1}/\sqrt{2} + \frac{\pi}{2}\right) \approx 0.2961$	$\frac{7}{24}$	$\frac{17}{96}$
$\int_{0}^{1/2} u^2\varphi(u)\, du$	$-\frac{1}{4\sqrt{2\pi}} - \frac{1}{2(\pi)^{3/2}}\left(arctan^{1}/\sqrt{2} + \frac{\pi}{2}\right) + \frac{1}{2\sqrt{\pi}} \approx -0.0141$	$-\frac{1}{24}$	$-\frac{1}{96}$

Normal: $f(x) = \frac{1}{\sqrt{2\pi}\sigma} e^{-\frac{v^2}{2\sigma^2}}, -\infty < x < \infty$
Double exponential: $f(x) = \frac{1}{2} e^{-|x|}, -\infty < x < \infty$
Logistic: $f(x) = e^{-x}\left(1 + e^{-x}\right)^{-2}, -\infty < x < \infty.$

the asymptotic equivalence of a linear rank statistic to a sum of independent random variables in the multi-sample case.

Theorem 9.4. *Let* $\{V_i\}_1^n$ *be i.i.d. uniform random variables on the interval* $(0,1)$. *Set*

$$T_1 = \sum_{i=1}^{N_r} \left(d_i - \bar{d}\right) \varphi(V_i, f),$$

$$T_2 = \sum_{i=1}^{N_r} \left(d_i - \bar{d}\right) \varphi\left(\frac{\mu(i)}{N_r + 1}, f\right).$$

Then under Theorem 9.2

$$E_0 (T_1 - T_2)^2 \to 0$$

as $\min\{n_1, \ldots, n_r\} \to \infty$ *and hence* $T_1 \sim T_2$ *under* H_0.

Proof. See Alvo and Pan (1997). □

Theorem 9.5. *Let*

$$T_1 = \sum_{i=1}^{N_r} \left(d_i - \bar{d}\right) \varphi(V_i, f)$$

and

$$\log L_n = \sum_{i=1}^{n} \log \frac{f(x-d_i)}{f(s-\bar{d})}$$

Then

$$\left(\log L_n - T_1 + \frac{b^2}{2}\right) \xrightarrow{P} 0$$

as $\min\{n_1, \ldots, n_r\} \to \infty$*. Moreover,* $\log L_n$ *is asymptotically* $\mathcal{N}\left(-\frac{b^2}{2}, b^2\right)$.

Proof. The asymptotic normality of T_1 follows from the Lindeberg condition. See Theorem VI.2.1 of Hájek and Sidak (1967) for details of the proof. □

The next theorem provides the general result which is useful for obtaining the non-null distribution of test statistics involving more general distance functions such as that of Hamming.

Theorem 9.6. *Consider a generalized linear rank statistic*

$$T = \sum_{i=1}^{N_r} a_{i\pi(i)},$$

and suppose that under H_0

$$T \sim \mathcal{N}(0,1).$$

Let

$$d(i,j) = a_{ij} - \bar{a}_{i.} - \bar{a}_{.j} + \bar{a}_{..}$$

with

$$\max_{1 \le i,j \le N_r} d(i,j) \approx O(N_r^p), p < 0$$

Assume as well $I(f) < \infty$ *holds. Then under the alternative* H_1 *the asymptotic distribution of* T *is* $\mathcal{N}(\sigma_{12}, 1)$ *where*

$$\sigma_{12} = E_0[T\, T_2]$$

and

$$T_2 = \sum_{i=1}^{N_r} (d_i - \bar{d})\, \varphi\left(\frac{\pi(i)}{N_r+1}, f\right).$$

Proof. The proof follows closely the proof in Theorem VI.2.4 of Hájek and Sidak (1967). Recall

$$T_1 = \sum_{i=1}^{N_r} (d_i - \bar{d})\, \varphi(V_i, f).$$

Under H_0 from Theorem 9.4, T_1 and T_2 are asymptotically equivalent. Under H_0, Theorem 9.5 then implies

$$\log L_{N_r} \sim T_1 - \frac{b^2}{2}.$$

It follows that under H_0

$$(T, \log L_{N_r}) \sim \left(T, T_1 - \frac{b^2}{2}\right) \sim \left(T, T_2 - \frac{b^2}{2}\right).$$

We can see that $T_2 \sim \mathcal{N}(0, b^2)$ and hence if

$$(T, T_2) \sim \mathcal{N}_2\left(\begin{pmatrix} 0 \\ 0 \end{pmatrix}, \begin{pmatrix} 1 & \sigma_{12} \\ \sigma_{12} & b^2 \end{pmatrix}\right),$$

then from Le Cam's third lemma, it will follow that under the alternative

$$T \sim \mathcal{N}(\sigma_{12}, 1).$$

In view of the Cramér-Wold device (Section 2.1.2), it remains to show that for arbitrary constants c_1, c_2

$$c_1 T + c_2 T_2 \sim Normal.$$

For that purpose we make use of Hoeffding's combinatorial (see Section 3.3) central limit theorem. Let

$$a_{ij}^* = c_1 a_{ij} + c_2 \left(d_i - \bar{d}\right) \varphi\left(\frac{j}{N_r + 1}, f\right).$$

Then

$$c_1 T + c_2 T_2 = \sum_{i=1}^{N_r} a_{i\pi(i)}^*.$$

Put

$$\begin{aligned} d^*(i,j) &= a_{ij}^* - \bar{a}_{i.}^* - \bar{a}_{.j}^* + \bar{a}_{..}^* \\ &= c_1 d(i,j) + c_2 \left(d_i - \bar{d}\right) \left[\varphi\left(\frac{j}{N_r+1}, f\right) - \bar{\varphi}(., f)\right]. \end{aligned}$$

Since $\varphi(v, f)$ is an integrable function, there exists a constant $M > 0$ such that $|\varphi(v, f)| \leq M$, *a.s.*. Also,

$$\begin{aligned} \left|d_i - \bar{d}\right| &\leq \max_{1 \leq k \leq r} \left|\Delta_k - \frac{\sum n_k \Delta_k}{N_r}\right| \\ &\approx N_r^{-1/2} \max_{1 \leq k \leq r} \left|\delta_k - \frac{\sum t\delta_k}{N_r}\right| \\ &\approx O\left(N_r^{-1/2}\right). \end{aligned}$$

Hence

$$\begin{aligned} |d^*(i,j)| &\leq |c_1 d(i,j)| + \left| c_2 \left(d_i - \bar{d}\right) \left[\varphi\left(\frac{j}{N_r+1}, f\right) - \bar{\varphi}(., f) \right] \right| \\ &\leq |c_1| \max_{1\leq i,j \leq N_r} |d(i,j)| + |c_2 M| \left\| d_i - \bar{d} \right\| \\ &\approx O(N_r^p) + O\left(N_r^{-1/2}\right) \approx o(1). \end{aligned}$$

On the other hand,

$$\begin{aligned} Var_0\{c_1 T + c_2 T_2\} &= \frac{1}{N_r - 1} \sum (d^*(i,j))^2 \\ &\to c^2 \end{aligned}$$

and

$$\frac{\max_{1\leq i,j\leq N_r} (d^*(i,j))^2}{\frac{1}{N_r-1}\sum (d^*(i,j))^2} \to 0$$

This completes the proof. □

Applying Theorem 9.6 with $p = -\frac{1}{2}$, we see that Hamming's statistic H_r is, under the alternative, asymptotically normal with mean

$$m_H = 1 + E_0[H_r T_2]$$

and variance, as in (6.5)

$$\sigma_H^2 = \frac{r-1}{N_r - 1}$$

The calculation of the mean is shown in the next lemma.

Lemma 9.1. *For the Hamming distance,*

$$E_0[H_r T_2] \approx N^{-1/2} \sum_{k=1}^{r} \delta_k \int_{W_{k-1}}^{W_k} [\varphi(v,f) - \bar{\varphi}(f)]\, dv$$

Proof. We may write

$$\begin{aligned} E_0[H_r T_2] &= E_0\left\{ \sum a_{i\mu(i)} \sum \left(d_i - \bar{d}\right) \varphi\left(\frac{\mu(i)}{N_r+1}, f\right) \right\} \\ &= \frac{1}{N_r - 1} \sum_{k=1}^{r} \sum_{i=N_{k-1}+1}^{N_k} \left\{ \sum_{j=1}^{N_r} a_{ij} \left(d_i - \bar{d}\right) \left[\varphi\left(\frac{j}{N_r+1}, f\right) - \bar{\varphi}(., f) \right] \right\} \\ &= \frac{1}{N_r - 1} \sum_{k=1}^{r} n_k \left\{ \sum_{j=N_{k-1}+1}^{N_{kr}} \frac{1}{n_k} \left(\Delta_k - \bar{d}\right) \left[\varphi\left(\frac{j}{N_r+1}, f\right) - \bar{\varphi}(., f) \right] \right\} \end{aligned}$$

$$
\begin{aligned}
&= \frac{1}{N_r - 1} \sum_{k=1}^{r} \left(\Delta_k - \bar{d} \right) \left\{ \sum_{j=N_{k-1}+1}^{N_{kr}} \left[\varphi\left(\frac{j}{N_r + 1}, f \right) - \bar{\varphi}(., f) \right] \right\} \\
&\approx N^{-1/2} \sum_{k=1}^{r} \left\{ \delta_k - \sum_{j=1}^{r} w_k \delta_k \right\} \int_{W_{k-1}}^{W_k} \left[\varphi(v, f) - \bar{\varphi}(f) \right] dv \\
&= N^{-1/2} \sum_{k=1}^{r} \delta_k \int_{W_{k-1}}^{W_k} \left[\varphi(v, f) - \bar{\varphi}(f) \right] dv
\end{aligned}
$$

We may specialize these results for the two-sample case. In particular, we may calculate for $r = 2$ and sample sizes n_1, n_2 with $\frac{n_1}{n_1+n_2} \to \lambda$,

$$
m_H = 1 + \frac{bI^{-1/2(f)}}{n_1 + n_2} \left\{ \left(\frac{\lambda}{1 - \lambda} \right)^{-1/2} \left[\int_{\lambda}^{1} \varphi(v, f)\, dv - \int_{0}^{\lambda} \varphi(v, f)\, dv \right] \right\},
$$

and

$$
\sigma_H^2 = \frac{1}{n_1 + n_2 - 1}.
$$

It now follows that the asymptotic relative efficiency for the Hamming distance is given by

$$
ARE = \frac{\left\{ \sum_{k=1}^{r} \delta_k \int_{W_{k-1}}^{W_k} \left[\varphi(v, f) - \bar{\varphi}(f) \right] dv \right\}^2}{(r - 1)\, I(f) \left\{ \sum_{k=1}^{r} w_k \left(\delta_k - \sum_{k=1}^{r} w_k \delta_k \right)^2 \right\}}. \tag{9.21}
$$

□

9.3. Asymptotic Efficiency in the Unordered Multi-Sample Test

In the unordered multi-sample problem, the test statistic for the Spearman distance has an asymptotic χ^2 distribution under both the null and alternative hypothesis. For that reason, it is necessary to redefine the notion of efficiency. Suppose that we are interested in testing

$$
H_0 : \theta = 0,
$$

against

$$
H_1 : \theta \neq 0.
$$

Consider two test statistics Q_1, Q_2 for which

$$
Q_i \xrightarrow{\mathcal{L}} \begin{cases} \chi_k^2 & \textit{under } H_0 \\ \chi_k^2\left(\delta_i^2\right) & \textit{under } H_1 \end{cases}
$$

where δ_i^2 is a noncentrality parameter for $i = 1, 2$ respectively. Suppose that Q_1 is the locally most powerful test. Then we may define the asymptotic efficiency of Q_2 relative to Q_1 to be the ratio

$$e = \left(\frac{\delta_2}{\delta_1}\right)^2.$$

Suppose instead that we have a situation where the degrees of freedom under the null and alternative hypotheses are different. In that case, we may use Theorem VI.4.6 of Hájek and Sidak (1967) which states that when the noncentrality parameter $\delta \to 0$, then the power function under the alternative is given by

$$P\left(\chi_k^2\left(\delta^2\right) \geq \chi_{\alpha,n}^2\right) - \alpha \;\approx\; \delta 2^{-(n+2)/2} \exp\left(-\frac{\chi_{\alpha,n}^2}{2}\right)\left[\Gamma\left(\frac{n+2}{2}\right)\right]^{-1}\left(\chi_{\alpha,n}^2\right)^{n/2}.$$

On the other hand, suppose that the maximum asymptotic power can be reached by the test based on $\chi_{r-1}^2\left(b^2\right)$. Hence the asymptotic relative efficiency can be defined to be approximately

$$\frac{P\left(\chi_k^2\left(\delta^2\right) \geq \chi_{\alpha,n}^2\right) - \alpha}{P\left(\chi_k^2\left(b^2\right) \geq \chi_{\alpha,n}^2\right) - \alpha} \approx 2^{(r-1-n)/2} \exp\left(\frac{\chi_{\alpha,r-1}^2 - \chi_{\alpha,n}^2}{2}\right) \frac{\Gamma\left(\frac{r+2}{2}\right)}{\Gamma\left(\frac{n+2}{2}\right)} \frac{\left(\chi_{\alpha,n}^2\right)^{n/2}}{\left(\chi_{\alpha,r-1}^2\right)^{(r-1)/2}} \frac{\delta}{b^2}. \tag{9.22}$$

In the case where $r = 2$,

$$b^2 = I\left(f\right) w_1 w_2 \left(\delta_1 - \delta_2\right)^2.$$

Theorem 9.7. *The Spearman statistic for the multi-sample unordered problem is asymptotically chi-square $\chi_{r-1}^2\left(\delta_S^2\right)$ with noncentrality parameter*

$$\delta_S^2 \;=\; 12 b^2 \left(\int_0^1 v \varphi\left(v, f\right) dv\right)^2 / I\left(f\right).$$

Proof. See Alvo and Pan (1997). □

Theorem 9.8. *The Hamming statistic for the multi-sample unordered problem is asymptotically chi-square $\chi_{(r-1)^2}^2\left(\delta_H^2\right)$ with noncentrality parameter*

$$\delta_H^2 \;=\; E\left[\bar{\mu}\right] \Sigma_H^{-1} E\left[\bar{\mu}\right],$$

where $\bar{\mu}$ was defined in Theorem 6.2.

Proof. See Alvo and Pan (1997). □

Theorems 9.7 and 9.8 permit us to calculate the asymptotic relative efficiencies in the unordered case. It can be shown that these are the same as for the ordered situation and hence we have the same results as above.

9.4. Exercises

Exercise 9.1. Show that for the two-sample problem the asymptotic relative efficiency for the normal density,

$$f(x) = \frac{1}{\sqrt{2\pi}\sigma} \exp\left(-\frac{x^2}{2\sigma^2}\right), -\infty < x < \infty, I(f) = \sigma^{-2}$$

is $ARE = \frac{3}{\pi} \approx 0.9554$.

Exercise 9.2. Show that for the two-sample problem the asymptotic relative efficiency for the double exponential,

$$f(x) = \frac{1}{2} \exp(-|x|), -\infty < x < \infty, I(f) = 1$$

is $ARE = \frac{3}{4} = 0.75$.

Example 9.4. Show that for the two-sample problem the asymptotic relative efficiency for the logistic distribution,

$$f(x) = e^{-x}\left(1 + e^{-x}\right)^{-2}, -\infty < x < \infty, I(f) = 1/3$$

is $ARE = 1$.

Part III.

Selected Applications

10. Multiple Change-Point Problems

10.1. Introduction

In the classical formulation of the single change-point problem, there is a sequence $X_1, \ldots, X_n$ of independent continuous random variables such that the X_i for $i \leq \tau$ have a common distribution function $F_1(x)$ and those for $i > \tau$ a common distribution $F_2(x)$. It is of interest to test the hypothesis of "no change," i.e., $\tau = n$ against the alternative of a change, $1 \leq \tau < n$.

We begin by formulating the problem in terms of a parametric framework and study the properties of the new model. We then construct a composite likelihood function which permits us to conduct tests of hypotheses based on a score statistic to assess the significance of the change-points. We demonstrate the consistency of the estimated change-point locations and present a binary segmentation algorithm to search for the multiple change-points. We then report on a number of simulation experiments in order to compare the performance of the proposed method with other methods in the literature. We apply the new method to detect the DNA copy number alterations in a human genomic data set and to identify the change-points on an interest rate time series. Our empirical results reveal that the proposed method is efficient for change-point detection even when the data are serially correlated.

10.2. Parametric Formulation for Change-Point Problems

Suppose that there exists a single change-point between two (not necessarily adjacent) segments in a sequence of independent random variables and let X and Y be any two random variables from the respective segments, X from the first and Y from the second. The null hypothesis states that there is no change-point from one segment to the next. Consider a kernel function $t = h(x, y)$ and let the density of $T = h(X, Y)$ be given by

$$\pi(t; \theta) = \exp[\theta t - K(\theta)] f_0(t) \tag{10.1}$$

M. Alvo, P. L. H. Yu, *A Parametric Approach to Nonparametric Statistics*,
Springer Series in the Data Sciences, https://doi.org/10.1007/978-3-319-94153-0_10

where $f_0(t)$ is the density of T under the null hypothesis and $K(\theta)$ is a normalizing constant. This is an example of exponential tilting where the first factor in (10.1) represents the alternative to the null hypothesis. Consider the specific case where the kernel is given by

$$h(x, y) = sgn(x - y). \tag{10.2}$$

When $\theta = 0$, there is no change-point and $sgn(x - y) = \pm 1$ with equal probability $\frac{1}{2}$ irrespective of the underlying common distribution of X and Y. Hence

$$f_0(t) = \frac{1}{2}, t = \pm 1.$$

and the normalizing constant $K(\theta)$ is calculated to be

$$K(\theta) = \ln(\cosh(\theta)).$$

We may express the null hypothesis in terms of the parameter θ

$$H_0 : \theta = 0$$

and in that case $K(0) = 0$. The kernel in (10.2) appears in the Mann–Whitney statistic when testing for a change in mean between two distributions. The use of different kernel functions allows us flexibility in measuring the change between two segments.

10.2.1. Estimating the Location of a Change-Point using a Composite Likelihood Approach

Consider a sequence of N independent observations $Z_1, \ldots, Z_N$ where there is a change-point that breaks the sequence into two segments: observations $\boldsymbol{X}_\tau = \{Z_1, Z_2, \ldots, Z_\tau\}$, and $\boldsymbol{Y}_\tau = \{Z_{\tau+1}, Z_{\tau+2}, \ldots, Z_N\}$. Instead of setting up the multivariate distribution of all N observations, we here adopt the composite likelihood approach based on the kernel defined in (10.2). Let $\{t_{ij}\}$ be the collection of kernel values

$$t_{ij} = \text{sgn}(z_j - z_i), i = 1, \ldots, \tau, j = \tau + 1, \ldots, N.$$

We define the composite likelihood as

$$L(\theta; \boldsymbol{X}_\tau, \boldsymbol{Y}_\tau) = \prod_{i=1}^{\tau} \prod_{j=\tau+1}^{N} f_T(t_{ij}; \theta)$$

using the density in (10.1). Hence the composite log-likelihood, apart from a constant, is given by

$$\ell(\theta; \boldsymbol{X}_\tau, \boldsymbol{Y}_\tau) = \theta \sum_{i=1}^{\tau} \sum_{j=\tau+1}^{N} \operatorname{sgn}(z_i - z_j) - \tau(N-\tau)K(\theta). \tag{10.3}$$

Given τ, maximizing $\ell(\theta; \boldsymbol{X}_\tau, \boldsymbol{Y}_\tau)$ with respect to θ leads to the estimate of θ:

$$\hat{\theta}(\tau) = \tanh^{-1}\left[\frac{1}{\tau(N-\tau)}\left(\sum_{i=1}^{\tau} \sum_{j=\tau+1}^{N} \operatorname{sgn}(z_i - z_j)\right)\right]. \tag{10.4}$$

A change-point location $\hat{\tau}$ is then estimated as

$$\hat{\tau} = \operatorname*{argmax}_{\tau} \ell(\hat{\theta}(\tau); \boldsymbol{X}_\tau, \boldsymbol{Y}_\tau).$$

In the following lemma, we prove the almost sure convergence of this statistic.

Lemma 10.1. Consider the model (10.1) and let $X_i, i = 1, \ldots, m$ and $Y_j, j = 1, \ldots, n$ be two sequences of independent random variables. Let $U_{m,n} = \sum_{i=1}^{m} \sum_{j=1}^{n} S_{ij}$, where $S_{ij} = \operatorname{sgn}(X_i - Y_j)$. Let $\mu_{XY} = E[\operatorname{sgn}(X - Y)]$. Then as $\min\{m, n\} \longrightarrow \infty$,

$$\frac{U_{m,n}}{mn} \xrightarrow{\text{a.s.}} \mu_{XY}.$$

Proof. Using the result in (Lehmann (1975), p. 335),

$$\begin{aligned} Var(U_{m,n}) &= \sum_{i,j} Var(S_{ij}) + \sum_{ijk\ell} Cov(S_{ij}, S_{kl}) \\ &= mnVar(S_{ij}) + mn\,(m-1)\,Cov(S_{ij}, S_{kj}) + mn\,(n-1)\,Cov(S_{ij}, S_{i\ell}) \\ &\leq mn(m+n-1)M, \end{aligned}$$

where M is an upper bound for $Var(S_{ij})$. It follows from Chebyshev's inequality that for each $\epsilon > 0$ we have as $\min\{m, n\} \longrightarrow \infty$,

$$P\left(|U_{m,n} - E(U_{m,n})| > mn\epsilon\right) \leq \frac{1}{m^2n^2\epsilon^2} Var(U_{m,n}) = O\left(\frac{2M}{\min\{m,n\}\epsilon^2}\right) \rightarrow 0$$

and hence $U_{m,n} - E(U_{m,n}) \xrightarrow{P} 0$. We have for subsequences $\{m^2\}, \{n^2\}$,

$$\sum_m \sum_n P\left(|U_{m^2,n^2} - E(U_{m^2,n^2})| > m^2n^2\epsilon\right) \leq \sum_m \sum_n O\left(\frac{2M}{mn\,\min\{m,n\}\epsilon^2}\right) < \infty.$$

Hence by the Borel-Cantelli Lemma 2.2, we have

$$P\left(|U_{m^2,n^2} - E(U_{m^2,n^2})| > m^2n^2\epsilon \text{ i.o.}\right) = 0$$

and for the subsequences $\{m^2\}, \{n^2\}$,

$$\frac{U_{m^2,n^2} - E(U_{m^2,n^2})}{m^2n^2} \xrightarrow{a.s.} 0. \tag{10.5}$$

To show there is little difference between the sequences and the subsequences, let

$$D_{m,n} = \max |U_{k_1,k_2} - E(U_{k_1,k_2}) - (U_{m^2,n^2} - E(U_{m^2,n^2}))|,$$

where the maximum is taken over $m^2 \le k_1 < (m+1)^2$,$n^2 \le k_2 < (n+1)^2$. From Chebyshev's inequality it also follows similarly that

$$P\left(D_{m,n} > m^2n^2\epsilon \text{ i.o.}\right) \le \frac{1}{m^4n^4\epsilon^2}Var(D_{m,n}) \le O\left(\frac{64M}{(\min\{m,n\})^2\epsilon^2}\right),$$

and hence

$$\frac{D_{m,n}}{m^2n^2} \xrightarrow{a.s.} 0. \tag{10.6}$$

Note that for $m^2 \le k_1 < (m+1)^2$ and $n^2 \le k_2 < (n+1)^2$,

$$\frac{|U_{k_1,k_2} - E(U_{k_1,k_2})|}{k_1k_2} \le \frac{|U_{m^2,n^2} - E(U_{m^2,n^2})| + D_{m,n}}{m^2n^2},$$

and the proof of almost sure convergence then follows by using (10.5) and (10.6). □

10.2.2. Estimation of Multiple Change-Points

The estimation of multiple change-points can be done by applying the above method recursively via a binary segmentation algorithm. The main idea is to recursively detect a new change-point in one of the segments generated from those change-points estimated previously. Consider we currently have $k-1$ change-points estimated at locations $\hat{\tau}_1, \hat{\tau}_2, \ldots, \hat{\tau}_{k-1}$ with $0 = \hat{\tau}_0 < \hat{\tau}_1 < \ldots, < \hat{\tau}_{k-1} < \hat{\tau}_k = N$, resulting in a partition of the N observations into k segments $S_1, \ldots, S_k$, where $S_i = \{Z_{\hat{\tau}_{i-1}+1}, \ldots, Z_{\hat{\tau}_i}\}$. Now we can apply the composite likelihood method of estimating a single change-point mentioned in Section 10.2.1 to the observations in each of the k segments. Suppose that in the i th segment, the composite likelihood (10.3) is maximized at a proposed change-point location $\hat{\tau}(i)$, which partitions the segment into two other segments denoted by $\boldsymbol{X}_{\hat{\tau}(i)}$ and $\boldsymbol{Y}_{\hat{\tau}(i)}$ with sizes $m_i = \hat{\tau}(i) - \hat{\tau}_{i-1}$ and $n_i = \hat{\tau}_i - \hat{\tau}(i)$. The location of the k th estimated change-point is the one with the largest scaled composite likelihood:

$$\hat{\tau}(i^*) = \underset{\hat{\tau}(i), i=1,\ldots,k}{\text{argmax}} \; Q(\boldsymbol{X}_{\hat{\tau}(i)}, \boldsymbol{Y}_{\hat{\tau}(i)}), \tag{10.7}$$

$$Q(\boldsymbol{X}_{\hat{\tau}(i)}, \boldsymbol{Y}_{\hat{\tau}(i)}) = \ell(\hat{\theta}(\hat{\tau}(i)); \boldsymbol{X}_{\hat{\tau}(i)}, \boldsymbol{Y}_{\hat{\tau}(i)})/(m_i + n_i). \tag{10.8}$$

The purpose of the scaling in (10.7) is to give consistent estimates of the change-point locations (see Section 10.3). In searching for K change-points, this binary segmentation procedure has a computation cost of order $O(KN \ln N)$ as compared to a standard grid search procedure with a computation cost of $O(K2^N)$. We can further speed up the procedure by searching for the change-points for the segments in parallel.

10.2.3. Testing the Significance of a Change-Point

The binary segmentation algorithm proposed in the previous section requires specifying in advance the number of change-points which is generally unknown in practice. It is suspected that some estimated change-points are not significant and can be removed. In this section, a score test is proposed to determine the statistical significance of a change-point. The test can also serve as a stopping rule in searching for a new change-point in the binary segmentation algorithm.

Suppose that $k-1$ change-points have been identified. We wish to examine the significance of the k th estimated change-point, as defined by (10.7). This problem is equivalent to performing a test for $H_0 : \theta = 0$ under model (10.1) based on the observations in the i^* th segment. We make use of the test statistic:

$$\frac{U(\boldsymbol{X}_{\hat{\tau}(i^*)}, \boldsymbol{Y}_{\hat{\tau}(i^*)})^2}{I(0)},$$

where $U(\boldsymbol{X}_{\hat{\tau}(i^*)}, \boldsymbol{Y}_{\hat{\tau}(i^*)})$ and $I(0)$ are the score function and Fisher information respectively evaluated at $\theta = 0$:

$$U(\boldsymbol{X}_{\hat{\tau}(i^*)}, \boldsymbol{Y}_{\hat{\tau}(i^*)}) = \left.\frac{\partial \ell(\theta(\hat{\tau}(i^*)); \boldsymbol{X}_{\hat{\tau}(i^*)}, \boldsymbol{Y}_{\hat{\tau}(i^*)})}{\partial \theta}\right|_{\theta=0} = \sum_{i=\hat{\tau}_{i^*-1}+1}^{\hat{\tau}(i^*)} \sum_{j=\hat{\tau}(i^*)+1}^{\hat{\tau}_{i^*}} \text{sgn}(z_i - z_j),$$

$$I(0) = -E\left(\frac{\partial^2 \ell(\theta(\hat{\tau}(i^*)); \boldsymbol{X}_{\hat{\tau}(i^*)}, \boldsymbol{Y}_{\hat{\tau}(i^*)})}{\partial \theta^2}\right)\Bigg|_{\theta=0} = m_{i^*} n_{i^*} K''(0) = m_{i^*} n_{i^*}.$$

We reject H_0 for large values of the test statistic. For a fixed change-point location, the test statistic has an asymptotic chi-square distribution with 1 d.f. under H_0. However, the maximum of the test statistics among all the possible change-point locations does not follow a chi-square distribution. We may instead compute the p-value of the test statistic through a permutation test. Since under H_0, the $X's$ and $Y's$ are independent and identically distributed, we can permute the observations in the segment and calculate the test statistic for each permutation sample. Usually we can calculate the exact p-value if we consider all possible permutations. However, listing all possible permutations would be time-consuming for large samples. Instead, we may consider 1000 random permutations to obtain an approximate p-value.

We can conduct the permutation test at each step of including a new change-point. If the p-value of the test is larger than a prespecified level of significance, the binary segmentation algorithm stops. Otherwise, we continue to search for another new change-point.

10.3. Consistency of the Estimated Change-Point Locations

10.3.1. The Case of Single Change-Point

In the case of a single point and a given τ, the scaled composite log-likelihood in (10.7) evaluated at the estimate $\hat{\theta}(\tau)$ stated in (10.4) can be written as

$$\mathcal{Q}(\boldsymbol{X}_\tau, \boldsymbol{Y}_\tau) = \frac{\tau(N-\tau)}{N}\left(\hat{w}\tanh^{-1}(\hat{w}) + \frac{1}{2}\ln(1-\hat{w}^2)\right), \quad \hat{w} = \frac{1}{\tau(N-\tau)}\sum_{i=1}^{\tau}\sum_{j=\tau+1}^{N}\operatorname{sgn}(z_i - z_j).$$

We define the estimate $\hat{\tau}_N$ of the change-point location as

$$\hat{\tau}_N = \arg\max_{\tau} \mathcal{Q}(\boldsymbol{X}_\tau, \boldsymbol{Y}_\tau).$$

The following lemma shows that $\hat{\tau}_N$ is a strongly consistent estimator for a single change-point location. Here we assume that $\mu_{XY} = E\left(\operatorname{sgn}(X-Y)\right) < 1$. Otherwise, it is trivial that X is always greater than Y.

Lemma 10.2. *Suppose there is a single change-point. Let γ be the true proportion of observations belonging to the segment defined under model (10.1). Let $\{\delta_N\}$ be a sequence of positive numbers such that $\delta_N \to 0, N\delta_N \to \infty$, as $N \to \infty$. Then for N large enough, $\gamma \in [\delta_N, 1-\delta_N]$ and for all $\epsilon > 0$,*

$$P\left(\lim_{N\to\infty}\left|\gamma - \frac{\hat{\tau}_N}{N}\right| < \epsilon\right) = 1.$$

Proof. For any $\tilde{\gamma} \in [\delta_N, 1-\delta_N]$, let $\boldsymbol{X}(\tilde{\gamma}) = \{Z_1, \ldots, Z_{\lfloor\tilde{\gamma}N\rfloor}\}$ and $\boldsymbol{Y}(\tilde{\gamma}) = \{Z_{\lfloor\tilde{\gamma}N\rfloor+1}, \ldots, Z_N\}$. Then as $N \longrightarrow \infty$,

$$\hat{w} \xrightarrow{\text{a.s.}} \left[\frac{\gamma}{\tilde{\gamma}}I(\tilde{\gamma} \geq \gamma) + \frac{1-\gamma}{1-\tilde{\gamma}}I(\tilde{\gamma} < \gamma)\right]\mu_{XY} = w(\tilde{\gamma}),$$

uniformly in $\tilde{\gamma}$. As Q is a continuous function of $\tilde{\gamma}$, it can be shown that as $N \longrightarrow \infty$,

$$\frac{1}{N} Q(\boldsymbol{X}(\tilde{\gamma}), \boldsymbol{Y}(\tilde{\gamma})) \xrightarrow{\text{a.s.}} \tilde{\gamma}(1-\tilde{\gamma}) h(w(\tilde{\gamma})),$$

uniformly in $\tilde{\gamma}$, where $h(a) = a \tanh^{-1}(a) + \frac{1}{2}\ln(1-a^2)$. Notice $|w(\tilde{\gamma})| < 1$ and applying the Taylor series expansion of h at $a = 0$, there exists a large K such that with $a = w(\tilde{\gamma})$,

$$h(a) = \frac{a^2}{2} + \frac{a^4}{12} + \cdots + \frac{a^{2K}}{2K(2K-1)},$$

and hence we have

$$\tilde{\gamma}(1-\tilde{\gamma}) h(w(\tilde{\gamma})) = \sum_{k=1}^{K} \frac{\mu_{XY}^{2k}}{2k(2k-1)} \left[\frac{\gamma^{2k}(1-\tilde{\gamma})}{\tilde{\gamma}^{2k-1}} I(\tilde{\gamma} \geq \gamma) + \frac{(1-\gamma)^{2k}\tilde{\gamma}}{(1-\tilde{\gamma})^{2k-1}} I(\tilde{\gamma} < \gamma) \right]. \tag{10.9}$$

It is easy to show that for $k \geq 1$, $\frac{(1-\tilde{\gamma})}{\tilde{\gamma}^{2k-1}}$ is monotonic decreasing in $\tilde{\gamma}$ for $\tilde{\gamma} \geq \gamma$ while $\frac{\tilde{\gamma}}{(1-\tilde{\gamma})^{2k-1}}$ is monotonic increasing in $\tilde{\gamma}$ for $\tilde{\gamma} < \gamma$. Therefore, $\tilde{\gamma}(1-\tilde{\gamma})h(w(\tilde{\gamma}))$ attains its maximum when $\tilde{\gamma} = \gamma$. The rest of the proof proceeds analogously to the proof of Theorem 1 of Matteson and James (2014). □

10.3.2. The Case of Multiple Change-Points

Let us first consider that there are two change-points, $\boldsymbol{\gamma} = \{\gamma^{(1)}, \gamma^{(2)}\}$ such that the sequence of N observations is partitioned into 3 segments $\boldsymbol{X}_{\boldsymbol{\gamma}} = \{Z_1, \ldots, Z_{\lfloor \gamma^{(1)} N \rfloor}\}$, $\boldsymbol{W}_{\boldsymbol{\gamma}} = \{Z_{\lfloor \gamma^{(1)} N \rfloor + 1}, \ldots, Z_{\lfloor \gamma^{(2)} N \rfloor}\}$ and $\boldsymbol{Y}_{\boldsymbol{\gamma}} = \{Z_{\lfloor \gamma^{(2)} N \rfloor + 1}, \ldots, Z_N\}$ and any two random variables obtained from two distinct segments satisfy model (10.1). Let $\mu_{XY} = E(\text{sgn}(X - Y))$. Similarly, we can define μ_{XW} and μ_{WY}. Assume that at least one of μ_{XY}, μ_{XW} and μ_{WY} has its absolute value less than one.

Lemma 10.3. *Consider any change-point $\tilde{\gamma}$ such that $\gamma^{(1)} \leq \tilde{\gamma} \leq \gamma^{(2)}$. Then the sequence of N observations is partitioned into 2 segments $\boldsymbol{X}(\tilde{\gamma}) = \{Z_1, \ldots, Z_{\lfloor \tilde{\gamma} N \rfloor}\}$ and $\boldsymbol{Y}(\tilde{\gamma}) = \{Z_{\lfloor \tilde{\gamma} N \rfloor + 1}, \ldots, Z_N\}$. Let $\hat{w} = \frac{1}{\lfloor \tilde{\gamma} N \rfloor (N - \lfloor \tilde{\gamma} N \rfloor)} \sum_{i=1}^{\lfloor \tilde{\gamma} N \rfloor} \sum_{j=\lfloor \tilde{\gamma} N \rfloor + 1}^{N} \text{sgn}(z_i - z_j)$. As $N \longrightarrow \infty$,*

$$\sup_{\tilde{\gamma} \in [\gamma^{(1)}, \gamma^{(2)}]} |\hat{w} - p(\tilde{\gamma})| \xrightarrow{a.s.} 0,$$

where

$$p(\tilde{\gamma}) = \frac{\gamma^{(1)}(\gamma^{(2)} - \tilde{\gamma})}{\tilde{\gamma}(1-\tilde{\gamma})} \mu_{XW} + \frac{\gamma^{(1)}(1-\gamma^{(2)})}{\tilde{\gamma}(1-\tilde{\gamma})} \mu_{XY} + \frac{(\tilde{\gamma} - \gamma^{(1)})(1-\gamma^{(2)})}{\tilde{\gamma}(1-\tilde{\gamma})} \mu_{WY}. \tag{10.10}$$

The proof is similar to Lemma 10.2 and is therefore omitted.

As Q is a continuous function of $\tilde{\gamma}$, it can be shown that as $N \longrightarrow \infty$,

$$Q(\boldsymbol{X}(\tilde{\gamma}), \boldsymbol{Y}(\tilde{\gamma}))/N \stackrel{\text{a.s.}}{\longrightarrow} \tilde{\gamma}(1-\tilde{\gamma})h(p(\tilde{\gamma})) = q(\tilde{\gamma}),$$

uniformly in $\tilde{\gamma}$.

Theorem 10.1. *For $\mathcal{A}_N \subset (\delta_N, 1-\delta_N)$ and real x, define*

$$d(x, \mathcal{A}_N) = inf\left\{|x-y| : y \in \mathcal{A}_N\right\}.$$

Let $\hat{\tau}_N$ be the estimated change-point and set

$$\mathcal{A}_N = \left\{y \in [\delta_N, 1-\delta_N] : q(y) \geq q(\gamma), \forall \gamma\right\}.$$

Then

$$d\left(\frac{\hat{\tau}_N}{N}, \mathcal{A}_N\right) \stackrel{a.s.}{\longrightarrow} 0, \text{ as } N \to \infty.$$

Proof. Note that $p(\tilde{\gamma})$ in (10.10) can be rewritten in a form: $p(\tilde{\gamma}) = \frac{a\tilde{\gamma}+b}{\tilde{\gamma}(1-\tilde{\gamma})}$ for some a and b, and $|p(\tilde{\gamma})| < 1$. Applying the Taylor series expansion of h at $a = 0$, there exists a large K such that

$$q(\tilde{\gamma}) = \tilde{\gamma}(1-\tilde{\gamma})h(p(\tilde{\gamma}))$$

can be approximated by

$$q(\tilde{\gamma}) = \sum_{k=1}^{K} \frac{\mu_{XY}^{2k}}{2k(2k-1)} \frac{(a\tilde{\gamma}+b)^{2k}}{(\tilde{\gamma}(1-\tilde{\gamma}))^{2k-1}}.$$

It is easy to show that $q(\tilde{\gamma})$ is continuously differentiable and strictly convex as the sums and products of convex functions are also convex.

Hence, for any two points $\tilde{\gamma}_1, \tilde{\gamma}_2 \in [\gamma^{(1)}, \gamma^{(2)}]$, there exists a $c > 0$ such that

$$|q(\tilde{\gamma}_1) - q(\tilde{\gamma}_2)| > c\,|\tilde{\gamma}_1 - \tilde{\gamma}_2| + o\left(|\tilde{\gamma}_1 - \tilde{\gamma}_2|\right).$$

The rest of the proof proceeds analogously to the proof of Theorem 2 of Matteson and James (2014). □

Finally, the extension of the consistency proof for multiple change-points is straightforward by noting that in the case of multiple change-points, $p(\tilde{\gamma})$ in (10.10) is still represented in the form: $\frac{a\tilde{\gamma}+b}{\tilde{\gamma}(1-\tilde{\gamma})}$.

10.4. Simulation Experiments

In this section, we report on the simulation experiments conducted to study the performance of the proposed algorithm.

10.4.1. Model Setup

Consider the Blocks data sets (Donoho and Johnstone, 1995) which are generally considered difficult for multiple change-point detection problems in view of the highly heterogeneous segment levels and lengths. We generate the Blocks data sets with 11 change-points and N samples:

$$
\begin{aligned}
Z_i &= \sum_{j=1}^{11} h_j J(Nt_i - \tau_j) + \sigma\varepsilon_i \qquad\qquad J(x) = \{1 + \mathrm{sgn}(x)\}/2 \\
\{\tau_j/N\} &= \{0.1, 0.13, 0.15, 0.23, 0.25, 0.40, 0.44, 0.65, 0.76, 0.78, 0.81\} \\
\{h_j\} &= \{2.01, -2.51, 1.51, -2.01, 2.51, -2.11, 1.05, 2.16, -1.56, 2.56, -2.11\}
\end{aligned}
$$

where the t_i's are equally spaced in $[0, 1]$ and $\{\tau_j/N\}$ marks the segment's information. The position of each change-point is τ_j and the mean difference is controlled by h_j.

Various i.i.d. distributions for ε_i are considered in our simulation experiments: $\mathcal{N}(0, 1)$: standard normal distribution, $t(2)$: Student's t distribution with two degrees of freedom, $\chi^2(3)$: chi-squared distribution with three degrees of freedom, $Cauchy(0, 1)$: Cauchy distribution with location 0 and scale 1, $Pareto(\alpha = 0.5)$, $Pareto(\alpha = 1.5)$: Pareto distributions with $\alpha = 0.5$, and 1.5 and $LN(0, 1)$: log-normal distribution with location 0 and scale 1. In our simulation, N is set to be 500 and 1000 with $\sigma = 0.5$. Examples of simulated model with various error distributions can be found in Figure 10.1.

10.4.1.1. Performance Measures

In order to compare the performances of various change-point methods, we consider two distance measures between the estimated change-point set $\hat{\Gamma}_{est}$ and the true change-point set Γ_{true}:

$$
\begin{aligned}
\xi_1(\hat{\Gamma}_{est}, \Gamma_{true}) &= \max_{b \in \Gamma_{true}} \min_{a \in \hat{\Gamma}_{est}} |a - b|, \\
\xi_2(\Gamma_{true}, \hat{\Gamma}_{est}) &= \max_{b \in \hat{\Gamma}_{est}} \min_{a \in \Gamma_{true}} |a - b|.
\end{aligned}
$$

Note that $\xi_1(\hat{\Gamma}_{est}, \Gamma_{true})$ measures the under-segmentation error and $\xi_2(\Gamma_{true}, \hat{\Gamma}_{est})$ measures the over-segmentation error. We also calculate the sum of ξ_1 and ξ_2 as a measure of the total change-point location error.

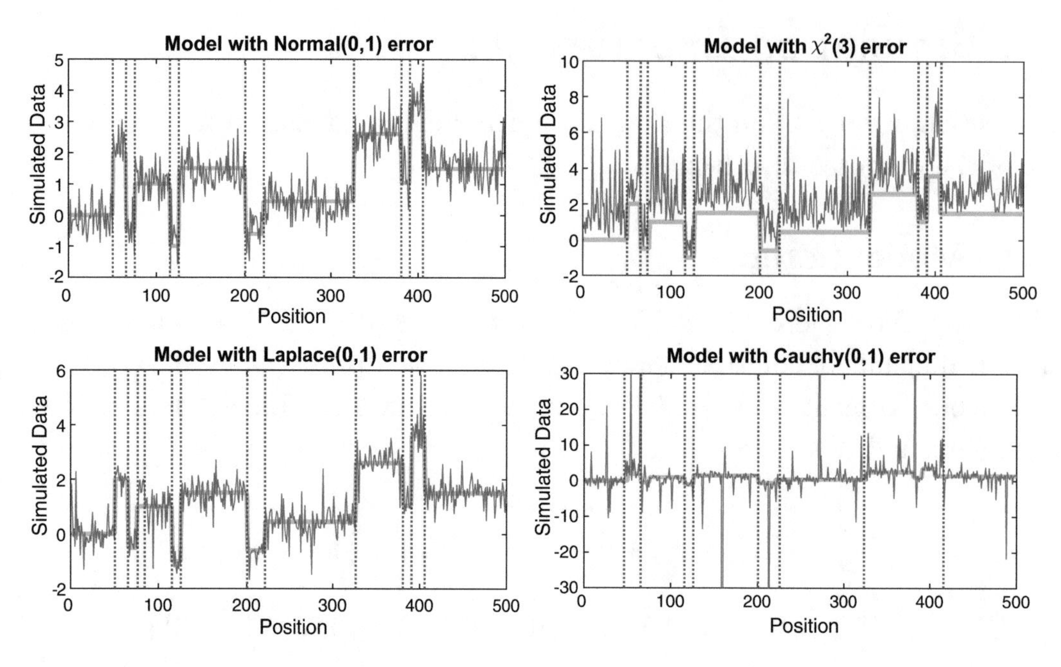

Figure 10.1.: Examples of simulated independence model introduced by Donoho and Johnstone (1995) with different error terms. The green line showed the true mean of each segments. The red dotted lines indicate the location of the change-points found by our method when the number of change-points is unknown.

10.4.2. Simulation Results

10.4.2.1. Known Number of Change-Points

We first simulate data under the model described in Section 10.4.1 with known number of change-points, i.e., 11. We compare our method with the traditional parametric likelihood ratio approach (Hinkley, 1970). Note that there are two ways to minimize the objective function in the parametric likelihood ratio (PLR) method which can be estimated using the binary segmentation (BS) algorithm and dynamic programming with the PELT algorithm (Killick et al., 2012). These two algorithms are implemented using the "change-point" R package. We also compete with the nonparametric methods called e.divisive which is implemented using the "ecp" R package with $\alpha = 1$ and $k = 11$. The simulation results are shown in Table 10.1.

As expected, parametric methods perform better under normal assumption. However, our method and e.divisive outperform the parametric likelihood ratio approach and provide a robust inference for the change-points when the error term does not follow normal or Laplace distribution. We can also see that the standard deviation increases substantially for PLR methods when the normality assumption is violated. In the cases of $Laplace(0,1)$ $Cauchy(0,1)$, $Pareto(\alpha = 0.5)$, $Pareto(\alpha = 1.5)$, our method has some advantages over the e.divisive method.

Table 10.1.: Simulation Results for known number of change-points (11). Standard deviations are given in parentheses. Number in bold represents the method with the smallest total error.

Error	$\mathcal{N}$	$\xi_1(\hat{\Gamma}_{est}, \Gamma_{true})$				$\xi_2(\Gamma_{true}, \hat{\Gamma}_{est})$				$\xi_1 + \xi_2$			
		Our method	PLR(BS)	PLR(PELT)	e.divisive	Our method	PLR(BS)	PLR(PELT)	e.divisive	Our method	PLR(BS)	PLR(PELT)	e.divisive
$\mathcal{N}(0,1)$	500	2.30(1.87)	1.90(2.05)	1.10(0.79)	2.00(0.86)	2.30(1.87)	1.80(1.64)	1.10(0.79)	2.00(0.86)	4.60	3.70	**2.20**	4.00
	1000	4.10(3.99)	2.40(2.84)	1.10(1.41)	2.55(2.91)	4.10(3.99)	2.40(2.84)	1.10(1.41)	2.55(2.91)	8.20	4.80	**2.20**	5.10
$\chi^2_{(3)}$	500	17.00(13.73)	19.50(17.99)	20.55(10.62)	9.15(8.41)	12.70(10.57)	13.80(10.46)	18.25(15.36)	16.15(17.21)	29.70	33.3	38.80	**25.30**
	1000	14.45(15.91)	25.20(22.38)	27.00(20.88)	9.70(7.33)	15.40(15.30)	30.80(53.03)	30.10(50.10)	9.00(7.32)	29.85	56.00	57.10	**18.70**
$Laplace(0,1)$	500	2.00(2.05)	1.40(1.39)	0.90(1.37)	2.10(0.97)	2.00(2.05)	1.40(1.39)	0.90(1.37)	2.10(0.97)	4.00	2.80	**1.80**	4.20
	1000	1.60(1.78)	1.05(0.76)	0.75(0.64)	1.65(0.67)	1.70(1.78)	1.05(0.76)	0.75(0.64)	1.65(0.67)	3.30	2.10	**1.50**	3.30
$Cauchy(0,1)$	500	24.40(17.59)	94.80(52.51)	94.30(47.92)	18.80(7.69)	23.45(18.67)	46.10(22.36)	47.75(23.57)	29.30(19.67)	**47.85**	140.90	142.05	48.10
	1000	16.00(18.55)	206.95(115.54)	169.30(91.42)	35.00(19.73)	18.50(35.10)	107.10(51.04)	96.15(44.65)	55.60(44.75)	**34.50**	314.05	265.45	90.60
$t(2)$	500	8.30(7.83)	23.95(11.46)	40.65(23.31)	4.85(4.36)	8.75(9.88)	20.15(15.76)	48.65(26.19)	5.65(6.18)	17.05	44.10	89.30	**10.50**
	1000	5.95(5.28)	31.40(23.22)	77.95(42.44)	3.75(2.02)	5.95(5.28)	43.40(59.78)	81.60(44.38)	3.75(2.02)	11.90	74.80	159.55	**7.50**
$Pareto(\alpha = 0.5)$	500	11.40(11.31)	75.65(50.02)	81.30(25.61)	16.25(10.50)	11.10(9.91)	45.25(27.30)	53.20(24.22)	28.85(26.82)	**22.5**	120.9	134.5	45.1
	1000	7.80(5.35)	169.05(105.86)	188.35(117.21)	28.15(28.93)	10.95(20.20)	104.70(43.37)	106.90(40.50)	43.30(40.39)	**18.75**	273.75	295.25	71.45
$Pareto(\alpha = 1.5)$	500	29.10(18.73)	98.95(44.31)	90.90(43.25)	42.45(16.83)	28.00(23.51)	47.90(21.54)	50.05(19.96)	44.40(17.14)	**57.10**	146.85	140.95	86.85
	1000	24.30(14.69)	244.65(121.40)	210.30(97.97)	71.60(40.23)	23.05(24.01)	97.55(51.27)	102.20(44.52)	74.80(36.84)	**47.35**	342.20	312.50	146.40
$LN(0,1)$	500	5.60(5.61)	18.75(12.00)	32.45(17.98)	3.65(3.50)	10.00(17.63)	12.85(13.81)	32.30(19.03)	8.75(19.84)	15.60	31.60	64.75	**12.40**
	1000	4.05(3.17)	20.00(19.46)	53.00(27.08)	3.20(1.70)	4.05(3.17)	7.15(6.30)	65.80(52.44)	3.20(1.70)	8.10	27.15	118.80	**6.40**

10.4.2.2. Unknown Number of Change-Points

Simulation experiments were also conducted when the number of change-points is unknown. In determining the number of change-points, we adopt the permutation test with 1% significance level in our proposed method while the PLR(PELT) method applies BIC, and the e.divisive method is implemented using the "ecp" R package with $\alpha = 1$ and 1% significance level. We also consider the e.cp3o method which is implemented using the "ecp" R package with $\delta = 2$, $K = 15$ and $\alpha = 1$. Table 10.2 displays the simulation results.

Based on the total change-point location error $\xi_1 + \xi_2$, we can see from Table 10.2 that the PLR(PELT) method performs poorly when the error distribution is not normal or Laplace as it poorly estimates the number of change-points. The e.cp3o method also cannot estimate the number of change-points well. The e.divisive method outperforms our model in some error terms such as $\mathcal{N}(0,1)$, $Laplace(0,1)$, and $t(2)$. However, the e.divisive method cannot find any change-point when the error distributions are $Cauchy(0,1)$, $Pareto(\alpha = 0.5)$, and $Pareto(\alpha = 1.5)$. Note that the e.divisive method requires the existence of the first two moments of the underlying distributions because of the divergence measure used in e.divisive. $Cauchy(0,1)$, $Pareto(\alpha = 0.5)$, and $Pareto(\alpha = 1.5)$ do not have both first two moments being finite, and the e.divisive method may greatly underestimate the number of change-points. However, our method provides a robust estimation in these situations. Based on our parametric embedding, our method is capable of detecting various types of changes even for some distributions without finite first two moments.

10.5. Applications

10.5.1. Array CGH Data

DNA copy numbers refer to the number of copies of genomic DNA in a human. The usual number is two for normal cells for all the non-sex chromosomes. Variations are indicative of disease such as cancer. Hence, there is a need to produce copy number maps. The array data used consist of the $\log_2$-ratio of normalized intensities of the red/green channels indexed by marker locations on a chromosome where the red and green channels measure the intensities of the cancer and normal samples respectively. There have been a number of different techniques proposed for analyzing copy number variation (CNV) data. It is known that CNVs account for an abundance of genetic variation and may influence phenotypic differences.

The change-point detection method above was applied to the array CGH data (Array Comparative Genomic Hybridization) with experimentally tested DNA copy number alterations by Snijders et al. (2001). This data can be downloaded from http://www.nature.com/ng/journal/v29/n3/suppinfo/ng754_S1.html. These array data sets consist

Table 10.2.: Simulation results for unknown number of change-points(11). Number in bold represents the method with the smallest total error. "-" indicates that no change-point is identified.

Error	$\mathcal{N}$	$\xi_1(\hat{\Gamma}_{est}, \Gamma_{true})$				$\xi_2(\Gamma_{true}, \hat{\Gamma}_{est})$			
		our method	PLR(PELT)	e.cp3o	e.divisive	our method	PLR(PELT)	e.cp3o	e.divisive
$\mathcal{N}(0,1)$	500	1.80(1.11)	5.55(8.94)	32.65(20.65)	2.00(0.86)	2.70(2.72)	1.10(0.72)	2.20(9.37)	2.00(0.86)
	1000	1.55(1.05)	1.10(1.41)	86.20(67.45)	1.70(0.86)	23.55(50.01)	1.10(1.41)	6.60(16.37)	10.50(36.01)
$\chi^2_{(3)}$	500	27.40(19.67)	9.45(8.61)	71.60(32.83)	29.60(19.29)	7.60(5.54)	39.15(27.86)	30.50(17.03)	5.40(5.04)
	1000	16.75(20.84)	10.00(6.15)	199.75(101.38)	16.20(15.17)	16.55(16.66)	91.75(47.19)	55.40(23.64)	8.95(8.27)
$Laplace(0,1)$	500	2.95(6.00)	2.30(3.71)	33.20(21.75)	2.00(1.17)	4.00(9.53)	1.00(1.52)	7.15(13.35)	2.95(3.56)
	1000	1.05(0.69)	0.75(0.64)	72.00(48.18)	1.65(0.67)	14.20(40.59)	0.75(0.64)	12.55(21.85)	3.10(6.37)
$Cauchy(0,1)$	500	42.75(26.73)	43.80(17.99)	121.90(59.00)	-	9.30(6.94)	72.55(16.68)	31.35(21.10)	-
	1000	23.80(25.80)	83.60(23.47)	202.55(74.73)	-	12.75(9.51)	138.60(39.64)	62.70(36.70)	-
$t(2)$	500	12.70(10.56)	10.40(6.49)	102.50(51.54)	7.70(6.45)	6.55(4.74)	51.65(26.07)	30.70(19.08)	5.60(4.79)
	1000	8.15(16.32)	12.35(13.98)	206.55(115.11)	3.60(1.88)	11.10(22.46)	102.20(40.93)	65.50(33.41)	3.75(2.02)
$Pareto(\alpha = 0.5)$	500	13.20(12.83)	36.80(12.95)	124.70(52.27)	-	9.35(6.84)	66.80(19.67)	27.60(16.27)	-
	1000	6.55(4.51)	95.70(40.05)	259.90(115.63)	-	13.05(20.95)	124.25(35.78)	70.20(34.80)	-
$Pareto(\alpha = 1.5)$	500	60.95(25.47)	46.60(16.00)	138.80(60.35)	-	10.60(6.89)	60.80(19.40)	31.95(17.69)	-
	1000	39.05(22.59)	98.45(51.44)	322.75(195.95)	-	19.25(21.19)	119.60(42.57)	73.90(35.24)	-
$\mathcal{LN}(0,1)$	500	6.65(5.78)	8.20(7.85)	91.25(51.07)	7.25(7.22)	4.60(7.19)	42.65(24.10)	31.60(16.09)	8.50(19.32)
	1000	3.55(1.88)	5.00(5.42)	207.35(96.31)	3.20(1.70)	10.75(21.88)	107.00(43.62)	66.25(30.46)	6.05(12.58)

Error	$\mathcal{N}$	Error of the Number of change-point				$\xi_1 + \xi_2$			
		our method	PLR(PELT)	e.cp3o	e.divisive	our method	PLR(PELT)	e.cp3o	e.divisive
$\mathcal{N}(0,1)$	500	0.20(0.52)	0.30(0.66)	6.20(0.41)	0.00(0.00)	4.5	6.65	34.85	**4**
	1000	0.70(0.73)	0.00(0.00)	6.55(1.00)	0.15(0.37)	25.1	**2.2**	92.8	12.2
$\chi^2_{(3)}$	500	2.30(0.98)	5.40(3.66)	6.95(1.05)	3.05(1.73)	**35**	48.6	102.1	**35**
	1000	0.00(0.86)	9.35(2.70)	8.00(1.21)	0.35(0.99)	33.3	101.75	255.15	**25.15**
$Laplace(0,1)$	500	0.00(0.56)	0.15(0.37)	6.15(0.67)	0.20(0.52)	6.95	**3.3**	40.35	4.95
	1000	0.40(0.60)	0.00(0.00)	6.30(0.86)	0.05(0.22)	15.25	**1.5**	84.55	4.75
$Cauchy(0,1)$	500	3.40(1.93)	13.00(0.00)	7.85(1.50)	9.90(0.97)	**52.05**	116.35	153.25	-
	1000	0.45(1.00)	13.00(0.00)	7.80(1.36)	9.65(1.04)	**36.55**	222.2	265.25	-
$t(2)$	500	0.80(1.06)	7.05(3.09)	7.75(1.02)	0.40(0.94)	19.25	62.05	133.2	**13.3**
	1000	0.10(0.64)	12.55(1.00)	7.60(1.35)	0.05(0.22)	19.25	114.55	272.05	**7.35**
$Pareto(\alpha = 0.5)$	500	0.40(0.99)	13.00(0.00)	8.00(1.38)	8.65(1.81)	**22.55**	103.6	152.3	-
	1000	0.25(0.55)	13.00(0.00)	8.25(1.07)	8.75(1.83)	**19.6**	219.95	330.1	-
$Pareto(\alpha = 1.5)$	500	4.50(1.32)	13.00(0.00)	7.75(1.21)	10.70(0.66)	**71.55**	107.4	170.75	-
	1000	1.10(1.02)	13.00(0.00)	8.00(1.34)	10.65(0.59)	**58.3**	218.05	396.65	-
$\mathcal{LN}(0,1)$	500	0.40(0.68)	6.65(3.25)	7.35(1.23)	0.35(0.93)	**11.25**	50.85	122.85	15.75
	1000	0.30(0.66)	11.80(1.47)	7.75(1.37)	0.10(0.45)	14.3	112	273.6	**9.25**

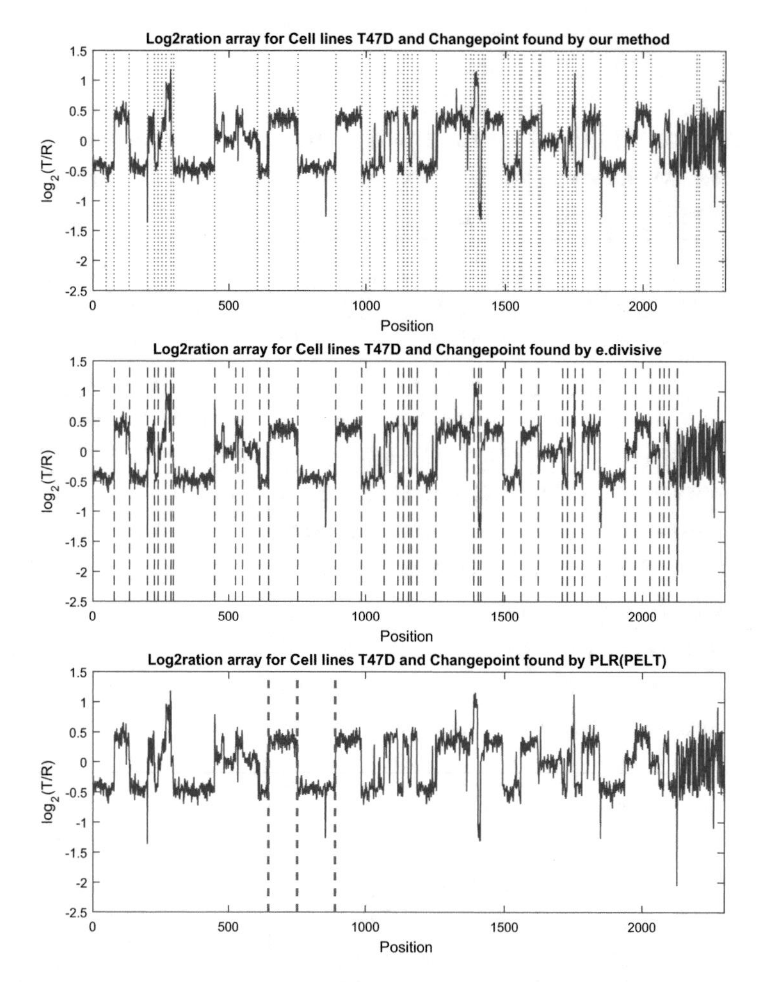

Figure 10.2.: Log2ration array for Cell lines T47D and change-points found by our method, e.divisive and PLR(PRLT). Vertical dotted lines are the change-points found by these method.

of single experiments on 15 human cell strains and 9 other cell lines. Each cell line is divided by chromosomes. Each array contains measurements for approximately 2200 BACs (bacterial artificial chromosomes) spotted in triplicates. The variable used for analysis is the normalized average of the log_2 test over reference ratio.

We first apply our method, as well as the e.divisive and PRL(PELT) methods to one of the cell lines labeled T47D with sample size $N = 2295$. To determine the number of change-points, our method uses the 5% significance level. The parametric likelihood approach (PLR) with PELT algorithm uses BIC penalty in the R package "change-point." The e.divisive method uses $\alpha = 1$, minimum segment length=2, and 5% significance level. Figure 10.2 flags the change-points found by these three methods. We can see that our method successfully locates most of the change-points especially those short but significantly distinct segments. Note that our algorithm won't flag any single extreme data point as a segment although our minimum length of a segment is set to be 1. Comparing with the e.divisive method, our method found 51 change-points while e.divisive found 43 change-points. Looking at the additional change-points found by our method, some of them such as three change-points after position 2200 are hard to find. In terms of the computational time spent in searching for the change-points, e.divisive took about 15 minutes while our method took about 2 minutes only in a PC with the same 3.6GHz. Surprisingly, the PLR(RELT) only finds 3 change-points and misses all of the important change-points with BIC penalty.

Continuing, we applied our method as well as the e.divisive method to 15 chromosomes in the data and all of which except GM07081 chromosome 15 are identified as having DNA copy number alterations by a method called spectral karyotyping. Figure 10.3 illustrates all change-points identified by our method and e.divisive based on 5% significance level. The results reveal that our method successfully finds those identified change-points in all chromosomes. Such was not the case for the e.divisive method. See, for example, GM03563 Chromosome 3, GM01750 chromosome 9, GM01535 chromosome 5, and GM0781 chromosome 7. Sometimes change-points can appear in the narrow regions at the beginnings or ends of the data such as GM01535 Chromosome 12. However, e.divisive, HMM, and CBS could not detect these kind of change-points at the beginnings or ends of the data (Chromosome 9 on GM03563 and Chromosome 12 on GM01535) while our method could. The LB method could not find the change-points in Chromosome 15 on GM07081 while our method found a relatively reasonable segment in the sequence. It should be noted that LB, HMM, and CBS are parametric methods relying on certain distributional assumptions. Our method is entirely nonparametric though embedded in a parametric setting. In addition, our method can find chromosomes with DNA copy number alterations which have not been confirmed by spectral karyotyping. They may either be false positives or represent real DNA copy number alterations which are undetectable due of the low resolution of spectral karyotyping.

10.5.2. Interest Rate Data

The monthly real interest rate data over the period from January 1980 to June 2015 ($N = 426$) are studied. Following Garcia and Perron (1996); Bai and Perron (2003), we focus on a more recent three-month treasury bill rate data deflated by the CPI. The data are downloaded from the US Board of Governors of the Federal Reserve System (http://www.federalreserve.gov/econresdata/default.htm) and US Bureau of Labor Statistics website (http://www.bls.gov/cpi/#tables).

We applied our method based on a 1% significance level to the real interest rate data set over 1980:1–2015:6 ($N = 426$) shown in Figure 10.4. To compare the methods, four economic recession periods and the positions of change-point found by our method, e.divisive and PLR(PLET) are shown in Figure 10.5. A summary of the change-point positions found by three methods for the four recession periods is also presented in Table 10.3.

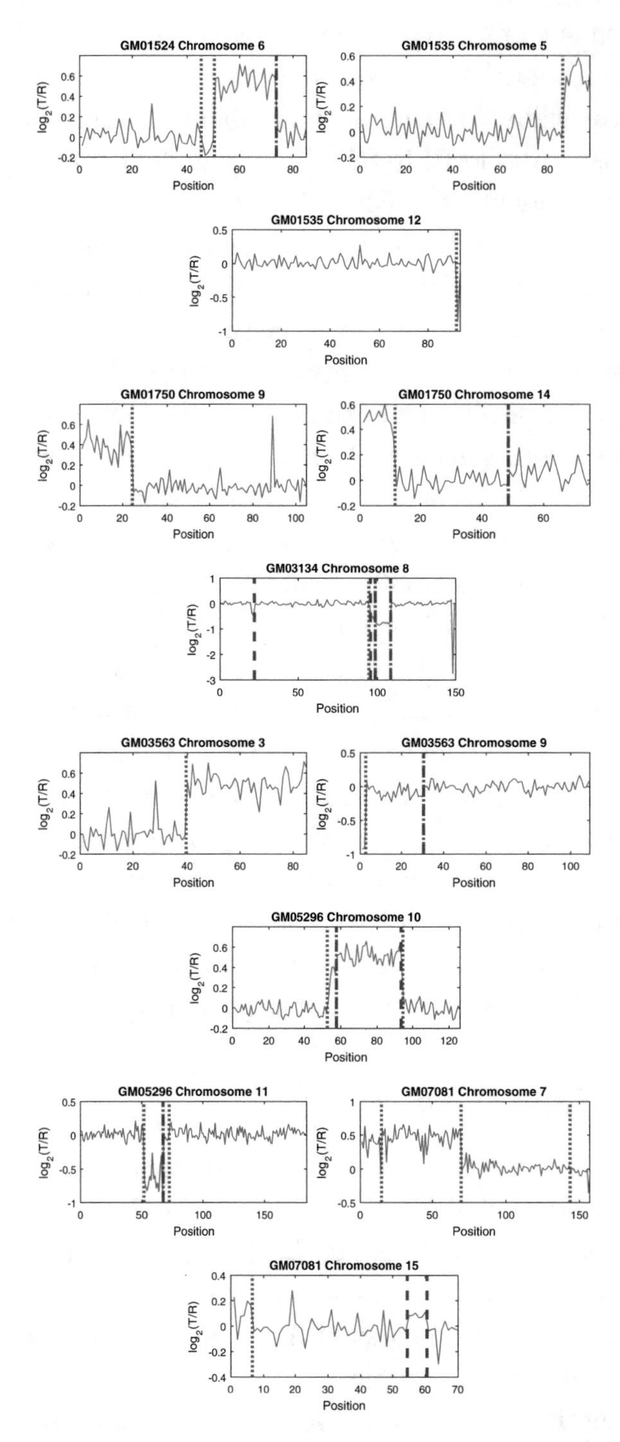

Figure 10.3.: The chromosomes with identified alterations and the change-points found by our method and e.divisive. Our results are shown by red dotted line. The results of e.divisive are shown by blue dashed line. The blue-red dash-dot line means the two methods find same change-point.

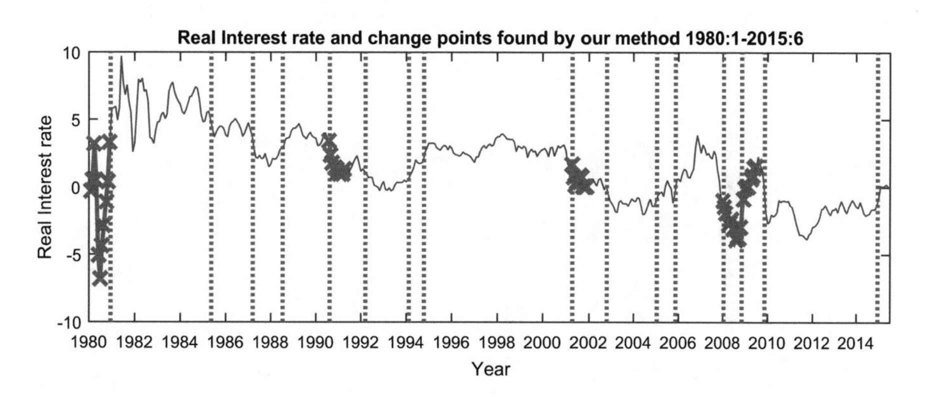

Figure 10.4.: The real interest rate data and change-points found by our method

Table 10.3.: Four recession periods and the positions of change-point found by our method, e.divisive and PLR(PLET). Numbers in bold refer to the method which successfully finds the starting point or the end point of the recession period.

Four recessions period	Position of change-point found		
	Our method	e.divisive	PLR(PLET)
1980:1–1980:11	**1980:11**	**1980:11**	1980:10
1990:7–1991:3	**1990:7**	1990:8	**1990:7**
2001:3–2001:11	**2001:3**	**2001:3**	**2001:3**
2007:12–2009:6	**2007:12**	2007:10	2007:10

From the first subgraph (1980:1–1984:12) in Figure 10.5, our method and e.divisive both successfully find one change-point at the end of the recession period in 1980:11. However, PLR(PELT) method falsely overestimates the number of change-points and falsely detects many change-points at the beginning of the data (during the period 1980–1981). This may be caused by the great variation and outliers at the beginning of the data set. From the second subgraph (1989:1–1993:12) in Figure 10.5, our method and PLR(PELT) successfully locate the change-point at the beginning of the recession period in 1990:7, while e.divisive method locates the change-point a bit late in 1990–1998. From the third subgraph (2000:1–2004:12) in Figure 10.5, all three methods successfully locate the change-point at the beginning of the recession period in 2001:03. From the fourth subgraph (2007:1–2011:12) in Figure 10.5, only our method successfully locates the change-point at the beginning of the recession period in 2007:12, while the other two methods find the change-point two months early due to the sudden fall of the interest rate data. Finally, we conclude that our method successfully locates the change-points over all four recession periods.

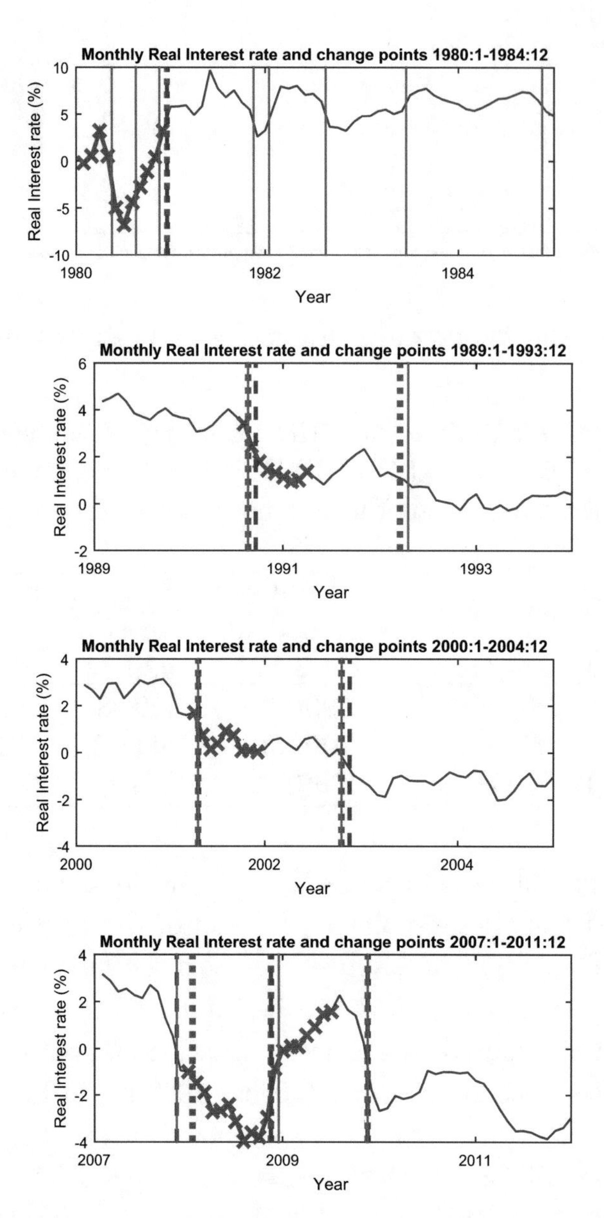

Figure 10.5.: The real interest rate data around four recession periods. Vertical lines are the change-points found by our method, e.divisive and PLR(PRLT). Red star points are the periods when U.S. business Recessions actually happened defined by US National Bureau of Economic Research. Our results are shown by red dotted line. The results of e.divisive are shown by blue dashed line. The results of PLR(PRLT) are shown by purple thin line. When different type lines coincide, it means different methods find same change-point. Details of the position found can be seen in Table 10.3.

Chapter Notes

Although our method assumes independence, we also conducted a series of simulation experiments where this assumption is violated (see Alvo et al. (2017)). We slightly changed the structure of the error terms (ε_i's) in the simulation setting to test the performance in this situation. The simulation results revealed that our method outperforms the other methods when the normality assumption is violated, particularly in the cases of heavy-tailed distributions. We also checked the independence assumption in the two real data-sets studied in Section 10.5. We first computed the residuals after removing the mean of each segment estimated by our method and then applied the Ljung–Box Q test to the residuals with maximum lag of 20. The p-value of the test is 0.0016 for the T47D cell line and 0.0001 for the real interest rate data, indicating that the independence assumptions for both data sets may not be valid. We conclude that our method is less sensitive to departures from serial independence.

11. Bayesian Models for Ranking Data

Ranking data are often encountered in practice when judges (or individuals) are asked to rank a set of t items, which may be political goals, candidates in an election, types of food, etc. We see examples in voting and elections, market research, and food preference just to name a few. By studying ranking data, we can understand the judges' perception and preferences on the ranked alternatives.

Let $\boldsymbol{R} = (R(1), \ldots, R(t))'$ be ranking t items, labeled $1, \ldots, t$. It will be more convenient to standardize the rankings as:

$$\boldsymbol{y} = \left(\boldsymbol{R} - \frac{t+1}{2}\mathbf{1}\right) / \sqrt{\frac{t(t^2-1)}{12}},$$

where $\boldsymbol{y}$ is the $t \times 1$ vector with $\|\boldsymbol{y}\| \equiv \sqrt{\boldsymbol{y}'\boldsymbol{y}} = 1$. We consider the following ranking model:

$$\pi(\boldsymbol{y}|\kappa, \boldsymbol{\theta}) = C(\kappa, \boldsymbol{\theta}) \exp\left\{\kappa\boldsymbol{\theta}'\boldsymbol{y}\right\},$$

where the parameter $\boldsymbol{\theta}$ is a $t \times 1$ vector with $\|\boldsymbol{\theta}\| = 1$, parameter $\kappa \geq 0$, and $C(\kappa, \boldsymbol{\theta})$ is the normalizing constant. In the case of the distance-based models (Alvo and Yu, 2014), the parameter $\boldsymbol{\theta}$ can be viewed as if a modal ranking vector. In fact, if $\boldsymbol{R}$ and $\boldsymbol{\mu}_0$ represent an observed ranking and the modal ranking of t items respectively, then the probability of observing $\boldsymbol{R}$ under the Spearman distance-based model is proportional to

$$\begin{aligned} \exp\left\{-\lambda\left(\frac{1}{2}\sum_{i=1}^{t}\left(R(i) - \boldsymbol{\mu}_0(i)\right)^2\right)\right\} &= \exp\left\{-\lambda\left(\frac{t(t+1)(2t+1)}{12} - \boldsymbol{\mu}_0'\boldsymbol{R}\right)\right\} \\ &\propto \exp\left\{\kappa\boldsymbol{\theta}'\boldsymbol{y}\right\}, \end{aligned}$$

where $\kappa = \lambda\frac{t(t^2-1)}{12}$, and $\boldsymbol{y}$ and $\boldsymbol{\theta}$ are the standardized rankings of $\boldsymbol{R}$ and $\boldsymbol{\mu}_0$, respectively. However, the $\boldsymbol{\mu}_0$ in the distance-based model is a discrete permutation vector of integers $\{1, 2, \ldots, t\}$ but the $\boldsymbol{\theta}$ in our model is a real-valued vector, representing a consensus view of the relative preference of the items from the individuals. Since both $\|\boldsymbol{\theta}\| = 1$ and $\|\boldsymbol{y}\| = 1$, the term $\boldsymbol{\theta}'\boldsymbol{y}$ can be seen as $\cos\phi$ where ϕ is the angle between the

M. Alvo, P. L. H. Yu, *A Parametric Approach to Nonparametric Statistics*,
Springer Series in the Data Sciences, https://doi.org/10.1007/978-3-319-94153-0_11

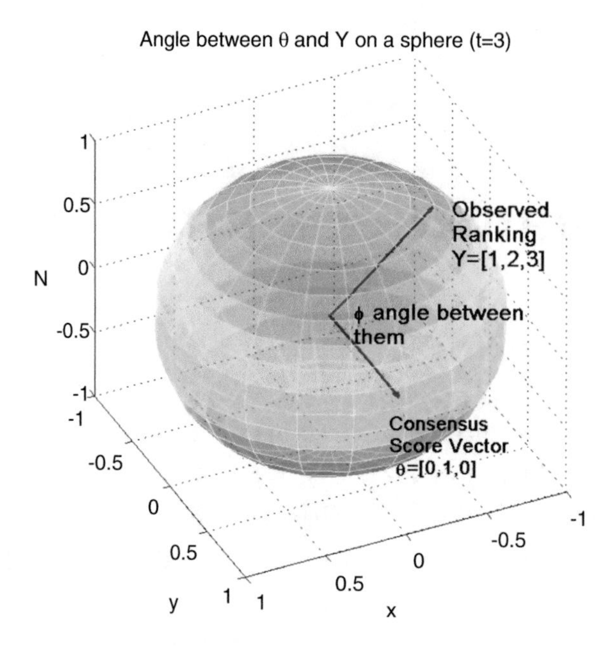

Figure 11.1.: Illustration for the angle between the consensus score vector $\boldsymbol{\theta} = (0, 1, 0)'$ and the standardized observation of $(1, 2, 3)'$ on the sphere when $t = 3$.

consensus score vector $\boldsymbol{\theta}$ and the observation $\boldsymbol{y}$. Figure 11.1 illustrates an example of the angle between the consensus score vector $\boldsymbol{\theta} = (0, 1, 0)'$ and the standardized observation of $\boldsymbol{R} = (1, 2, 3)'$ on the sphere for $t = 3$. The probability of observing a ranking is proportional to the cosine of the angle from the consensus score vector. The parameter κ can be viewed as a concentration parameter. For small κ, the distribution of rankings will appear close to a uniform whereas for larger values of κ, the distribution of rankings will be more concentrated around the consensus score vector. We call this new model as *angle-based ranking model.*

To compute the normalizing constant $C(\kappa, \boldsymbol{\theta})$, let P_t be the set of all possible permutations of the integers $1, \ldots, t$. Then

$$(C(\kappa, \boldsymbol{\theta}))^{-1} = \sum_{\boldsymbol{y} \in \mathrm{P}_t} \exp\left\{\kappa \boldsymbol{\theta}^T \boldsymbol{y}\right\}. \tag{11.1}$$

Notice that the summation is over the $t!$ elements in P_t. When t is large, says greater than 15, the exact calculation of the normalizing constant is prohibitive. Using the fact that the set of $t!$ permutations lie on a sphere in $(t-1)$-space, our model resembles the continuous von Mises-Fisher distribution, abbreviated as $vMF(\boldsymbol{x}|\boldsymbol{m}, \kappa)$, which is defined on a $(p-1)$ unit sphere with mean direction $\boldsymbol{m}$ and concentration parameter κ:

$$p(\boldsymbol{x}|\kappa, \boldsymbol{m}) = V_p(\kappa) \exp(\kappa \boldsymbol{m}' \boldsymbol{x}),$$

where

$$V_p(\kappa) = \frac{\kappa^{\frac{p}{2}-1}}{(2\pi)^{\frac{p}{2}} I_{\frac{p}{2}-1}(\kappa)},$$

and $I_a(\kappa)$ is the modified Bessel function of the first kind with order a. Consequently, we may approximate the sum in (11.1) by an integral over the sphere:

$$C(\kappa, \boldsymbol{\theta}) \simeq C_t(\kappa) = \frac{\kappa^{\frac{t-3}{2}}}{2^{\frac{t-3}{2}} t! I_{\frac{t-3}{2}}(\kappa)\Gamma(\frac{t-1}{2})},$$

where $\Gamma(.)$ is the gamma function. Table 11.1 shows the error rate of the approximate log-normalizing constant as compared to the exact one computed by direct summation. Here, κ is chosen to be 0.01 to 2 and t ranges from 4 to 11. Note that the exact calculation of the normalizing constant for $t = 11$ requires the summation of $11! \approx 3.9 \times 10^7$ permutation. The computer ran out of memory (16GB) beyond $t = 11$. This approximation seems to be very accurate even when $t = 3$. The error drops rapidly as t increases. Note that this approximation allows us to approximate the first and second derivatives of $\log C$ which can facilitate our computation in what follows.

Notice that κ may grow with t as $\boldsymbol{\theta}'\boldsymbol{y}$ is a sum of t terms. It can be seen from the applications in Section 11.4 that in one of the clusters for the APA data ($t = 5$), κ is $7.44 (\approx 1.5t)$ (see Table 11.4). We thus compute the error rate for $\kappa = t$ and $\kappa = 2t$ as shown in Figure 11.2. It is found that the approximation is still accurate with error rate of less than 0.5% for $\kappa = t$ and is acceptable for large t when $\kappa = 2t$ as the error rate decreases in t.

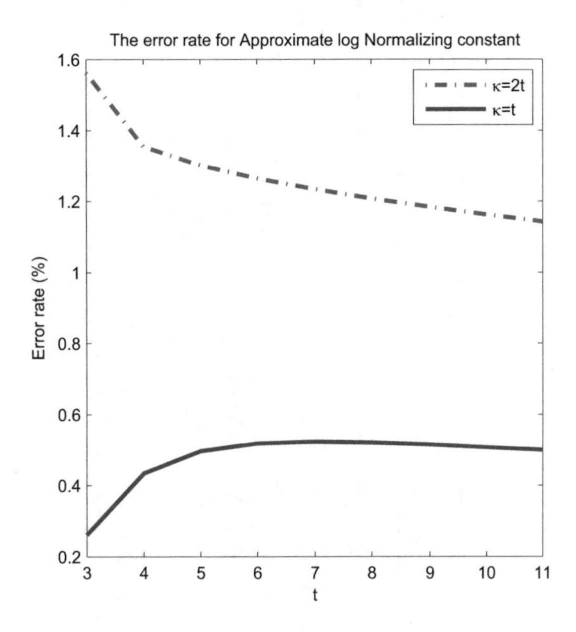

Figure 11.2.: The error rate of the approximate log-normalizing constant as compared to the exact one computed by direct summation for $\kappa = t$ and $\kappa = 2t$.

Table 11.1.: The error rate of the approximate log-normalizing constant as compared to the exact one computed by direct summation.

κ	t								
	3	4	5	6	7	8	9	10	11
0.01	$<$0.00001%	$<$0.00001%	$<$0.00001%	$<$0.00001%	$<$0.00001%	$<$0.00001%	$<$0.00001%	$<$0.00001%	$<$0.00001%
0.1	$<$0.00001%	$<$0.00001%	$<$0.00001%	$<$0.00001%	$<$0.00001%	$<$0.00001%	$<$0.00001%	$<$0.00001%	$<$0.00001%
0.5	0.00003%	0.00042%	0.00024%	0.00013%	0.00007%	0.00004%	0.00003%	0.00002%	0.00001%
0.8	0.00051%	0.00261%	0.00150%	0.00081%	0.00046%	0.00027%	0.00017%	0.00011%	0.00008%
1	0.00175%	0.00607%	0.00354%	0.00194%	0.00110%	0.00066%	0.00041%	0.00027%	0.00018%
2	0.05361%	0.06803%	0.04307%	0.02528%	0.01508%	0.00932%	0.00598%	0.00398%	0.00273%

11.1. Maximum Likelihood Estimation (MLE) of Our Model

Let $\boldsymbol{Y} = \{\boldsymbol{y}_1, \ldots, \boldsymbol{y}_N\}$ be a random sample of N standardized rankings drawn from $p(\boldsymbol{y}|\kappa, \boldsymbol{\theta})$. The log-likelihood of $(\kappa, \boldsymbol{\theta})$ is then given by

$$l(\kappa, \boldsymbol{\theta}) = N \log C_t(\kappa) + \sum_{i=1}^{N} \kappa \boldsymbol{\theta}' \boldsymbol{y}_i \tag{11.2}$$

Maximizing (11.2) subject to $\|\boldsymbol{\theta}\| = 1$ and $\kappa \geq 0$, we find that the maximum likelihood estimator of $\boldsymbol{\theta}$ is given by $\hat{\boldsymbol{\theta}}_{MLE} = \frac{\sum_{i=1}^{N} \boldsymbol{y}_i}{\|\sum_{i=1}^{N} \boldsymbol{y}_i\|}$, and $\hat{\kappa}$ is the solution of

$$A_t(\kappa) \equiv \frac{-C_t'(\kappa)}{C_t(\kappa)} = \frac{I_{\frac{t-1}{2}}(\kappa)}{I_{\frac{t-3}{2}}(\kappa)} = \frac{\left\|\sum_{i=1}^{N} \boldsymbol{y}_i\right\|}{N} \equiv r. \tag{11.3}$$

A simple approximation to the solution of (11.3) following Banerjee et al. (2005) is given by

$$\hat{\kappa}_{MLE} = \frac{r(t-1-r^2)}{1-r^2}.$$

A more precise approximation can be obtained from a few iterations of Newton's method. Using the method suggested by Sra (2012), starting from an initial value κ_0, we can recursively update κ by iteration:

$$\kappa_{i+1} = \kappa_i - \frac{A_t(\kappa_i) - r}{1 - A_t(\kappa_i)^2 - \frac{t-2}{\kappa_i} A_t(\kappa_i)}, \quad i = 0, 1, 2, \ldots.$$

11.2. Bayesian Method with Conjugate Prior and Posterior

Taking a Bayesian approach, we consider the following conjugate prior for $(\kappa, \boldsymbol{\theta})$ as

$$p(\kappa, \boldsymbol{\theta}) \propto [C_t(\kappa)]^{\nu_0} \exp\{\beta_0 \kappa \boldsymbol{m}_0' \boldsymbol{\theta}\}, \tag{11.4}$$

where $\|\boldsymbol{m}_0\| = 1$, $\nu_0, \beta_0 \geq 0$. Given $\boldsymbol{Y}$, the posterior density of $(\kappa, \boldsymbol{\theta})$ can be expressed by

$$p(\kappa, \boldsymbol{\theta}|\boldsymbol{Y}) \propto \exp\{\beta \kappa \boldsymbol{m}' \boldsymbol{\theta}\} V_t(\beta\kappa) \cdot \frac{[C_t(\kappa)]^{N+\nu_0}}{V_t(\beta\kappa)},$$

where $\boldsymbol{m} = \left(\beta_0 \boldsymbol{m_0} + \sum_{i=1}^N \boldsymbol{y}_i\right)\beta^{-1}$, $\beta = \left\|\beta_0 \boldsymbol{m}_0 + \sum_{i=1}^N \boldsymbol{y}_i\right\|$. The posterior density can be factored as

$$p(\kappa, \theta|\boldsymbol{Y}) = p(\theta|\kappa, \boldsymbol{Y})p(\kappa|\boldsymbol{Y}) \tag{11.5}$$

where $p(\boldsymbol{\theta}|\kappa, \boldsymbol{Y}) \sim vMF(\boldsymbol{\theta}|\boldsymbol{m}, \beta\kappa)$ and

$$p(\kappa|\boldsymbol{Y}) \propto \frac{[C_t(\kappa)]^{N+\nu_0}}{V_t(\beta\kappa)} = \frac{\kappa^{\frac{t-3}{2}(\upsilon_0+N)} I_{\frac{t-2}{2}}(\beta\kappa)}{\left[I_{\frac{t-3}{2}}(\kappa)\right]^{\nu_0+N} (\beta\kappa)^{\frac{t-2}{2}}}.$$

The normalizing constant for $p(\kappa|\boldsymbol{Y})$ is not available in closed form. Nunez-Antonio and Gutiérrez-Pena (2005) suggested using a sampling-importance-resampling (SIR) procedure with a proposal density chosen to be the gamma density with mean $\hat{\kappa}_{MLE}$ and variance equal to some prespecified number such as 50 or 100. However, in a simulation study, it was found that the choice of this variance is crucially related to the performance of SIR. An improper choice of variance may lead to slow or unsuccessful convergence. Also the MCMC method leads to intensive computational complexity. Furthermore, when the sample size N is large, $\beta\kappa$ can be very large which complicates the computation of the term $I_{\frac{t-2}{2}}(\beta\kappa)$ in $V_t(\beta\kappa)$. Thus the calculation of the weights in the SIR method will fail when N is large. We conclude that in view of the difficulties for directly sampling from $p(\kappa|\boldsymbol{Y})$, it may be preferable to approximate the posterior distribution with an alternative method known as variational inference (abbreviated VI from here on).

11.3. Variational Inference

Variational inference provides a deterministic approximation to an intractable posterior distribution through optimization. We first adopt a joint vMF-Gamma distribution as the prior for $(\kappa, \boldsymbol{\theta})$:

$$\begin{aligned} p(\kappa, \boldsymbol{\theta}) &= p(\boldsymbol{\theta}|\kappa)p(\kappa) \\ &= vMF(\boldsymbol{\theta}|\boldsymbol{m}_0, \beta_0\kappa)Gamma(\kappa|a_0, b_0), \end{aligned}$$

where $Gamma(\kappa|a_0, b_0)$ is the Gamma density function with shape parameter a_0 and rate parameter b_0 (i.e., mean equal to $\frac{a_0}{b_0}$), and $p(\boldsymbol{\theta}|\kappa) = vMF(\boldsymbol{\theta}|\boldsymbol{m}_0, \beta_0\kappa)$. The choice of $Gamma(\kappa|a_0, b_0)$ for $p(\kappa)$ is motivated by the fact that for large values of κ, $p(\kappa)$ in (11.4) tends to take the shape of a Gamma density. In fact, for large values of κ, $I_{\frac{t-3}{2}}(\kappa) \simeq \frac{e^\kappa}{\sqrt{2\pi\kappa}}$, and hence $p(\kappa)$ becomes the Gamma density with shape $(\nu_0 - 1)\frac{t-2}{2} + 1$

and rate $\nu_0 - \beta_0$:

$$p(\kappa) \propto \frac{[C_t(\kappa)]^{\nu_0}}{V_t(\beta_0\kappa)} \propto \kappa^{(\nu_0-1)\frac{t-2}{2}} \exp(-(\nu_0 - \beta_0)\kappa).$$

11.3.1. Optimization of the Variational Distribution

In the variational inference framework, we aim to determine q so as to minimize the Kullback-Leibler (KL) divergence between $p(\boldsymbol{\theta}, \kappa|\boldsymbol{Y})$ and $q(\boldsymbol{\theta}, \kappa)$:

$$KL\,(q|p) = E_q\left[\ln \frac{q\,(\boldsymbol{\theta}, \kappa)}{p\,(\boldsymbol{\theta}, \kappa|\boldsymbol{Y})}\right].$$

This can be shown to be equivalent to maximizing the evidence lower bound (ELBO) (Blei et al., 2017). So the optimization of the variational factors $q(\boldsymbol{\theta}|\kappa)$ and $q(\kappa)$ is performed by maximizing the evidence lower bound $\mathcal{L}(q)$ with respect to q on the log-marginal likelihood, which in our model is given by

$$\begin{aligned}\mathcal{L}(q) &= E_{q(\boldsymbol{\theta},\kappa)}\left[\ln \frac{p(\boldsymbol{Y}|\kappa, \boldsymbol{\theta})p(\boldsymbol{\theta}|\kappa)p(\kappa)}{q(\boldsymbol{\theta}|\kappa)q(\kappa)}\right] \qquad (11.6)\\ &= E_{q(\boldsymbol{\theta},\kappa)}\left[f(\boldsymbol{\theta}, \kappa)\right] - E_{q(\boldsymbol{\theta},\kappa)}\left[\ln q(\boldsymbol{\theta}|\kappa)\right] - E_{q(\kappa)}\left[\ln q(\kappa)\right] + constant,\end{aligned}$$

where all the expectations are taken with respect to $q(\boldsymbol{\theta}, \kappa)$ and

$$\begin{aligned}f(\boldsymbol{\theta}, \kappa) = &\sum_{i=1}^{N} \kappa\boldsymbol{\theta}'\boldsymbol{y}_i + N\left(\frac{t-3}{2}\right)\ln\kappa - N\ln I_{\frac{t-3}{2}}(\kappa) + \kappa\beta_0\boldsymbol{m}_0'\boldsymbol{\theta}\\ &+ \left(\frac{t-2}{2}\right)\ln\kappa - \ln I_{\frac{t-2}{2}}(\kappa\beta_0) + (a_0 - 1)\ln\kappa - b_0\kappa.\end{aligned}$$

For fixed κ, the optimal posterior distribution ln $q^*(\boldsymbol{\theta}|\kappa)$ is ln $q^*(\boldsymbol{\theta}|\kappa) = \kappa\beta_0\boldsymbol{m}_0'\boldsymbol{\theta} + \sum_{i=1}^{N} \kappa\boldsymbol{\theta}'\boldsymbol{y}_i + constant$. We recognize $q^*(\boldsymbol{\theta}|\kappa)$ as a von Mises-Fisher distribution $vMF(\boldsymbol{\theta}|\boldsymbol{m}, \kappa\beta)$ where

$$\beta = \left\|\beta_0\boldsymbol{m}_0 + \sum_{i=1}^{N}\boldsymbol{y}_i\right\| \quad \text{and} \quad \boldsymbol{m} = \left(\beta_0\boldsymbol{m}_0 + \sum_{i=1}^{N}\boldsymbol{y}_i\right)\beta^{-1}.$$

Let $g(\kappa)$ denote the remaining terms in $f(\boldsymbol{\theta}, \kappa) - \ln q(\boldsymbol{\theta}|\kappa)$ which only involve κ:

$$g(\kappa) = \left[N\left(\frac{t-3}{2}\right) + a_0 - 1\right]\ln\kappa - b_0\kappa - N\ln I_{\frac{t-3}{2}}(\kappa) - \ln I_{\frac{t-2}{2}}(\kappa\beta_0) + \ln I_{\frac{t-2}{2}}(\kappa\beta).$$

It is still difficult to maximize $E_{q(\kappa)}\left[g(\kappa)\right] - E_{q(\kappa)}\left[\ln q(\kappa)\right]$ since it involves the evaluation of the expected modified Bessel function. Following the similar idea in Taghia et al.

(2014), we first find a tight lower bound $\underline{g(\kappa)}$ for $g(\kappa)$ so that

$$\mathcal{L}(q) \geq \underline{\mathcal{L}(q)} = E_{q(\kappa)}\left[\underline{g(\kappa)}\right] - E_{q(\kappa)}\left[\ln q(\kappa)\right] + constant.$$

From the properties of the modified Bessel function of the first kind, it is known that the function $\ln I_\nu(x)$ is strictly concave with respect to x and strictly convex relative to $\ln x$ for all $\nu > 0$. Then, we can have the following two inequalities:

$$\ln\, I_\nu(x) \leq \ln\, I_\nu(\bar{x}) + \left(\frac{\partial}{\partial x}\ln I_\nu(\bar{x})\right)(x - \bar{x}), \tag{11.7}$$

$$\ln\, I_\nu(x) \geq \ln\, I_\nu(\bar{x}) + \left(\frac{\partial}{\partial x}\ln I_\nu(\bar{x})\right)\bar{x}(\ln x - \ln \bar{x}), \tag{11.8}$$

where $\frac{\partial}{\partial x}\ln I_\nu(\bar{x})$ is the first derivative of $\ln I_\nu(x)$ evaluated at $x = \bar{x}$. Applying inequality (11.7) for $\ln I_{\frac{t-3}{2}}(\kappa)$ and inequality (11.8) for $\ln I_{\frac{t-2}{2}}(\kappa\beta_0)$, we have

$$\begin{aligned}
g(\kappa) \geq \underline{g(\kappa)} = &\left[N\left(\frac{t-3}{2}\right) + a_0 - 1\right]\ln\kappa - b_0\kappa + \ln I_{\frac{t-2}{2}}(\beta\bar{\kappa}) \\
&+ \frac{\partial}{\partial\beta\kappa}\ln I_{\frac{t-2}{2}}(\beta\bar{\kappa})\beta\bar{\kappa}\left(\ln\beta\kappa - \ln\beta\bar{\kappa}\right) - N\ln I_{\frac{t-3}{2}}(\bar{\kappa}) \\
&- N\frac{\partial}{\partial\kappa}\ln I_{\frac{t-3}{2}}(\bar{\kappa})\left(\kappa - \bar{\kappa}\right) - \ln I_{\frac{t-2}{2}}(\beta_0\bar{\kappa}) - \frac{\partial}{\partial\beta_0\kappa}\ln I_{\frac{t-2}{2}}(\beta_0\bar{\kappa})\beta_0\left(\kappa - \bar{\kappa}\right).
\end{aligned}$$

Since the equality holds when $\kappa = \bar{\kappa}$, we see that the lower bound of $\mathcal{L}(q)$ is tight. Rearranging the terms, we have the approximate optimal solution as $\ln q^*(\kappa) = (a - 1)\ln\kappa - b\kappa + constant$, where

$$a = a_0 + N\left(\frac{t-3}{2}\right) + \beta\bar{\kappa}\left[\frac{\partial}{\partial\beta\kappa}\ln I_{\frac{t-2}{2}}(\beta\bar{\kappa})\right], \tag{11.9}$$

$$b = b_0 + N\frac{\partial}{\partial\kappa}I_{\frac{t-3}{2}}(\bar{\kappa}) + \beta_0\left[\frac{\partial}{\partial\beta_0\kappa}\ln I_{\frac{t-2}{2}}(\beta_0\bar{\kappa})\right]. \tag{11.10}$$

We recognize $q^*(\kappa)$ to be a $Gamma(\kappa|a, b)$ with shape a and rate b. Finally, the posterior mode $\bar{\kappa}$ can be obtained from the previous iteration as:

$$\bar{\kappa} = \begin{cases} \frac{a-1}{b} & \text{if } a > 1, \\ \frac{a}{b} & \text{otherwise.} \end{cases} \tag{11.11}$$

11.3.2. Comparison of the True Posterior Distribution and Its Approximation Obtained by Variational Inference

Since we use a factorized approximation for the posterior distribution in the variational inference approach, it is of interest to compare the true posterior distribution with its approximation obtained using the variational inference approach. We simulated two data sets with $\kappa = 1$, $\boldsymbol{\theta} = (-0.71, 0, 0.71)'$, $t = 3$ and different data sizes of $N = 20$ and $N = 100$. We generated samples from the posterior distribution by SIR method in Section 11.2 using a gamma density with mean $\hat{\kappa}_{MLE}$ and variance equal to 0.2 as the proposal density. We then applied the above variational inference to generate samples from the posterior distribution. Figure 11.3 exhibits the histograms and box-plots for the posterior distributions of κ and $\boldsymbol{\theta}$ for different settings.

From Figure 11.3, we see that the posterior distribution using the Bayesian-VI is very close to the posterior distribution obtained by the Bayesian-SIR method. When the sample size is small ($N = 20$), there are more outliers for the Bayesian-SIR method while the posterior κ for the Bayesian-VI method seems to be more concentrated. When the sample size is large, the posterior estimates of $\boldsymbol{\theta}$ and κ become more accurate and Bayesian-VI is closer to the posterior distribution obtained by the Bayesian-SIR method.

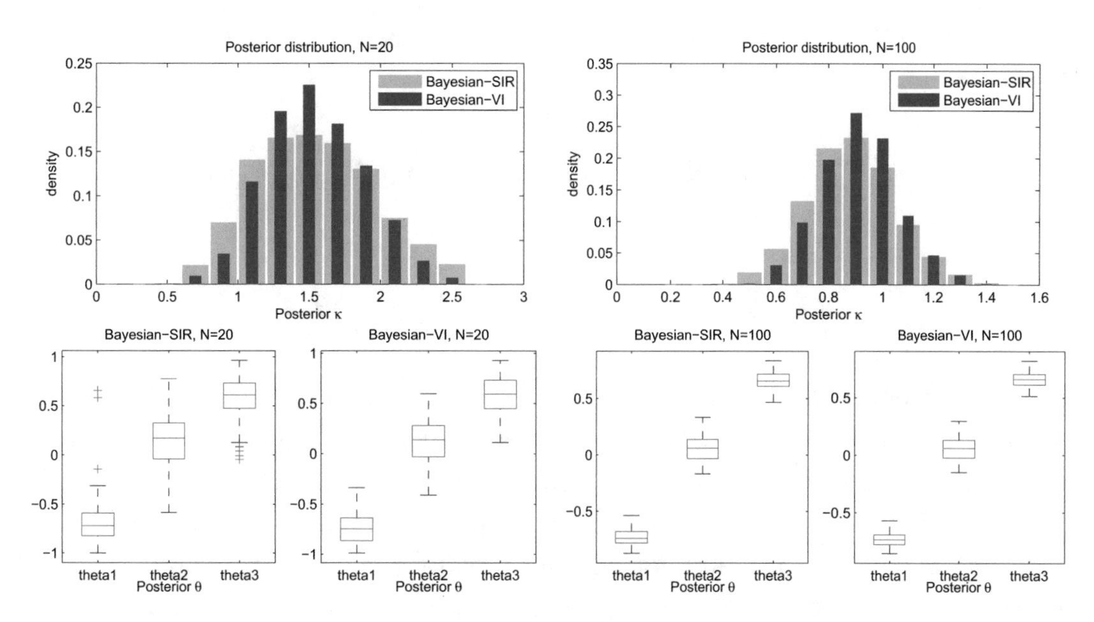

Figure 11.3.: Comparison of the posterior distribution obtained by Bayesian SIR method and the approximate posterior distribution by variational inference approach. The comparison is illustrated for different data sizes of $N = 20$ (left) and $N = 100$ (right).

11.3.3. Angle-Based Model for Incomplete Rankings

A judge may rank a set of items in accordance with some criteria. However, in real life, some of the ranking data may be missing either at random or by design. For example, in the former case, some of the items may not be ranked due to the limited knowledge of the judges. In this kind of incomplete ranking data, a missing item could have any rank and this is called subset rankings. In another instance called top-k rankings, the judges may only rank the top 10 best movies among several recommended. The unranked movies would in principle receive ranks larger than 10. In those cases, the notation $\boldsymbol{R}^I = (2, -, 3, 4, 1)'$ refers to a subset ranking with item 2 unranked while $\boldsymbol{R}^I = (2, *, *, *, 1)'$ represents a top two ranking with item 5 ranked first and item 1 ranked second.

In the usual Bayesian framework, missing data problems can be resolved by appealing to Gibbs sampling and data augmentation methods. Let $\{\boldsymbol{R}_1^I, \ldots, \boldsymbol{R}_N^I\}$ be a set of N observed incomplete rankings, and let $\{\boldsymbol{R}_1^*, \ldots, \boldsymbol{R}_N^*\}$ be their unobserved complete rankings. We want to have the following posterior distribution:

$$p(\boldsymbol{\theta}, \kappa | \boldsymbol{R}_1^I, \ldots, \boldsymbol{R}_N^I) \propto p(\boldsymbol{\theta}, \kappa) p(\boldsymbol{R}_1^I, \ldots, \boldsymbol{R}_N^I | \boldsymbol{\theta}, \kappa),$$

which can be achieved by Gibbs sampling based on the following two full conditional distributions:

$$p(\boldsymbol{R}_1^*, \ldots, \boldsymbol{R}_N^* | \boldsymbol{R}_1^I, \ldots, \boldsymbol{R}_N^I, \boldsymbol{\theta}, \kappa) = \prod_{i=1}^{N} p(\boldsymbol{R}_i^* | \boldsymbol{R}_i^I, \boldsymbol{\theta}, \kappa),$$

$$p(\boldsymbol{\theta}, \kappa | \boldsymbol{R}_1^*, \ldots, \boldsymbol{R}_N^*) \propto p(\boldsymbol{\theta}, \kappa) \prod_{i=1}^{N} p(\boldsymbol{R}_i^* | \boldsymbol{\theta}, \kappa).$$

Sampling from $p(\boldsymbol{R}_1^*, \ldots, \boldsymbol{R}_N^* | \boldsymbol{R}_1^I, \ldots, \boldsymbol{R}_N^I, \boldsymbol{\theta}, \kappa)$ can be generated by using the Bayesian SIR method or the Bayesian VI method which have been discussed in the previous sections. More concretely, we need to fill in the missing ranks for each observation and for that we appeal to the concept of compatibility described in Alvo and Yu (2014) which considers for an incomplete ranking, the class of complete order preserving rankings. For example, suppose we observe one incomplete subset ranking $\boldsymbol{R}^I = (2, -, 3, 4, 1)'$. The set of corresponding compatible rankings is $\{(2,5,3,4,1)', (2,4,3,5,1)', (2,3,4,5,1)', \ (3,2,4,5,1)', (3,1,4,5,2)'\}$.

Generally speaking, let $\Omega(\boldsymbol{R}_i^I)$ be the set of complete rankings compatible with $\boldsymbol{R}_i^I$. For an incomplete subset ranking with k out of t items being ranked, we will have a total $t!/k!$ complete rankings in its compatible set. Note that $p(\boldsymbol{R}_i^* | \boldsymbol{R}_i^I, \boldsymbol{\theta}, \kappa) \propto p(\boldsymbol{R}_i^* | \boldsymbol{\theta}, \kappa)$, $\boldsymbol{R}_i^* \in \Omega(\boldsymbol{R}_i^I)$. Obviously, direct sampling from this distribution will be tedious for large t. Instead, we use the Metropolis-Hastings algorithm to draw samples from this distribution with the proposed candidates generated uniformly from $\Omega(\boldsymbol{R}_i^I)$. The idea of introducing compatible rankings allows us to treat different kinds of incomplete rankings easily. It is easy to sample uniformly from the compatible rankings since we just need to fill-in the missing ranks under different situations. In the case of top-k rankings, the compatibility set will be defined to ensure that the unranked items receive rankings larger than k. Note that it is also possible to use Monte Carlo EM approach to handle incomplete rankings under a maximum likelihood setting where the Gibbs sampling is used in the E-step (see Yu et al. (2005)).

11.4. Applications

11.4.1. Sushi Data Sets

We investigate the two data sets of Kamishima (2003) for finding the difference in food preference patterns between eastern and western Japan. Historically, western Japan has been mainly affected by the culture of the Mikado emperor and nobles, while eastern Japan has been the home of the Shogun and Samurai warriors. Therefore, the preference patterns in food are different between these two regions (Kamishima, 2003).

The first data set consists of complete rankings of $t = 10$ different kinds of sushi given by 5000 respondents according to their preference. The region of respondents is also recorded ($N = 3285$ for eastern Japan, 1715 for western Japan). We apply the MLE, Bayesian-SIR, and Bayesian-VI on both eastern and western Japan data. We chose non-informative priors for both Bayesian-SIR and Bayesian-VI. Specifically, the prior parameter $\boldsymbol{m}_0$ is chosen uniformly whereas β_0, a_0, and b_0 are chosen to be small numbers close to zero. Since the sample size N is quite large compared to t, the estimated models for all three methods are almost the same. Figure 11.4 compares the posterior means of $\boldsymbol{\theta}$ between eastern Japan (blue bar) and western Japan (red bar) obtained by Bayesian-VI method. Note that the more negative value of θ_i means that the more preferable sushi i is. From Figure 11.4, we see that the main difference for sushi preference between eastern and western Japan occurs in salmon roe, squid, sea eel, shrimp, and tuna. People in eastern Japan have a greater preference for salmon roe and tuna than the western Japanese. On the other hand, the latter have a greater preference for squid, shrimp, and sea eel. Table 11.2 shows the posterior parameter obtained by Bayesian-VI. It can be seen that the eastern Japanese are slightly more cohesive than western Japanese since the posterior mean of κ is larger.

The second data set contains incomplete rankings given by 5000 respondents who were asked to pick and rank some of the $t = 100$ different kinds of sushi according to their preference, and most of them only selected and ranked the top 10 out of 100 sushi. Figure 11.5 compares the box-plots of the posterior means of $\boldsymbol{\theta}$ between eastern Japan (blue box) and western Japan (red box) obtained by Bayesian-VI. The posterior

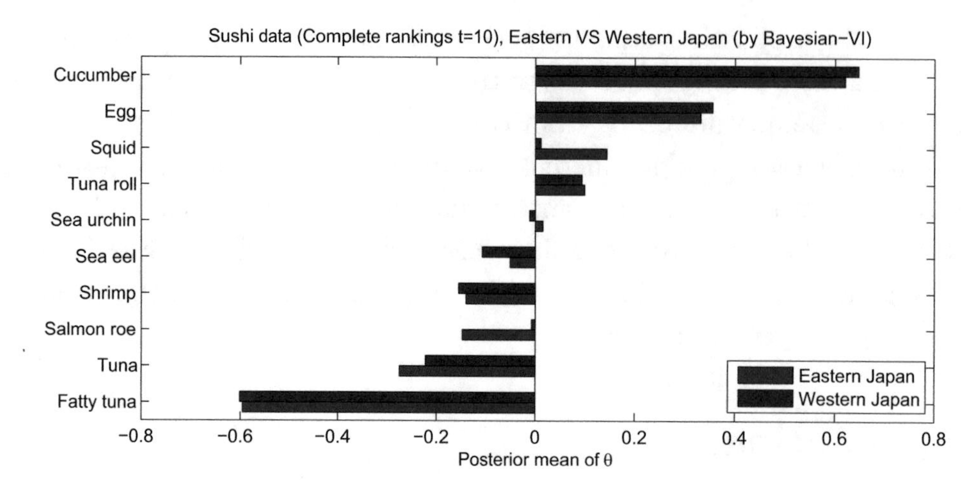

Figure 11.4.: Posterior means of $\boldsymbol{\theta}$ for the sushi complete ranking data ($t = 10$) in eastern Japan (blue bar) and western Japan (red bar) obtained by Bayesian-VI.

Table 11.2.: Posterior parameters for the sushi complete ranking data ($t = 10$) in eastern Japan and western Japan obtained by Bayesian-VI.

Posterior Parameter	Eastern Japan	Western Japan
β	1458.85	741.61
a	18509.84	9462.70
b	3801.57	2087.37
Posterior Mean of κ	4.87	4.53

distribution of $\boldsymbol{\theta}$ is based on the Gibbs samplings after dropping the first 200 samples during the burn-in period. Since there are too many kinds of sushi, this graph doesn't allow us to show the name of each sushi. However, we can see that about one-third of the 100 kinds of sushi have fairly large posterior means of θ_i and their values are pretty close to each other. This is mainly because these sushi are less commonly preferred by Japanese and the respondents hardly chose these sushi in their list. As these sushi are usually not ranked as top 10, it is natural to see that the posterior distributions of their θ_i's tend to have a larger variance.

From Figure 11.5, we see that there exists a greater difference between eastern and western Japan for small θ_i's. Figure 11.6 compares the box-plots of the top 10 smallest posterior means of $\boldsymbol{\theta}$ between eastern Japan (blue box) and western Japan (red box). The main difference for sushi preference between eastern and western Japan appears to be in sea eel, salmon roe, tuna, sea urchin, and sea bream. The eastern Japanese prefer salmon roe, tuna, and sea urchin sushi more than the western Japanese, while the latter like sea eel and sea bream more than the former. Generally speaking, tuna and sea urchin are more oily food, while salmon roe and tuna are more seasonal food. So from the analysis of both data sets, we can conclude that the eastern Japanese usually prefer more oily and seasonal food than the western Japanese (Kamishima, 2003).

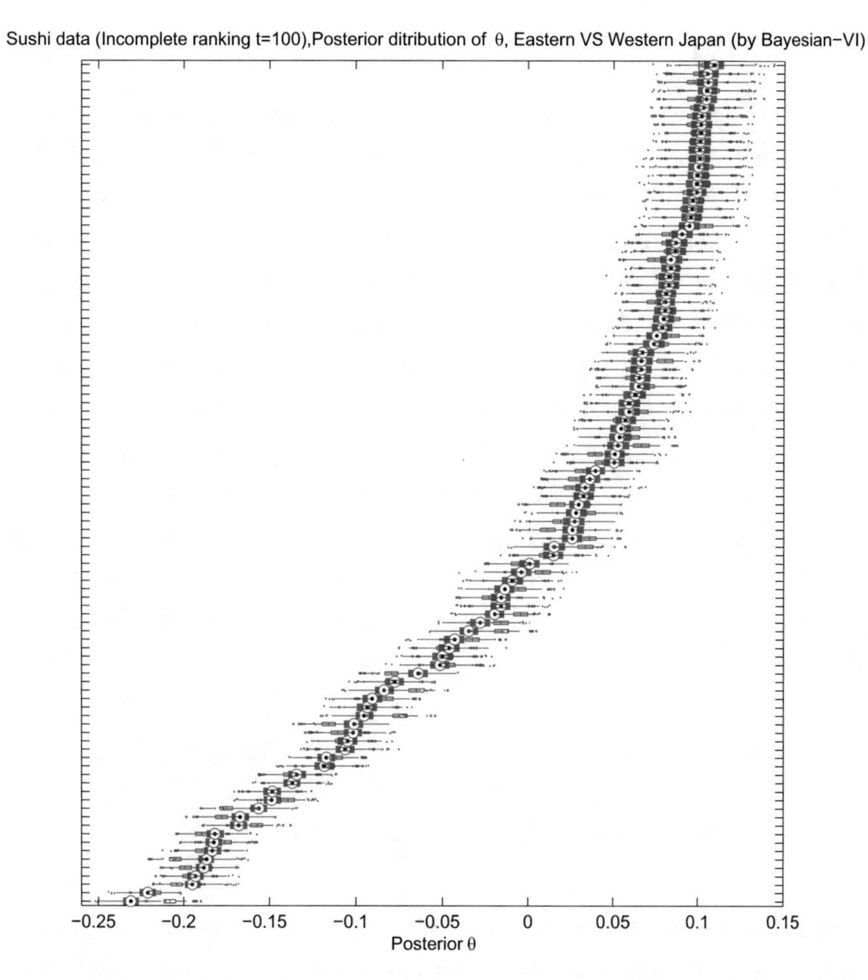

Figure 11.5.: Box-plots of the posterior means of $\boldsymbol{\theta}$ for the sushi incomplete rankings ($t = 100$) in eastern Japan (blue box-plots) and western Japan (red box-plots) obtained by Bayesian-VI.

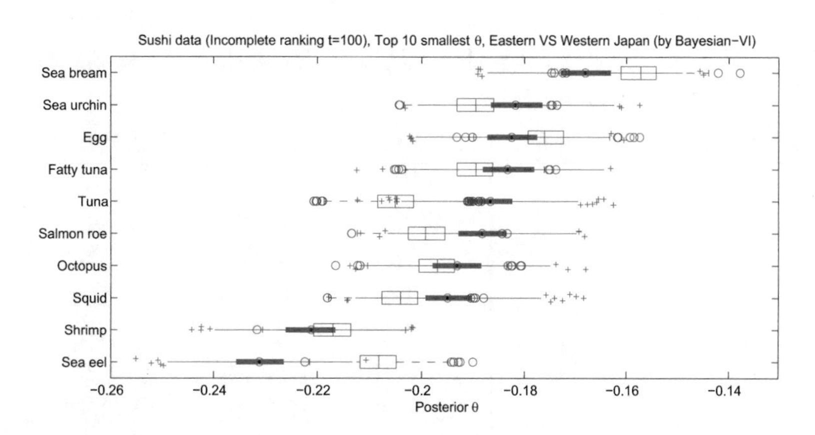

Figure 11.6.: Box-plots of the top 10 smallest posterior means of $\boldsymbol{\theta}$ for the sushi incomplete rankings ($t = 100$) in eastern Japan (blue box-plots and blue circles for outliers) and western Japan (red box-plots and red pluses for outliers) obtained by Bayesian-VI.

11.4.2. APA Data

We revisit the well-known APA data set of Diaconis (1988a) which contains 5738 full rankings of 5 candidates for the presidential election of the American Psychological Association (APA) in 1980. For this election, members of APA had to rank five candidates {A,B,C,D,E} in the order of their preference. Candidates A and C are research psychologists, candidates D and E are clinical psychologists, and candidate B is a community psychologist. This data set has been studied by Diaconis (1988a) and Kidwell et al. (2008) who found that the voting population was divided into 3 clusters.

We fit the data using a mixture of G angle-based models, see Xu et al. (2018) for the details. We chose a non-informative prior for the Bayesian-VI method for a different number of clusters $G = 1$ to 5. Specifically, the prior parameter $\boldsymbol{m}_{0g}$ is a randomly chosen unit vector whereas β_{0g}, d_{0g}, a_{0g}, and b_{0g} are chosen as random numbers close to zero. The p_{ig} are initialized as $\frac{1}{G}$. Table 11.3 shows the Deviance information criterion (DIC) for $G = 1$ to 5. It can be seen that the mixture model with $G = 3$ clusters attains the smallest DIC.

Table 11.3.: Deviance information criterion (DIC) for the APA ranking data.

G	1	2	3	4	5
DIC	54827	53497	53281	53367	53375

Table 11.4 indicates the posterior parameters for the three-cluster solution and Figure 11.7 exhibits the posterior means of $\boldsymbol{\theta}$ for the three clusters obtained by Bayesian-VI. It is very interesting to see that Cluster 1 votes clinical psychologists D and E as their first and second choices and dislike especially the research psychologist C. Cluster 2

Table 11.4.: Posterior parameters for the APA ranking data ($t = 5$) for three clusters obtained by Bayesian-VI.

Posterior Parameter	Cluster 1	Cluster 2	Cluster 3
$\boldsymbol{m}$	0.06	-0.44	0.26
	0.02	0.19	0.14
	0.78	-0.64	-0.75
	-0.54	0.49	0.55
	-0.33	0.39	-0.19
β	1067.10	1062.34	414.74
d	3231.09	1317.21	1189.72
a	4756.33	9224.97	1821.73
b	3330.45	1239.41	1197.80
Posterior mean of $\boldsymbol{\kappa}$	1.43	7.44	1.52
Posterior mean of $\boldsymbol{\tau}$	56.31%	22.96%	20.73%

prefers research psychologists A and C but dislikes the others. Cluster 3 prefers research psychologist C. From Table 11.4, Cluster 1 represents the majority (posterior mean of $\tau_1 = 56.31\%$). Cluster 2 is small but more cohesive since the posterior mean of κ_2 is larger. Cluster 3 has a posterior mean of $\tau_3 = 20.73\%$ and κ_3 is 1.52. The preferences of the five candidates made by the voters in the three clusters are heterogeneous and the mixture model enables us to draw further inference from the data.

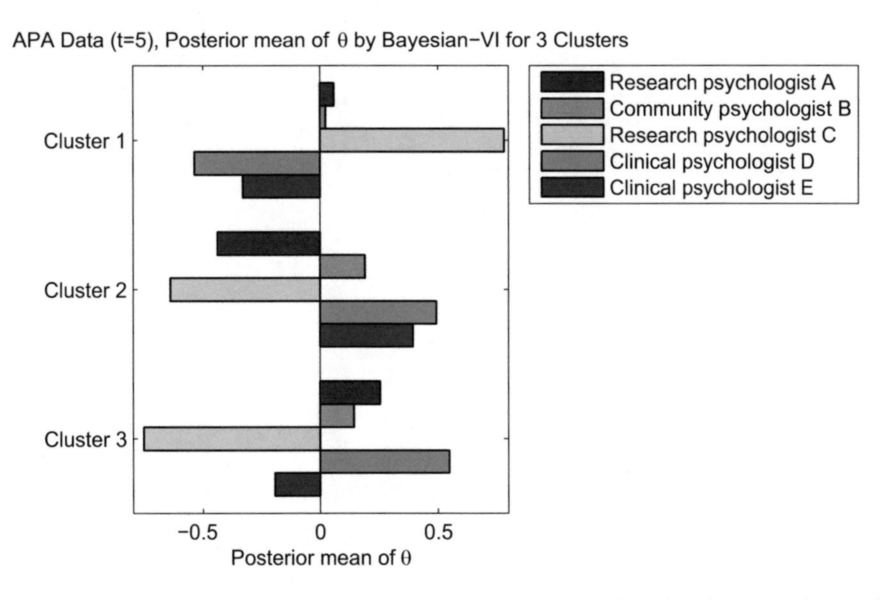

Figure 11.7.: Comparison of the posterior mean of θ for the APA data ($t = 5$) for three clusters obtained by Bayesian-VI.

Chapter Notes

We proposed a new class of general exponential ranking model called angle-based ranking models. The model assumed a consensus score vector $\boldsymbol{\theta}$ where the rankings reflect the rank-order preference of the items. The probability of observing a ranking is proportional to the cosine of the angle from the consensus score vector. Unlike distance-based models, the consensus score vector $\boldsymbol{\theta}$ proposed exhibits detailed information on item preferences while distance-based model only provide equal-spaced modal ranking. We applied the method to sushi data and concluded that certain types of sushi are seldom eaten by the Japanese. Our consensus score vector $\boldsymbol{\theta}$ defined on a unit sphere can be easily re-parameterized to incorporate additional arguments or covariates in the model. The judge-specific covariates could be age, gender, and income, the item-specific covariates could be prices, weights, and brands, and the judge-item-specific covariates could be some personal experience on using each phone or brand. Adding those covariates into the model could greatly improve the power of prediction of our model. We could also develop Bayesian inference methods to facilitate the computation. Further details are available in Xu et al. (2018).

12. Analysis of Censored Data

Censored data occur when the value of an observation is only partially known. For example, it may be known that someone's exact wealth is unknown but it may be known that their wealth exceeds one million dollars. In left censoring, the data may fall below a certain value whereas in right censoring, it may be above a certain value. Type I censoring occurs when the subjects of an experiment are right censored. Type II censoring occurs when the experiment stops after a certain number of subjects have failed; the remaining subjects are then right censored. Truncated data occur when observations never lie outside a given range. For example, all data outside the unit interval is discarded. A good example to illustrate the ideas occurs in insurance companies. Left truncation occurs when policyholders are subject to a deductible whereas right censoring occurs when policyholders are subject to an upper pay limit.

A major theme of this chapter is to demonstrate that the key to deriving fundamental results with the embedding approach for censored data lies in an appropriate choice of the parametric family. We first briefly introduce survival analysis and then provide an overview of the developments of rank tests for censored data, highlighting the difficulties caused by ranking incomplete data and describing important landmarks in overcoming these difficulties. Next, we generalize the parametric embedding approach to give a new derivation of what these landmark results have finally led to. More importantly, coupled with the LAN and local minimaxity results, the approach introduced yields asymptotically optimal tests for local alternatives in the embedded parametric family. Since the actual alternatives are unknown, the problem of adaptive (data-dependent) choice of the score function for rank tests has witnessed important developments. We provide a brief review of this topic and its implications on the choice of the parametric family in parametric embedding.

12.1. Survival Analysis

We recall some notions from survival analysis. Let T be a random variable which denotes the survival time. The survival function denoted $S(t)$ is the probability that an individual survives up to time t :

$$S(t) = P(T > t) = 1 - F(t),$$

M. Alvo, P. L. H. Yu, *A Parametric Approach to Nonparametric Statistics*,
Springer Series in the Data Sciences, https://doi.org/10.1007/978-3-319-94153-0_12

where $F(t)$ is the distribution function of T assumed continuous. The survival function is a nonincreasing function of time with the properties

$$S(0) = 1, S(\infty) = 0.$$

A useful property of the mean is

$$\mu = E[T] = \int_0^\infty S(t)\, dt.$$

Next we define the hazard function.

Definition 12.1. The hazard function is defined as the limit of the probability that an individual fails in a very short time interval given that he has survived up to time t:

$$h(t) = \lim_{\Delta t \to 0} \frac{P(t < T < t + \Delta t | T > t)}{\Delta t}.$$

The hazard function can be expressed in terms of the survival function and the probability density function $f(t)$

$$h(t) = \frac{f(t)}{S(t)} = -\frac{S'(t)}{S(t)} = -\frac{d}{dt}\log\{S(t)\}. \tag{12.1}$$

The hazard function is also known as the instantaneous failure rate. One of the characteristics of survival data is the existence of incomplete data, the most common type of which is left truncation and right censoring.

Definition 12.2. Left truncation occurs when subjects enter a study at a specific time and are followed henceforth until the event occurs or the subject is censored. Right censoring occurs when a subject leaves the study before the event occurs or the study ends before the event has occurred.

In right censoring, the observations can be represented by a random vector (T, δ) where δ indicates if the survival time is observed $(\delta = 1)$ or not $(\delta = 0)$.

12.1.1. Kaplan-Meier Estimator

The Kaplan-Meier estimator is a nonparametric estimate of the survival function using lifetime data. Suppose that there are n individuals in a cohort and that $t_1 < t_2 < \ldots$ are the actual times of death. Let $d_1, d_2, \ldots$ denote the number of deaths that occur at each of these times and let $n_1, n_2, \ldots$ denote the corresponding number of patients remaining in the cohort. Hence,

$$n_{i+1} = n_i - d_i, i = 1, 2, ..$$

Suppose that

$$t \in [t_j, t_{j-1}), j = 1, 2, \dots$$

Then the probability of surviving beyond time t is

$$\hat{S}(t) = \prod_{i=1}^{j} \left(1 - \frac{d_i}{n_i}\right) \tag{12.2}$$

Equation 12.2 is known as the Kaplan-Meier estimator of the survival function. It is a step-wise function. Confidence intervals can be constructed using Greenwood's formula for the standard error Rodriguez (2005).

Assuming no censored observations, it can be shown by induction that

$$\hat{S}(t) = \frac{n_{i+1}}{n}, i = 0, 1, 2, \dots$$

In the case of censored observations, we may assume that

$$n_{i+1} = n_i - d_i - c_i, i = 0, 1, 2, ..$$

12.1.2. Locally Most Powerful Tests

We recall from Chapter 8 Hoeffding's change of measure formula

$$P\{R_1 = r_1, \dots, R_m = r_m\} = E_g \left[\frac{f(V_{(r_1)})}{g(V_{(r_1)})} \cdots \frac{f(V_{(r_m)})}{g(V_{(r_m)})}\right] \Big/ \binom{n}{m}, \tag{12.3}$$

where E_g denotes expectation with respect to the probability measure under which the n observations are i.i.d. with common density function g, assuming that g is positive whenever f is. In particular, consider testing $H_0 : f = g$ versus the location alternative $f(x) = g(x - \theta)$ for small positive values of θ. In this case, differentiating both sides of (4) with respect to θ and letting $\theta \downarrow 0$ yield

$$\frac{\partial}{\partial \theta} P\{R_1 = r_1, \dots, R_m = r_m\}\Big|_{\theta=0} = -\sum_{i=1}^{m} E_g \left[\frac{g'(V_{(r_i)})}{g(V_{(r_i)})}\right] \Big/ \binom{n}{m}. \tag{12.4}$$

Hence by an extension of the Neyman-Pearson lemma, the derivative of the power function at $\theta = 0$ is maximized by a test that rejects H_0 when the right-hand side of (4) exceeds some threshold C, which is chosen so that the test has type I error α when $\theta = 0$. This test, therefore, is locally most powerful, for testing alternatives of the form $f(x) = g(x - \theta)$, with $\theta \downarrow 0$, and examples include the Fisher-Yates test when g is standard normal and the Wilcoxon test when $g(x) = e^x/(1 + e^x)^2$ is the logistic density.

A parametric embedding argument similar to (7.3) can be used to give an alternative derivation of the local optimality of the Fisher-Yates and Wilcoxon tests. Define

$$\pi\left(\boldsymbol{x}_{1j},\boldsymbol{x}_{2j};\boldsymbol{\theta}_1,\boldsymbol{\theta}_2\right)=\exp\left\{\sum_{l=1}^{2}\boldsymbol{\theta}_l'\boldsymbol{x}_{lj}-K\left(\boldsymbol{\theta}_1,\boldsymbol{\theta}_2\right)\right\}p_{0j}, j=1,\ldots,t!, \tag{12.5}$$

where $\boldsymbol{\theta}_\ell=\left(\theta_{\ell 1},\ldots,\theta_{\ell k}\right)'$ represents the parameter vector for sample $\ell(=1,2)$ and x_{1j}, x_{2j} are the data from sample 1 and sample 2 with respective sizes m and $n-m$ that are associated with the ranking (permutation) ν_j, $j=1,\ldots,n!$. Under the null hypothesis $H_0:\boldsymbol{\theta}_1=\boldsymbol{\theta}_2$, we can assume without loss of generality that the underlying $V_1,\ldots,V_n$ from the combined sample are i.i.d. uniform (by considering $G(V_i)$, where G is the common distribution function, assumed to be continuous, of the (V_i) and that all rankings of the V_i are equally likely. Hence the above model represents an exponential family constructed by exponential tilting of the baseline measure (i.e., corresponding to H_0) on the rank-order data. This has the same spirit as Neyman's smooth test of the null hypothesis that the data are i.i.d. uniform against alternatives in the exponential family. The parametric embedding makes these results directly applicable to the rank-order statistics as was discussed in Chapter 7. In particular, this shows that the two-sample Wilcoxon test of H_0 is locally powerful for testing the uniform distribution against the truncated exponential distribution for which the $x_{\ell j}$ are constrained to lie in the range $(0,1)$ of the uniform distribution. Note that these exponential tilting alternatives differ from the location alternatives in the preceding paragraph not only in their distributional form (truncated exponential instead of logistic) but also in avoiding the strong assumption of the preceding paragraph that the data have to be generated from the logistic distribution even under the null hypothesis.

12.2. Local Asymptotic Normality, Hajek-Le Cam Theory, and Stein's Least Favorable Parametric Submodels

The local alternatives in Section 12.1 refer to θ near the value(s) θ_0 assumed by the null hypothesis. The sample size n is not involved in the analysis of local power. On the other hand, the central limit theorem has played a major role in the development of rank tests, as asymptotic normality is used to provide approximate critical values under the null hypothesis and to approximate the power function under alternatives within $O(n^{-1/2})$ from θ_0. Hajek (1962) applied Le Cam's contiguity theory to rank tests of the null hypothesis $H_0 : \Delta = 0$ in the simple regression model $Y_i = \alpha + \Delta c_i + \varepsilon_i$, in which ε_i are i.i.d. with common density function f, using linear rank statistics of the form

$$S_n = \sum_{i=1}^{n} (c_i - \bar{c})\varphi\left(\frac{R_i}{n+1}\right). \tag{12.6}$$

He derived the asymptotic normality of S_n under the null hypothesis and contiguous alternatives, and showed the test to have asymptotically maximum power uniformly for these alternatives if $\varphi = -(f' \circ F^{-1})/(f \circ F^{-1})$, where F is the distribution function with derivative f. Note that this result is consistent with the choice of the score function given by (12.4) for locally most powerful tests. Hajek (1968) subsequently introduced the projection method to extend these results to local alternatives that need not be contiguous to the null.

The rank tests in the preceding paragraph deal with the regression setting, which is related to the location alternatives in Section 12.1. If we focus on k-sample problems, then parametric embedding as in the second paragraph of that section can be applied and the idea of *local asymptotic normality* (LAN), which was also introduced by Le Cam in conjunction with contiguity, can be applied to derive the LAN property of the embedded family. As pointed out in Chapter 7 of van der Vaart (2007), a sequence of parametric models is LAN if asymptotically (as $n \to \infty$) their likelihood ratio processes behave like those for the normal mean model via a quadratic expansion of the log-likelihood function. Hajek (1970, 1972) and Le Cam made use of the LAN property to derive asymptotic optimality in parametric estimation and testing via convolution theorems and local asymptotic minimax bounds; see van der Vaart (2007), Chapter 8. The Hajek-Le Cam theory was originally introduced to resolve the long-standing problem concerning the efficiency of the maximum likelihood estimator in a parametric model. For the problem of estimating a location parameter or more general regression parameters, there is a corresponding asymptotic minimax theory introduced by Huber (1964, 1972, 1973, 1981) associated with robust estimators which consist of three types: the maximum likelihood type M-estimators, the R-estimators which are derived from the rank statistics, and the L-estimators which are linear combinations of order statistics. See Chapters 5, 13, and 22 of van der Vaart (2007). Although "it is customary to treat nonparametric statistical theory as a subject completely different from parametric theory," Stein developed the least favorable parametric subfamilies for nonparametric testing and estimation as "one of the obvious connections between the two subjects." The implication of Stein's idea on our parametric embedding theme is the possibility of establishing full asymptotic efficiency of a nonparametric test by using a "least favorable" parametric family of densities for parametric embedding (see Bickel (1982)).

12.3. Parametric Embedding with Censored and Truncated Data

In this section we generalize parametric embedding to the much more complicated setting of censored and truncated data, and use the generalization of parametric embedding to revisit a number of major developments for these data. An extension of rank tests to censored data began with Gehan's (1965) extension of the Wilcoxon test and Mantel's logrank test (Mantel, 1966). An idea similar to Gehan's was extended to truncated data by Bhattacharya et al. (1983). Lai and Ying (1991, 1992) gave a unified treatment of rank statistics for left-truncated and right-censored (LTRC) data. In section 12.3.1 we generalize the parametric embedding approach to censored data. Section 12.3.2 gives an overview of the development of hazard-induced rank tests, highlighting the difficulties caused by ranking incomplete data and describing important landmarks in overcoming these difficulties. Coupled with the LAN and local minimaxity results of Section 12.3.3, the approach introduced in Section 12.3.1 yields asymptotically optimal tests for local alternatives in the embedded family.

12.3.1. Extension of Parametric Embedding to Censored Data

We begin with the right-censored case for which our basic idea of using the hazard function instead of the density function for exponential tilting can become transparent. Recall that for complete data $V_1, \ldots, V_n$, the parametric embedding

$$\pi\left(\boldsymbol{x}_j; \boldsymbol{\theta}\right) = \exp\left\{\boldsymbol{\theta}' \boldsymbol{x}_j - K\left(\boldsymbol{\theta}\right)\right\} p_{0j}, j = 1, \ldots, t!$$

for one sample or

$$\pi\left(\boldsymbol{x}_{1j}, \boldsymbol{x}_{2j}; \boldsymbol{\theta}_1, \boldsymbol{\theta}_2\right) = \exp\left\{\sum_{l=1}^{2} \boldsymbol{\theta}_l' \boldsymbol{x}_{lj} - K\left(\boldsymbol{\theta}_1, \boldsymbol{\theta}_2\right)\right\} p_{0j}, j = 1, \ldots, t!$$

for two samples assumes (a) equally likely rankings that give rise to p_{0j} and i.i.d. uniform $G(V_1), \ldots, G(V_n)$ under the null hypothesis, and (b) exponential tilting via distinct values of $\boldsymbol{x}_j$ or $\boldsymbol{x}_{ij}$ that are functions of the ranks. Here, $\boldsymbol{\theta}_l = (\theta_{l1}, \ldots, \theta_{lk})'$ represents the parameter vector for sample l and $\boldsymbol{x}_{1j}, \boldsymbol{x}_{2j}$are the data for samples 1 and 2 respectively with respective sizes $m, n-m$. Under the null hypothesis $H_0 : \boldsymbol{\theta}_1 = \boldsymbol{\theta}_2$ and we can assume without loss in generality that the underlying V_i are i.i.d. uniform. The V_i are not completely observable when the data are censored so that the observations are $(\tilde{V}_i, \delta_i)$, where $\tilde{V}_i = \min(V_i, c_i)$ and $\delta_i = I_{\{V_i \leq c_i\}}$. Since the rank assigned to V_i for complete data is the empirical distribution function evaluated to V_i, the analog for censored data is $\hat{G}(\tilde{V}_i)$, where $\hat{G}$ is the Kaplan-Meier estimator which is the nonparametric MLE of G for censored data. Hence the model under the null hypothesis is that of i.i.d. uniform random variables censored by $G(c_i)$, providing a partial analog of (a). Since $\hat{G}$ puts all its mass at the uncensored observations (with $\delta = 1$), this causes some difficulty in generalizing (b) because the sample also contains censored observations. We note that at each uncensored observation $\tilde{V}_i$, the information in the ordered sample conveys not only the value of V_i but also how many observations $\tilde{V}_j$ in the sample are $\geq \tilde{V}_i$. When the V_i denote failure times in survival analysis, this means the size of the risk set, that is, the number of subjects who are at risk at an observed failure time V_i. This resolves the inherent difficulty of ordering the censored observations whose actual failure times are unknown except for their exceedance over c_i. To rank the data, we need to have a total order of the sample space, but the subset consisting of censored observations cannot be totally ordered because the underlying failure times are unknown. Using the observed failure time and the risk set size at each uncensored observations gives a partial analog of the ranking for complete data. To be at risk at an observed failure time V_i, the subject cannot fail prior to V_i. The jump in $\hat{G}(\tilde{V}_i)$ basically measures the conditional probability of failing in an infinitesimal interval around $\tilde{V}_i$ given that failure has not occurred prior to $\tilde{V}_i$. This means that we should think of hazard functions instead of density functions and perform exponential tilting using the hazard functions rather density functions.

Consider the two-sample problem with censored data. Let $V_{(1)} < \cdots < V_{(k)}$ denote the ordered uncensored observations of the combined sample, N_j (resp. M_j) denote the number of observations in the combined sample (resp. in sample 1) that are $\geq V_{(j)}$, and $u_j = 1$ (resp. 0) if $V_{(j)}$ comes from sample 1 (resp. sample 2). Note that $\{(1), \ldots, (k), M_1, N_1, \ldots, M_k, N_k\}$ is invariant under the group of strictly increasing transformations on the testing problem. We now introduce embedding of the null model into a smooth parametric family that also consists of alternatives. Instead of tilting the density functions as before, we define the change of measures via intensity (hazard) functions, as in Section II.7 of Andersen et al. (1993). Because the normalizing constant $e^{-K(\boldsymbol{\theta})}$gets canceled in the numerator and denominator, it does not appear in the likelihood ratio statistic. On the other hand, the denominator of (12.1) will induce a function $\lambda_0(t)$, which can be chosen as the baseline (or null hypothesis) hazard function, in the likelihood ratio. The analog therefore for one sample takes the proportional hazards form

$$\pi(\boldsymbol{x}_j;\boldsymbol{\theta}, t) = h_0(t) \exp(\boldsymbol{\theta}'\boldsymbol{x}_j). \tag{12.7}$$

We discuss below the choice of $\boldsymbol{x}_j$ that extends $\boldsymbol{x}_j = X(\nu_j)$ to LTRC data, for which we also define the hazard-induced rank statistics.

12.3.2. From Gehan and Bhattacharya et al. to Hazard-Induced Rank Tests

In this section we first focus on some landmark developments of two-sample rank tests for censored data in the literature and then show how the x_j can be chosen on the basis of the insights provided by these developments. We next show how these two-sample rank statistics can be extended to the k-sample and regression settings, and then further extend them for left-truncated and LTRC data.

The first landmark development was Gehan's extension of the Mann-Whitney version of the Wilcoxon test to censored data. Let T_{1i} (resp. T_{2j}) denote the actual failure times of sample 1 (resp. sample 2), and $(\tilde{T}_{1i}, \delta_{1i})$ and $(\tilde{T}_{2j}, \delta_{2j})$ be the corresponding observations. For complete data, the Mann-Whitney statistic is $W = \sum_{i=1}^{m} \sum_{j=1}^{n-m} w(T_{1i}, T_{2j})$, where $w(t_1, t_2) = 1$ (resp. -1) if $t_1 > t_2$ (resp. $t_1 < t_2$), and $w(t_1, t_2) = 0$ if $t_1 = t_2$. For censored data, Gehan replaced $w(T_{1i}, T_{2j})$ by

$$w(\tilde{T}_{1i}, \delta_{1i}; \tilde{T}_{2j}, \delta_{2j}) = \begin{cases} -1 & \text{if } \tilde{T}_{1i} \leq \tilde{T}_{2j} \text{ and } \delta_{1i} = 1 \\ 1 & \text{if } \tilde{T}_{1i} \geq \tilde{T}_{2j} \text{ and } \delta_{2j} = 1 \\ 0 & \text{otherwise,} \end{cases} \tag{12.8}$$

noting that comparisons can be made if the smaller of $\tilde{T}_i$ and $\tilde{T}'_j$ is uncensored.[1] Breslow (1970) subsequently extended this to the k-sample case and expressed W in the counting process form

$$W = \int Y(s)\, dN'(s) - \int Y'(s)\, dN(s), \tag{12.9}$$

where $N_1(s) = \sum_{i=1}^{m} I_{\{\tilde{T}_{1i} \leq s, \delta_{1i}=1\}}$, $N_2(s) = \sum_{j=1}^{n-m} I_{\{\tilde{T}_{2j} \leq s, \delta_{2j}=1\}}$ and $Y_1(s) = \sum_{i=1}^{m} I_{\{\tilde{T}_{1i} \geq s\}}$ and $Y_2(s) = \sum_{j=1}^{n-m} I_{\{\tilde{T}_{2j} \geq s\}}$ are the corresponding risk set sizes.

Instead of the weight processes Y and Y' that depend on both failures and censoring, Prentice (1978) suggested that a better alternative should depend on the survival experience in the combined sample. For complete data the classical two-sample rank statistics have the form $S_n = \sum_{i=1}^{m} a_n(R_i)$, where the scores $a_n(j)$ are obtained from a score function φ on $(0, 1]$ by $a_n(j) = \varphi(j/n)$ so that $S_n = \sum_{i=1}^{m} \varphi(G_n(T_i))$, where G_n is the distribution of the combined sample, or by some asymptotically equivalent variant such as the expected value of φ evaluated at the jth uniform order statistic from a sample of size n. As pointed out, the counterpart of $G_n(T_i)$ for censored data is $\hat{G}_n(\tilde{T}_i)$, where $\hat{G}_n$ is the Kaplan-Meier estimate based on the combined sample. If $\delta_i = 1$, $\tilde{T}_i$ is the actual failure time and has score $\varphi(\hat{G}_n(\tilde{T}_i))$. On the other hand, if $\delta_i = 0$, then the

[1] In fact, Gehan introduced a further refinement depending on whether the larger observation is censored or not.

actual failure time T_i is unknown, other than that it exceeds $\tilde{T}_i$ and therefore has score $\Phi(\hat{G}_n(\tilde{T}_i))$, where

$$\Phi(t) = \int_t^1 \varphi(v)\,dv/(1-t), \quad 0 \le t < 1, \tag{12.10}$$

represents the average of scores $\varphi(u)$ with $u \ge t$. This leads to the following extension of the classical rank statistic $\sum_{i=1}^m \varphi(G_n(T_i))$ to censored data:

$$S_n^* = \sum_{i=1}^m \left\{ \delta_i \varphi(\tilde{T}_i) + (1-\delta_i)\Phi(\tilde{T}_i) \right\}. \tag{12.11}$$

Prentice (1978) conjectured the asymptotic equivalence of (12.11) to another class of rank statistics that he proposed for censored data based on the generalized rank vector, which is a permutation of $\{1, \dots, n\}$ of the form

$$R = \left[(1), \dots, (k); \{(i\,1), \dots, (i\,\nu_i)\}_{i=0,\dots,k} \right], \tag{12.12}$$

where $V_{(1)} < \cdots < V_{(k)}$ are the ordered uncensored observations of the combined sample and $\{\tilde{V}_{(i\,1)}, \dots, \tilde{V}_{(i\,\nu_i)}\}$ is the unordered set of censored observations between $V_{(i)}$ and $V_{(i+1)}$, setting $V_{(0)} = 0$. Cuzick (1985) proved this conjecture under some smoothness assumptions on φ and also extended the proof to show in his Section 3 the asymptotic equivalence of (12.11) and

$$S_n = \sum_{j=1}^k \psi\left(\hat{G}_n(V_{(j)}) \right) \left(c_j - \frac{M_j}{N_j} \right), \quad \text{where } \psi = \varphi - \Phi. \tag{12.13}$$

This form of rank statistics for censored data dated back to Mantel (1966) with $\psi = 1$. As shown by Gu et al. (1991), there is a one-to-one correspondence between φ and ψ:

$$\varphi(t) = \psi(t) - \int_0^t \frac{\psi(s)}{1-s}\,ds, \quad 0 < t < 1,$$

and rank statistics of the form (12.13) can be expressed in the form of generalized Mann-Whitney statistics $W = \sum_{i=1}^m \sum_{j=1}^{n-m} w(\tilde{T}_i, \delta_i; \tilde{T}'_j, \delta'_j)$ with

$$w(\tilde{T}_{1i}, \delta_{1i}; \tilde{T}_{2j}, \delta_{2j}) = \begin{cases} -n\psi(\hat{G}_n(\tilde{T}_{1i}))/Y_\bullet(\tilde{T}_{1i}) & \text{if } \tilde{T}_{1i} \le \tilde{T}_{2j} \text{ and } \delta_{1i} = 1 \\ n\psi(\hat{G}_n(\tilde{T}_{2j}))/Y_\bullet(\tilde{T}_{2j}) & \text{if } \tilde{T}_{1i} \ge \tilde{T}_{2j} \text{ and } \delta_{2j} = 1 \\ 0 & \text{otherwise,} \end{cases} \tag{12.14}$$

where $Y_\bullet(s) = \sum_{i=1}^m I_{\{\tilde{T}_{1i} \ge s\}} + \sum_{j=1}^{n-m} I_{\{\tilde{T}_{2j} \ge s\}}$ is the risk set size of the combined sample at s.

The representation (12.13) is convenient for extensions from two-sample to the regression setting in which c_j are the covariates in the regression model $V_i = \Delta c_i + \varepsilon_i$, as in (12.6). The M_j/N_j in (12.13) is now generalized to

$$\bar{c}_j = \left(\sum_{i=1}^{n} c_i I_{\{\tilde{V}_i \geq \tilde{V}_{(j)}\}} \right) \Big/ N_j, \tag{12.15}$$

which is the average value of the covariate associated with the risk set at the uncensored observation $\tilde{V}_{(j)}$. Lai and Ying (1991, Theorem 1) established the asymptotic normality of these rank statistics under the null hypothesis $H_0 : \Delta = 0$ and under local alternatives. Analogous to the complete data case, these tests are asymptotically efficient when $\psi = (\lambda' \circ F^{-1})/(\lambda \circ F^{-1})$, where F is the common distribution function and λ the hazard function of the ε_i. They proved this result when the data can also be subject to left truncation.

Suppose (c_i, V_i, δ_i) can be observed only when $\tilde{V}_i = \min(V_i, \xi) \geq \tau_i$, where (τ_i, ξ_i, c_i) are independent random vectors that are independent of the ε_i. The τ_i are left truncation variables and V_i is also subject to right censoring by ξ_i. The case $\xi_i \equiv \infty$ corresponds to the left-truncated model, for which multiplication of V_i and τ_i by -1 converts it into a right-truncated model. Motivated by a controversy in cosmology involving Hubble's Law and Chronometric Theory, Bhattacharya et al. (1983) introduced a Mann-Whitney-type statistic $W_n(\beta) = \sum\sum_{i \neq j} w_{ij}(\Delta)$ in the regression model $V_i = \Delta c_i + \varepsilon_i$, in which c_i represents log velocity and V_i the negative log of luminosity; moreover, (c_i, V_i) can only be observed if $V_i \leq v_0$. This is a right-truncated model with truncation variables $\tau_i \equiv v_0$, and letting $(V_i^*, c_i^*), i = 1, \ldots, n$ denote the observations, they defined $e_i(\beta) = V_i^* - \Delta c_i^*$ and

$$w_{ij}(\Delta) = \begin{cases} c_i^* - c_j^* & \text{if } e_j(\Delta) < e_i(\Delta) \leq v_0 - \Delta c_j^* \\ c_j^* - c_i^* & \text{if } e_i(\Delta) < e_j(\Delta) \leq v_0 - \Delta c_i^* \\ 0 & \text{otherwise} \end{cases} \tag{12.16}$$

since it is impossible to compare $e_i(\Delta)$ and $e_j(\Delta)$ if

$$V_i^* - \Delta c_j^* > v_0 - \Delta c_j^* or\ V_j^* - \Delta c_i^* > v_0 - \Delta c_j^*.$$

Note the similarity of this idea to that proposed by Gehan for censored data, and again it has the same drawbacks as before. In fact, as shown by Lai and Ying (1991), what we discussed in the preceding paragraph for censored data can be readily extended to LTRC data $(u_i^*, \tilde{V}_i^*, \delta_i^*), i = 1, \ldots, n$, that are generated from the larger sample consisting of $(V_i, c_i), i = 1, \ldots, m(n) \triangleq \inf\left\{m : \sum_{i=1}^{m} I_{\{\tau_i \leq \min(V_i, c_i)\}} = n\right\}$, with $(\tilde{V}_i, \delta_i)$ observable only when $\tilde{V}_i \geq \tau_i$. The risk set size at t in this case is $Y(t) = \sum_{i=1}^{m(n)} I_{\{\tau_i - \Delta c_i \leq t \leq \tilde{V}_i - \Delta c_i\}}$ and the nonparametric MLE of the common distribution function G of ε_i is the product-limit estimator

$$\hat{G}_n(t) = 1 - \prod_{s \le t} \left(1 - \frac{(N(s) - N(s-))}{Y(s)}\right),$$

where $N(s) = \sum_{i=1}^{m(n)} I_{\{\tau_i - \Delta c_i \le \tilde{V}_i - \Delta c_i \le s, \delta_i = 1\}}$ when the value of Δ is specified (e.g., $\Delta = 0$ under the null hypothesis). The counting process $N(s)$ plays a fundamental role in the martingale theory underlying the analysis of rank tests via $N(s)$ and $Y(s)$ by Aalen (1978),Gill (1980) and Andersen et al. (1993, Chapter 5) for censored data, and by Lai and Ying (1991, 1992) for LTRC data.

As pointed out in Section 12.3.1, the parametric embedding associated with these regression models is that of a location shift family. Parametric embedding via exponential tilting as in (12.7) is associated with another kind of regression models, called hazard regression models, which model how the hazard functions (rather than the means) of V_i vary with the covariates u_i. Seminal contributions to this problem were made by Cox (1972) who introduced the model (9) for censored survival data. Kalbfleisch and Prentice (1973) derived the marginal likelihood $L(\boldsymbol{\theta})$ of the rank vector R for this model:

$$L(\boldsymbol{\theta}) = \prod_{j=1}^{k} \left\{ e^{\boldsymbol{\theta}' \boldsymbol{x}_{(j)}} \Big/ \left(\sum_{i \in I_j} e^{\boldsymbol{\theta}' \boldsymbol{x}_{(i)}} \right) \right\}, \tag{12.17}$$

where $I_j = \left\{ i : \tilde{V}_i \ge \tilde{V}_{(j)} \right\}$ is the risk set at the ordered uncensored observation $\tilde{V}_{(j)}$, which is the same as that given by Cox using conditional arguments and later by Cox (1975) using partial likelihood. This can be readily extended to LTRC data by redefining the risk set at $\tilde{V}_{(j)}$ as $\left\{ i : \tilde{V}_i \ge \tilde{V}_{(j)} \ge \tau_i \right\}$. Basically, the regression model in the preceding paragraph considers the residuals $\tilde{V}_i - \Delta c_i$, whereas for hazard regression we consider $\tilde{V}_i$ instead.

12.3.3. Semi-parametric Efficiency via Least Favorable Parametric Submodels

The LAN property for the embedded families (exponential tilting and location shifts) associated with rank tests for complete data can be extended to those for LTRC data discussed in the preceding two sections; see Chapter 8 of Andersen et al. (1993) for censored data and Lai and Ying (1992) for LTRC data in the regression setting. For the embedded family (9), the well-known arguments for Cox regression extend readily to LTRC data if the x_i in (9) are the vector of covariates u_i. For the two-sample problem in which x_i depends on the generalized rank vector, we can choose $x_j = \psi(\hat{G}_n(\tilde{V}^*_{(j)}))$ to devise an asymptotically efficient rank statistic as in the censored case, where $\psi = \varphi - \Phi$.

The asymptotic efficiency of the rank tests depends on the class of alternatives in the embedded parametric family, which may not contain the actual alternative.

The problem of finding the parametric family that gives the best asymptotic minimax bound has been an active area of research since the seminal paper of Stein that describes a basic idea inherently related to the theme of this chapter:

> Clearly a nonparametric problem is at least as difficult as any of the parametric problems obtained by assuming we have enough knowledge of the unknown state of nature to restrict it to a finite-dimensional set. For a problem in which one wants to estimate a single real-valued function of the unknown state of nature it frequently happens that there is, through each state of nature, a one-dimensional problem which is, for large samples, at least as difficult (to a first approximation) as any other finite-dimensional problem at that point. If a procedure does essentially as well, for large samples, as one could do for each such one-dimensional problem, one is justified in considering the procedure efficient for large samples.

The implication of Stein's idea on our parametric embedding theme is the possibility of establishing full asymptotic efficiency of a nonparametric/semi-parametric test by using a "least favorable" parametric family of densities for parametric embedding. Lai and Ying (1992, Section 2) have shown how this can be done for regression models with i.i.d. additive noise ε_i. The least favorable parametric family has hazard functions of the form $\lambda(t) + \theta\eta(t)$, where η is an approximation to $-\lambda'\Gamma_1/\Gamma_0$, λ is the hazard function of ε_i and it is assumed that for $h = 0, 1, 2$,

$$\Gamma_h(s) = \lim_{m \to \infty} m^{-1} \sum_{i=1}^{m} E\{c_i^h I_{\{\tau_i - \Delta c_i \le s \le c_i - \Delta c_i\}} / (1 - F(\tau_i - \Delta c_i)\}$$

exists for every s with $F(s) < 1$, where F is the distribution function of ε_i. In particular, the technical details underlying the approximation are given in (2.26 a, b, c) of that paper. Lai and Ying (1991, 1992) have also shown how these semi-parametric information bounds can be attained by using a score function that incorporates adaptive estimation of λ. For a comprehensive overview of semi-parametric efficiency and adaptive estimation in other contexts, see Bickel et al. (1993). For further details on the analysis of censored data, see Alvo et al. (2018).

Appendices

A. Description of Data Sets

A.1. Sutton Leisure Time Data

Ranks on			Number of white females	Number of black females
A: Males	B: Females	C: Both Sexes		
1	2	3	0	1
1	3	2	0	1
2	1	3	1	0
2	3	1	0	5
3	1	2	7	0
3	2	1	6	6

A.2. Umbrella Alternative Data

The Wechsler Adult intelligence scale scores on males by age groups.

Age Group				
16–19	20–34	35–54	55–69	>70
8.62	9.85	9.98	9.12	4.80
9.94	10.43	10.69	9.89	9.18
10.06	11.31	11.40	10.57	9.27

A.3. Song Data

The Song data (t=5) is described in Critchlow et al. (1991). Ninety-eight students were asked to rank 5 words, (1) score, (2) instrument, (3) solo, (4) benediction, and (5) suit, according to the association with the word "song." Critchlow et al. (1991) reported that the average ranks for words (1) to (5) are 2.72, 2.27, 1.60, 3.71, and 4.69, respectively. However, the available data given is in grouped format and the ranking of 15 students are unknown and hence discarded, resulting in 83 rankings, as shown in below.

M. Alvo, P. L. H. Yu, *A Parametric Approach to Nonparametric Statistics*,
Springer Series in the Data Sciences, https://doi.org/10.1007/978-3-319-94153-0

Rankings	Observed frequency
(32145)	19
(23145)	10
(13245)	9
(42135)	8
(12345)	7
(31245)	6
(32154)	6
(52134)	5
(21345)	4
(24135)	3
(41235)	2
(43125)	2
(52143)	2
others	0

A.4. Goldberg Data

This data is due to Goldberg (1976) data (t=10). In the data, 143 graduates were asked to rank 10 occupations according to the degree of social prestige. These 10 occupations are: (i) Faculty member in an academic institution (Fac), (ii) Mechanical engineer (ME), (iii) Operation researcher (OR), (iv) Technician (Tech), (v) Section supervisor in a factory (Sup), (vi) Owner of a company employing more than 100 workers (Own), (vii) Factory foreman (For), (viii) Industrial engineer (IE), (ix) Manager of a production department employing more than 100 workers (Mgr), and (x) Applied scientist (Sci). The data are given in Cohen and Mallows (1980) and have been analyzed by many researchers.

Feigin and Cohen (1978) analyzed the Goldberg data and found three outliers due to the fact that the corresponding graduates wrongly presented rankings in reverse order. After reversing these 3 rankings, the average ranks received by the 10 occupations are 8.57, 4.90, 6.29, 1.90, 4.34, 8.13, 1.47, 6.27, 5.29, 7.85, with the convention that bigger rank means more prestige. Then the preference of graduates is in the order: Fac > Own > Sci > OR > IE > Mgr > ME > Sup > Tech > For.

A.5. Sushi Data set

We first investigate the two data sets of Kamishima (2003) for finding the difference in food preference patterns between eastern and western Japan. Historically, western Japan has been mainly affected by the culture of the Mikado emperor and nobles, while eastern Japan has been the home of the Shogun and Samurai warriors. Therefore, the preference patterns in food are different between these two regions (Kamishima, 2003).

The first data set is a complete ranking data with $t = 10$. 5000 respondents are asked to rank 10 different kinds of Sushi according to their preference. The region of respondents is also recorded (N=3285 for eastern Japan, 1715 for western Japan).

A.6. APA Data

We revisit the well-known APA data set of Diaconis (1988b) which contains 5738 full rankings resulting from the American Psychological Association (APA) presidential election of 1980. For this election, members of APA had to rank five candidates in order of preference. Candidates A and C were research psychologists, candidates D and E were clinical psychologists, and candidate B was a community psychologist. This data set has been studied by Diaconis (1988b) and Kidwell et al. (2008) which found that the voting population was divided into 3 clusters. We also fit the data using our mixture model stated in Section 11.4.2. See also Xu et al. (2018).

A.7. January Precipitation Data (in mm) for Saint John, New Brunswick, Canada

Year	Precipitation		Year	Precipitation		Year	Precipitation
1894	102.9		1927	117.6		1960	167.4
1895	112.3		1928	126.5		1961	82.3
1896	29.0		1929	119.4		1962	119.6
1897	100.6		1930	89.2		1963	169.2
1898	129.0		1931	112.5		1964	127.8
1899	95.5		1932	145.8		1965	87.9
1900	165.9		1933	112.8		1966	83.3
1901	121.9		1934	134.6		1967	88.9
1902	55.1		1935	300.0		1968	149.6
1903	101.3		1936	135.1		1969	115.8
1904	110.7		1937	138.7		1970	26.2
1905	142.0		1938	100.3		1971	82.8
1906	111.3		1939	85.3		1972	132.3
1907	83.1		1940	38.9		1973	153.4
1908	108.7		1941	57.2		1974	103.1
1909	144.8		1942	112.5		1975	168.9
1910	130.0		1943	28.2		1976	205.5
1911	81.3		1944	40.4		1977	102.5
1912	77.5		1945	127.3		1978	283.2
1913	108.7		1946	82.1		1979	296.2
1914	73.2		1947	110.4		1980	50.2
1915	147.1		1948	154.9		1981	134.3
1916	71.4		1949	156.7		1982	196.3
1917	109.7		1950	142.7		1983	75.3
1918	94.5		1951	161.3		1984	141.6
1919	142.5		1952	300.2		1985	44.0
1920	87.1		1953	141.0		1986	135.4
1921	80.3		1954	197.9		1987	135.0
1922	85.3		1955	156.0		1988	95.7
1923	141.7		1956	203.5		1989	101.0
1924	133.9		1957	152.7		1990	162.1
1925	119.9		1958	221.5		1991	102.2
1926	156.0		1959	147.3			

A.8. Annual Temperature Data (in °C) in Hong Kong

Year	max	mean	min		Year	max	mean	min
1961	34.2	22.9	7.3		1989	34.3	23	7.6
1962	35.5	22.7	6		1990	36.1	23.1	7
1963	35.6	23.3	7.1		1991	34.5	23.5	4.6
1964	33.9	22.9	7		1992	35	22.8	8.4
1965	33.4	23.1	7.3		1993	33.5	23.1	5.4
1966	34.7	23.8	7.5		1994	34.1	23.6	7.9
1967	34.4	22.9	4.6		1995	34.2	22.8	9.2
1968	35.7	22.9	5.7		1996	34.3	23.3	5.8
1969	34.7	22.7	4		1997	33.2	23.3	10.2
1970	33.6	22.8	7.6		1998	34.4	24	8.9
1971	33.7	22.7	5.5		1999	35.1	23.8	5.8
1972	34.7	22.8	3.8		2000	34.2	23.3	7.2
1973	33.1	23.3	7		2001	34	23.6	8.9
1974	34.3	22.8	4.2		2002	33.6	23.9	6.8
1975	33.9	22.8	4.3		2003	33.7	23.6	8.8
1976	35.2	22.5	5.7		2004	34.6	23.4	7.7
1977	34.9	23.3	6.2		2005	35.4	23.3	6.4
1978	34.2	22.8	6.9		2006	34	23.5	8
1979	33.8	23.1	6.1		2007	35.3	23.7	10.6
1980	35	23	5.5		2008	34.6	23.1	7.9
1981	33.3	23.1	9.5		2009	34.9	23.5	9.4
1982	34.8	22.9	8.9		2010	34.1	23.2	5.8
1983	33.9	23	6.1		2011	35	23	7.2
1984	34.4	22.5	7		2012	34.5	23.4	7.1
1985	33	22.6	8.8		2013	34.9	23.3	9.2
1986	34.8	22.8	4.8		2014	34.6	23.5	7.3
1987	34.2	23.4	7.6		2015	36.3	24.2	10.3
1988	33.8	22.8	9.5		2016	35.6	23.6	3.1

Bibliography

Aalen, O. (1978). Nonparametric estimation of partial transition probabilities in multiple decrement models. *Ann. Statist.*, 6(3):534–545.

Adkins, L. and Fligner, M. (1998). A non-iterative procedure for maximum likelihood estimation of the parameters of Mallows' model based on partial rankings. *Communications in Statistics: Theory and Methods*, 27(9):2199–2220.

Albert, J. (2008). *Bayesian Computation with R.* Springer, second edition.

Alvo, M. (2008). Nonparametric tests of hypotheses for umbrella alternatives. *Canadian Journal of Statistics*, 36:143–156.

Alvo, M. (2016). Bridging the gap: a likelihood: function approach for the analysis of ranking data. *Communications in Statistics - Theory and Methods, Series A*, 45:5835–5847.

Alvo, M. and Berthelot, M.-P. (2012). Nonparametric tests of trend for proportions. *International Journal of Statistics and Probability*, 1:92–104.

Alvo, M. and Cabilio, P. (1991). On the balanced incomplete block design for rankings. *The Annals of Statistics*, 19:1597–1613.

Alvo, M. and Cabilio, P. (1994). Rank test of trend when data are incomplete. *Environmetrics*, 5:21–27.

Alvo, M. and Cabilio, P. (1995). Testing ordered alternatives in the presence of incomplete data. *Journal of the American Statistical Association*, 90:1015–1024.

Alvo, M. and Cabilio, P. (1998). Applications of Hamming distance to the analysis of block data. In Szyszkowicz, B., editor, *Asymptotic Methods in Probability and Statistics: A Volume in Honour of Miklós Csörgõ*, pages 787–799. Elsevier Science, Amsterdam.

Alvo, M. and Cabilio, P. (1999). A general rank based approach to the analysis of block data. *Communications in Statistics: Theory and Methods*, 28:197–215.

M. Alvo, P. L. H. Yu, *A Parametric Approach to Nonparametric Statistics*,
Springer Series in the Data Sciences, https://doi.org/10.1007/978-3-319-94153-0

Alvo, M. and Cabilio, P. (2000). Calculation of hypergeometric probabilities using Chebyshev polynomials. *The American Statistician*, 54:141–144.

Alvo, M. and Cabilio, P. (2005). General scores statistics on ranks in the analysis of unbalanced designs. *The Canadian Journal of Statistics*, 33:115–129.

Alvo, M., Cabilio, P., and Feigin, P. (1982). Asymptotic theory for measures of concordance with special reference to Kendall's tau. *The Annals of Statistics*, 10:1269–1276.

Alvo, M., Lai, T. L., and Yu, P. L. H. (2018). Parametric embedding of nonparametric inference problems. *Journal of Statistical Theory and Practice*, 12(1):151–164.

Alvo, M. and Pan, J. (1997). A general theory of hypothesis testing based on rankings. *Journal of Statistical Planning and Inference*, 61:219–248.

Alvo, M. and Xu, H. (2017). The analysis of ranking data using score functions and penalized likelihood. *Austrian Journal of Statistics*, 46:15–32.

Alvo, M. and Yu, P. L. H. (2014). *Statistical Methods for Ranking Data.* Springer.

Alvo, M., Yu, P. L. H., and Xu, H. (2017). A semi-parametric approach to the multiple change-point problem. Working paper, The University of Hong Kong.

Andersen, P., Borgan, O., Gill, R., and Keiding, N. (1993). *Statistical Models Based on Counting Processes.* Springer: New York.

Anderson, R. (1959). Use of contingency tables in the analysis of consumer preference studies. *Biometrics*, 15:582–590.

Ansari, A. R. and Bradley, R. A. (1960). Rank-sum tests for dispersions. *Annals of Mathematical Statistics*, 31(4):1174–1189.

Asmussen, S., Jensen, J., and Rojas-Nandayapa, L. (2016). Exponential family techniques for the lognormal left tail. *Scandinavian Journal of Statistics*, 43:774–787.

Bai, J. and Perron, P. (2003). Computation and analysis of multiple structural change models. *Journal of Applied Econometrics*, 18(1):1–22.

Banerjee, A., Dhillon, I. S., Ghosh, J., and Sra, S. (2005). Clustering on the unit hypersphere using von Mises-Fisher distributions. *Journal of Machine Learning Research*, 6(Sep):1345–1382.

Beckett, L. A. (1993). Maximum likelihood estimation in Mallows' model using partially ranked data. In Fligner, M. A. and Verducci, J. S., editors, *Probability Models and Statistical Analyses for Ranking Data*, pages 92–108. Springer-Verlag.

Bhattacharya, P. K., Chernoff, H., and Yang, S. S. (1983). Nonparametric estimation of the slope of a truncated regression. *Ann. Statist.*, 11(2):505–514.

Bhattacharya, R., Lizhen, L., and Patrangenaru, V. (2016). *A Course in Mathematical Statistics and Large Sample Theory.* Springer.

Bickel, P. J. (1982). On adaptive estimation. *Annals of Statistics*, 10(3):647–671.

Bickel, P. J., Klaassen, C. A. J., Ritov, Y., and Wellner, J. A. (1993). *Efficient and Adaptive Estimation for Semiparametric Models.* The Johns Hopkins University Press.

Billingsley, P. (2012). *Probability and Measure.* John Wiley and Sons, anniversary edition.

Blei, D. M., Kucukelbir, A., and McAuliffe, J. D. (2017). Variational inference: A review for statisticians. *Journal of the American Statistical Association*, 112(518):859–877.

Box, George, E. and Tiao, George, C. (1973). *Bayesian Inference in Statistical Analysis.* Addison-Wesley Publishing Company.

Box, G. and Cox, D. (1964). An analysis of transformations. *Journal of the American Statistical Association*, 26:211–252.

Breslow, N. (1970). A generalized Kruskal-Wallis test for comparing K samples subject to unequal pattern of censorship. *Biometrika*, 57:579–594.

Busse, L. M., Orbanz, P., and Buhmann, J. M. (2007). Cluster analysis of heterogeneous rank data. In *Proceedings of the 24th International Conference on Machine Learning*, pages 113–120.

Cabilio, P. and Peng, J. (2008). Multiple rank-based testing for ordered alternatives with incomplete data. *Statistics and Probability Letters*, 78:2609–2613.

Casella, G. and Berger, R. L. (2002). *Statistical Inference.* Duxbury Press., second edition.

Casella, G. and George, E. I. (1992). Explaining the Gibbs sampler. *The American Statistician*, 46:167–174.

Cohen, A. and Mallows, C. (1980). Analysis of ranking data. Technical memorandum, AT&T Bell Laboratories, Murray Hill, N.J.

Conover, W. J. and Iman, R. L. (1981). Rank transformations as a bridge between parametric and nonparametric statistics. *The American Statistician*, 35(3):124–129.

Cox, D. and Hinkley, D. (1974). *Theoretical Statistics.* Chapman Hall, London.

Cox, D. R. (1972). Regression models and life-tables. *J. Roy. Statist. Soc. Ser. B.*, 34(2):187–220.

Cox, D. R. (1975). Partial likelihood. *Biometrika*, 62(2):269–276.

Critchlow, D. (1985). *Metric Methods for Analyzing Partially Ranked Data.* Springer-Verlag: New York.

Critchlow, D. (1986). A unified approach to constructing nonparametric rank tests. Technical Report 86–15, Department of Statistics, Purdue University.

Critchlow, D. (1992). On rank statistics: An approach via metrics on the permutation group. *Journal of Statistical Planning and Inference*, 32(325–346).

Critchlow, D. and Verducci, J. (1992). Detecting a trend in paired rankings. *Applied Statistics*, 41:17–29.

Critchlow, D. E., Fligner, M. A., and Verducci, J. S. (1991). Probability models on rankings. *Journal of Mathematical Psychology*, 35:294–318.

Cuzick, J. (1985). Asymptotic properties of censored linear rank tests. *Ann. Statist.*, 13(1):133–141.

Daniel, W. W. (1990). *Applied Nonparametric Statistics.* Duxbury, Wadsworth Inc., second edition.

Diaconis, P. (1988a). *Group Representations in Probability and Statistics.* Institute of Mathematical Statistics, Hayward.

Diaconis, P. (1988b). Group representations in probability and statistics. *Lecture Notes-Monograph Series*, 11:i–192.

Diaconis, P. and Graham, R. (1977). Spearman's footrule as a measure of disarray. *Journal of the Royal Statistical Society Series B*, 39:262–268.

Donoho, D. L. and Johnstone, I. M. (1995). Adapting to unknown smoothness via wavelet shrinkage. *Journal of the American Statistical Association*, 90(432):1200–1224.

Efron, B. (1981). Nonparametric standard errors and confidence intervals. *The Canadian Journal of Statistics*, 9(2):139–158.

Feigin, P. D. (1993). Modelling and analysing paired ranking data. In Fligner, M. A. and Verducci, J. S., editors, *Probability Models and Statistical Analyses for Ranking Data*, pages 75–91. Springer-Verlag.

Feigin, P. D. and Alvo, M. (1986). Intergroup diversity and concordance for ranking data: an approach via metrics for permutations. *The Annals of Statistics*, 14:691–707.

Feigin, P. D. and Cohen, A. (1978). On a model for concordance between judges. *Journal of the Royal Statistical Society Series B*, 40:203–213.

Feller, W. (1968). *An Introduction to Probability Theory and Its Applications*, volume I. John Wiley & Sons, Inc, New York, third edition.

Ferguson, T. (1967). *Mathematical Statistics: A Decision Theoretic Approach.* Academic Press, New York and London.

Ferguson, T. (1996). *A Course in Large Sample Theory.* John Wiley and Sons.

Fisher, R. (1935). *The Design of Experiments.* Oliver & Boyd, Edinburgh.

Fligner, M. A. and Verducci, J. S. (1986). Distance based ranking models. *Journal of the Royal Statistical Society Series B*, 48(3):359–369.

Fraser, D. (1957). *Non Parametric Methods in Statistics.* John Wiley and Sons., New York.

Friedman, M. (1937). The use of ranks to avoid the assumption of normality implicit in the analysis of variance. *Journal of the American Statistical Association*, 32:675–701.

Gao, X. and Alvo, M. (2005a). A nonparametric test for interaction in two-way layouts. *The Canadian Journal of Statistics*, 33:1–15.

Gao, X. and Alvo, M. (2005b). A unified nonparametric approach for unbalanced factorial designs. *Journal of the American Statistical Association*, 100:926–941.

Gao, X. and Alvo, M. (2008). Nonparametric multiple comparison procedures for unbalanced two-way layouts. *Journal of Statistical Planning and Inference*, 138:3674–3686.

Gao, X., Alvo, M., Chen, J., and Li, G. (2008). Nonparametric multiple comparison procedures for unbalanced one-way factorial designs. *Journal of Statistical Planning and Inference*, 138:2574–2591.

Garcia, R. and Perron, P. (1996). An analysis of the real interest rate under regime shifts. *The Review of Economics and Statistics*, pages 111–125.

Gehan, E. A. (1965). A generalized Wilcoxon test for comparing arbitrarily singly-censored samples. *Biometrika*, 52(1/2):203–223.

Geman, S. and Geman, D. (1984). Stochastic relaxation, Gibbs distributions and the Bayesian restoration of images. *IEEE Transactions on Pattern Analysis and Machine Intelligence*, 6:721–741.

Gibbons, J. D. and Chakraborti, S. (2011). *Nonparametric Statistical Inference*. Chapman Hall, New York, 5th edition.

Gill, R. D. (1980). *Censoring and Stochastic Integrals*. Mathematical Centre, Amsterdam.

Gotze, F. (1987). Approximations for multivariate U statistics. *Journal of Multivariate Analysis*, 22:212–229.

Gu, M. G., Lai, T. L., and Lan, K. K. G. (1991). Rank tests based on censored data and their sequential analogues. *Amer. J. Math. & Management Sci.*, 11(1–2):147–176.

Hajek, J. (1962). Asymptotically most powerful rank-order tests. *Ann. Math. Statist.*, 33(3):1124–1147.

Hajek, J. (1968). Asymptotic normality of simple linear rank statistics under alternatives. *Ann. Math. Statist.*, 39:325–346.

Hajek, J. (1970). A characterization of limiting distributions of regular estimates. *Z. fur Wahrsch. und Verw. Gebiete*, 14:323–330.

Hajek, J. (1972). Local asymptotic minimax and admissibility in estimation. In L. LeCam, J. N. and Scott, E., editors, *Proc. Sixth Berkeley Symp. Math. Statist. Prob.*, volume 1, pages 175–194. University of California Press, Berkeley.

Hastings, W. K. (1970). Monte Carlo sampling methods using Markov chains and their applications. *Biometrika*, 57:97–109.

Hettmansperger, Thomas, P. (1994). *Statistical Inference Based on Ranks*. John Wiley.

Hinkley, D. (1970). Inference about the change-point in a sequence of random variables. *Biometrika*, 57:1–17.

Hájek, J. and Sidak, Z. (1967). *Theory of Rank Tests*. Academic Press, New York.

Hoeffding, W. (1948). A class of statistics with asymptotically normal distribution. *Annals of Mathematical Statistics*, 19:293–325.

Hoeffding, W. (1951a). A combinatorial central limit theorem. *Annals of Mathematical Statistics*, 22:558–566.

Hoeffding, W. (1951b). Optimum non-parametric tests. In *Proceedings of the Second Berkeley Symposium on Mathematical Statistics and Probability*, pages 83–92, Berkeley, Calif. University of California Press.

Huber, P. J. (1964). Robust estimation of a location parameter. *Annals of Mathematical Statistics*, 35(1):73–101.

Huber, P. J. (1972). The 1972 Wald lecture robust statistics: A review. *Annals of Mathematical Statistics*, 43(4):1041–1067.

Huber, P. J. (1973). Robust regression: Asymptotics, conjectures and Monte Carlo. *Annals of Statistics*, 1(5):799–821.

Huber, P. J. (1981). *Robust statistics.* Wiley, New York.

Jarque, C. and Bera, A. (1987). A test of normality of observations and regression residuals. *International Statistical Review*, 55(2):163–172.

Jin, W. R., Riley, R. M., Wolfinger, R. D., White, K. P., Passador-Gundel, G., and Gibson, G. (2001). The contribution of sex, genotype and age to transcriptional variance in drosophila melanogaster. *Nature Genetics*, 29:389–395.

John, J. and Williams, E. (1995). *Cyclic Designs.* Chapman Hall, New York.

Kalbfleisch, J. (1978). Likelihood methods and nonparametric tests. *Journal of the American Statistical Association*, 73:167–170.

Kalbfleisch, J. D. and Prentice, R. L. (1973). Marginal likelihoods based on cox's regression and life model. *Biometrika*, 60(2):267–278.

Kamishima, T. (2003). Nantonac collaborative filtering: recommendation based on order responses. In *Proceedings of the ninth ACM SIGKDD international conference on Knowledge discovery and data mining*, pages 583–588. ACM.

Kannemann, K. (1976). An incidence test for k related samples. *Biometrische Zeitschrift*, 18:3–11.

Kendall, M. and Stuart, A. (1979). *The Advanced Theory of Statistics*, volume 2. Griffin, London, fourth edition.

Kidwell, P., Lebanon, G., and Cleveland, W. S. (2008). Visualizing incomplete and partially ranked data. *IEEE Transactions on Visualization and Computer Graphics*, 14(6):1356–1363.

Killick, R., Fearnhead, P., and Eckley, I. (2012). Optimal detection of changepoints with a linear computational cost. *Journal of the American Statistical Association*, 107(500):1590–1598.

Kruskal, W. H. (1952). A nonparametric test for the several sample problem. *Annals of Mathematical Statistics*, 23:525–540.

Kruskal, W. H. and Wallis, W. A. (1952). Use of ranks in one-criterion variance analysis. *Journal of the American Statistical Association*, 47(260):583–621.

Lai, T. L. and Ying, Z. (1991). Rank regression methods for left-truncated and right-censored data. *Ann. Statist.*, 19(2):531–556.

Lai, T. L. and Ying, Z. (1992). Asymptotically efficient estimation in censored and truncated regression models. *Statistica Sinica*, 2(1):17–46.

Lancaster, H. (1953). A reconciliation of chi square from metrical and enumerative aspects. *Sankhya*, 13:1–10.

Lee, A. (1990). *U-Statistics.* Marcel Dekker Inc., New York.

Lee, P. H. and Yu, P. L. H. (2012). Mixtures of weighted distance-based models for ranking data with applications in political studies. *Computational Statistics and Data Analysis*, 56:2486–2500.

Lehmann, E. (1975). *Nonparametrics: Statistical Methods Based on Ranks.* McGraw-Hill, New York.

Lehmann, E. and Stein, C. (1949). On the theory of some non-parametric hypotheses. *Ann. Math. Statist.*, 20(1):28–45.

Liang, F., Liu, C., and Carroll, J. D. (2010). *Advanced Markov Chain Monte Carlo Methods.* John Wiley & Sons.

Lindley, D. V. and Scott, W. F. (1995). *New Cambridge Statistical Tables.* Cambridge University Press, 2nd edition.

Lindsay, B. G. and Qu, A. (2003). Inference functions and quadratic score tests. *Statist. Sci.*, 18(3):394–410.

Mallows, C. L. (1957). Non-null ranking models. I. *Biometrika*, 44:114–130.

Mantel, N. (1966). Evaluation of survival data and two new rank order statistics arising in its consideration. *Cancer Chemotherapy Reports*, 50(3):163–170.

Marden, J. I. (1995). *Analyzing and Modeling Rank Data.* Chapman Hall, New York.

Matteson, D. S. and James, N. A. (2014). A nonparametric approach for multiple change point analysis of multivariate data. *Journal of the American Statistical Association*, 109:334–345.

McCullagh, P. (1993). Models on spheres and models for permutations. In Fligner, M. A. and Verducci, J. S., editors, *Probability Models and Statistical Analyses for Ranking Data*, pages 278–283. Springer-Verlag.

Metropolis, N., Rosenbluth, A. W., Rosenbluth, M. N., Teller, A. H., and Teller, E. (1953). Equations of state calculations by fast computing machines. *Journal of Chemical Physics*, 21:1087–1092.

Mielke, Paul W., J. and Berry, Kenneth, J. (2001). *Permutation Methods: A Distance Function Approach*. Springer.

Neyman, J. (1937). Smooth test for goodness of fit. *Skandinavisk Aktuarietidskrift*, 20:149–199.

Neyman, J. and Pearson, E. (1933). On the problem of the most efficient tests of statistical hypotheses. *Philo. Trans. Roy. Soc. A*, 231:289–337.

Nunez-Antonio, G. and Gutiérrez-Pena, E. (2005). A Bayesian analysis of directional data using the von Mises–Fisher distribution. *Communications in Statistics-Simulation and Computation*, 34(4):989–999.

Page, E. (1963). Ordered hypotheses for multiple treatments: a significance test for linear ranks. *Journal of the American Statistical Association*, 58:216–230.

Pearson, K. (1900). On the criterion that a given system of deviations from the probable in the case of a correlated system of variables is such that it can be reasonably supposed to have arisen from random sampling. *Philosophical Magazine*, pages 157–175.

Prentice, R. (1978). Linear rank tests with right censored data. *Biometrika*, 65:167–179.

Qian, Z. and Yu, P. L. H. (2018). Weighted distance-based models for ranking data using the r package rankdist. *Journal of Statistical Software*, page Forthcoming.

Ralston, A. (1965). *A First Course in Numerical Analysis*. McGraw Hill, New York.

Randles, Ronald, H. and Wolfe, Douglas, A. (1979). *Introduction to the Theory of Nonparametric Statistics*. John Wiley and Sons, Inc.

Rayner, J. C. W., Best, D. J., and Thas, O. (2009a). Generalised smooth tests of goodness of fit. *Journal of Statistical Theory and Practice*, pages 665–679.

Rayner, J. C. W., Thas, O., and Best, D. J. (2009b). *Smooth Tests of Goodness of Fit Using R.* John Wiley and Sons, 2nd edition.

Robert, C. and Casella, G. (2004). *Monte Carlo Statistical Methods.* Springer, New York, 2nd edition.

Rodriguez, G. (2005). *Nonparametric Survival Models.* Princeton University Press.

Royston, I. (1982). Expected normal order statistics (exact and approximate). *Journal of the Royal Statistical Society Series C*, 31(2):161–165. Algorithm AS 177.

Schach, S. (1979). An alternative to the Friedman test with certain optimality properties. *Ann. Statist.*, 7(3):537–550.

Sen, P. (1968a). Asymptotically efficient tests by the method of n rankings. *Journal of the Royal Statistical Society Series B*, 30:312–317.

Sen, P. (1968b). Asymptotically efficient tests by the method of n rankings. *Journal of the Royal Statistical Society Series B*, 30:312–317.

Serfling, Robert, J. (2009). *Approximating Theorems of Mathematical Statistics.* John Wiley and Sons.

Siegel, S. and Tukey, J. W. (1960). A nonparametric sum of ranks procedure for relative spread in unpaired samples. *Journal of the American Statistical Association*, 55:429–445.

Siegmund, D. (1976). Importance sampling in the Monte Carlo study of sequential tests. *The Annals of Statistics*, 4(4):673–684.

Snijders, A. M., Nowak, N., Segraves, R., Blackwood, S., Brown, N., Conroy, J., Hamilton, G., Hindle, A. K., Huey, B., Kimura, K., et al. (2001). Assembly of microarrays for genome-wide measurement of DNA copy number. *Nature genetics*, 29(3):263–264.

Sra, S. (2012). A short note on parameter approximation for von Mises-Fisher distributions: and a fast implementation of i s (x). *Computational Statistics*, 27(1):177–190.

Stein, C. (1956). Efficient nonparametric testing and estimation. In *Proceedings of the Third Berkeley Symposium on Mathematical Statistics and Probability, Volume 1: Contributions to the Theory of Statistics*, pages 187–195, Berkeley, Calif. University of California Press.

Taghia, J., Ma, Z., and Leijon, A. (2014). Bayesian estimation of the von-Mises Fisher mixture model with variational inference. *IEEE transactions on pattern analysis and machine intelligence*, 36(9):1701–1715.

Terry, M. (1952). Some rank order tests which are most powerful against specific parametric alternatives. *Ann. Math. Statist.*, 23(3):346–366.

Tierney, L. (1994). Markov chains for exploring posterior distributions (with discussion and rejoinder). *Annals of Statistics*, 22(4):1701–1762.

van der Vaart, A. (2007). *Asymptotic Statistics.* Cambridge University Press.

Varin, C., Reid, N., and Firth, D. (2011). An overview of composite likelihood methods. *Statistica Sinica*, 21:5–42.

Wald, A. (1949). Statistical decision functions. *Ann. Math. Statist.*, 22(2):165–205.

Xu, H., Alvo, M., and Yu, P. L. H. (2018). Angle-based models for ranking data. *Computational Statistics and Data Analysis*, 121:113–136.

Yu, P. L. H., Lam, K. F., and Alvo, M. (2002). Nonparametric rank test for independence in opinion surveys. *Austrian Journal of Statistics*, 31:279–290.

Yu, P. L. H., Lam, K. F., and Lo, S. M. (2005). Factor analysis for ranked data with application to a job selection attitude survey. *Journal of the Royal Statistical Society Series A*, 168(3):583–597.

Yu, P. L. H. and Xu, H. (2018). Rank aggregation using latent-scale distance-based models. *Statistics and Computing*, page Forthcoming.

Index

M. Alvo, P. L. H. Yu, *A Parametric Approach to Nonparametric Statistics*,
Springer Series in the Data Sciences, https://doi.org/10.1007/978-3-319-94153-0

Printed in the United States
By Bookmasters